Microengineering, MEMS, and Interfacing

A Practical Guide

MECHANICAL ENGINEERING
A Series of Textbooks and Reference Books

Founding Editor

L. L. Faulkner

Columbus Division, Battelle Memorial Institute
and Department of Mechanical Engineering
The Ohio State University
Columbus, Ohio

1. *Spring Designer's Handbook*, Harold Carlson
2. *Computer-Aided Graphics and Design*, Daniel L. Ryan
3. *Lubrication Fundamentals*, J. George Wills
4. *Solar Engineering for Domestic Buildings*, William A. Himmelman
5. *Applied Engineering Mechanics: Statics and Dynamics*, G. Boothroyd and C. Poli
6. *Centrifugal Pump Clinic*, Igor J. Karassik
7. *Computer-Aided Kinetics for Machine Design*, Daniel L. Ryan
8. *Plastics Products Design Handbook, Part A: Materials and Components; Part B: Processes and Design for Processes*, edited by Edward Miller
9. *Turbomachinery: Basic Theory and Applications*, Earl Logan, Jr.
10. *Vibrations of Shells and Plates*, Werner Soedel
11. *Flat and Corrugated Diaphragm Design Handbook*, Mario Di Giovanni
12. *Practical Stress Analysis in Engineering Design*, Alexander Blake
13. *An Introduction to the Design and Behavior of Bolted Joints*, John H. Bickford
14. *Optimal Engineering Design: Principles and Applications*, James N. Siddall
15. *Spring Manufacturing Handbook*, Harold Carlson
16. *Industrial Noise Control: Fundamentals and Applications*, edited by Lewis H. Bell
17. *Gears and Their Vibration: A Basic Approach to Understanding Gear Noise*, J. Derek Smith
18. *Chains for Power Transmission and Material Handling: Design and Applications Handbook*, American Chain Association
19. *Corrosion and Corrosion Protection Handbook*, edited by Philip A. Schweitzer
20. *Gear Drive Systems: Design and Application*, Peter Lynwander
21. *Controlling In-Plant Airborne Contaminants: Systems Design and Calculations*, John D. Constance
22. *CAD/CAM Systems Planning and Implementation*, Charles S. Knox
23. *Probabilistic Engineering Design: Principles and Applications*, James N. Siddall

24. *Traction Drives: Selection and Application*, Frederick W. Heilich III and Eugene E. Shube
25. *Finite Element Methods: An Introduction*, Ronald L. Huston and Chris E. Passerello
26. *Mechanical Fastening of Plastics: An Engineering Handbook*, Brayton Lincoln, Kenneth J. Gomes, and James F. Braden
27. *Lubrication in Practice: Second Edition*, edited by W. S. Robertson
28. *Principles of Automated Drafting*, Daniel L. Ryan
29. *Practical Seal Design*, edited by Leonard J. Martini
30. *Engineering Documentation for CAD/CAM Applications*, Charles S. Knox
31. *Design Dimensioning with Computer Graphics Applications*, Jerome C. Lange
32. *Mechanism Analysis: Simplified Graphical and Analytical Techniques*, Lyndon O. Barton
33. *CAD/CAM Systems: Justification, Implementation, Productivity Measurement*, Edward J. Preston, George W. Crawford, and Mark E. Coticchia
34. *Steam Plant Calculations Manual*, V. Ganapathy
35. *Design Assurance for Engineers and Managers*, John A. Burgess
36. *Heat Transfer Fluids and Systems for Process and Energy Applications*, Jasbir Singh
37. *Potential Flows: Computer Graphic Solutions*, Robert H. Kirchhoff
38. *Computer-Aided Graphics and Design: Second Edition*, Daniel L. Ryan
39. *Electronically Controlled Proportional Valves: Selection and Application*, Michael J. Tonyan, edited by Tobi Goldoftas
40. *Pressure Gauge Handbook*, AMETEK, U.S. Gauge Division, edited by Philip W. Harland
41. *Fabric Filtration for Combustion Sources: Fundamentals and Basic Technology*, R. P. Donovan
42. *Design of Mechanical Joints*, Alexander Blake
43. *CAD/CAM Dictionary*, Edward J. Preston, George W. Crawford, and Mark E. Coticchia
44. *Machinery Adhesives for Locking, Retaining, and Sealing*, Girard S. Haviland
45. *Couplings and Joints: Design, Selection, and Application*, Jon R. Mancuso
46. *Shaft Alignment Handbook*, John Piotrowski
47. *BASIC Programs for Steam Plant Engineers: Boilers, Combustion, Fluid Flow, and Heat Transfer*, V. Ganapathy
48. *Solving Mechanical Design Problems with Computer Graphics*, Jerome C. Lange
49. *Plastics Gearing: Selection and Application*, Clifford E. Adams
50. *Clutches and Brakes: Design and Selection*, William C. Orthwein
51. *Transducers in Mechanical and Electronic Design*, Harry L. Trietley
52. *Metallurgical Applications of Shock-Wave and High-Strain-Rate Phenomena*, edited by Lawrence E. Murr, Karl P. Staudhammer, and Marc A. Meyers
53. *Magnesium Products Design*, Robert S. Busk
54. *How to Integrate CAD/CAM Systems: Management and Technology*, William D. Engelke

55. *Cam Design and Manufacture: Second Edition; with cam design software for the IBM PC and compatibles*, disk included, Preben W. Jensen
56. *Solid-State AC Motor Controls: Selection and Application*, Sylvester Campbell
57. *Fundamentals of Robotics*, David D. Ardayfio
58. *Belt Selection and Application for Engineers*, edited by Wallace D. Erickson
59. *Developing Three-Dimensional CAD Software with the IBM PC*, C. Stan Wei
60. *Organizing Data for CIM Applications*, Charles S. Knox, with contributions by Thomas C. Boos, Ross S. Culverhouse, and Paul F. Muchnicki
61. *Computer-Aided Simulation in Railway Dynamics*, by Rao V. Dukkipati and Joseph R. Amyot
62. *Fiber-Reinforced Composites: Materials, Manufacturing, and Design*, P. K. Mallick
63. *Photoelectric Sensors and Controls: Selection and Application*, Scott M. Juds
64. *Finite Element Analysis with Personal Computers*, Edward R. Champion, Jr. and J. Michael Ensminger
65. *Ultrasonics: Fundamentals, Technology, Applications: Second Edition, Revised and Expanded*, Dale Ensminger
66. *Applied Finite Element Modeling: Practical Problem Solving for Engineers*, Jeffrey M. Steele
67. *Measurement and Instrumentation in Engineering: Principles and Basic Laboratory Experiments*, Francis S. Tse and Ivan E. Morse
68. *Centrifugal Pump Clinic: Second Edition, Revised and Expanded*, Igor J. Karassik
69. *Practical Stress Analysis in Engineering Design: Second Edition, Revised and Expanded*, Alexander Blake
70. *An Introduction to the Design and Behavior of Bolted Joints: Second Edition, Revised and Expanded*, John H. Bickford
71. *High Vacuum Technology: A Practical Guide*, Marsbed H. Hablanian
72. *Pressure Sensors: Selection and Application*, Duane Tandeske
73. *Zinc Handbook: Properties, Processing, and Use in Design*, Frank Porter
74. *Thermal Fatigue of Metals*, Andrzej Weronski and Tadeusz Hejwowski
75. *Classical and Modern Mechanisms for Engineers and Inventors*, Preben W. Jensen
76. *Handbook of Electronic Package Design*, edited by Michael Pecht
77. *Shock-Wave and High-Strain-Rate Phenomena in Materials*, edited by Marc A. Meyers, Lawrence E. Murr, and Karl P. Staudhammer
78. *Industrial Refrigeration: Principles, Design and Applications*, P. C. Koelet
79. *Applied Combustion*, Eugene L. Keating
80. *Engine Oils and Automotive Lubrication*, edited by Wilfried J. Bartz
81. *Mechanism Analysis: Simplified and Graphical Techniques, Second Edition, Revised and Expanded*, Lyndon O. Barton
82. *Fundamental Fluid Mechanics for the Practicing Engineer*, James W. Murdock
83. *Fiber-Reinforced Composites: Materials, Manufacturing, and Design, Second Edition, Revised and Expanded*, P. K. Mallick

84. *Numerical Methods for Engineering Applications*, Edward R. Champion, Jr.
85. *Turbomachinery: Basic Theory and Applications, Second Edition, Revised and Expanded*, Earl Logan, Jr.
86. *Vibrations of Shells and Plates: Second Edition, Revised and Expanded*, Werner Soedel
87. *Steam Plant Calculations Manual: Second Edition, Revised and Expanded*, V. Ganapathy
88. *Industrial Noise Control: Fundamentals and Applications, Second Edition, Revised and Expanded*, Lewis H. Bell and Douglas H. Bell
89. *Finite Elements: Their Design and Performance*, Richard H. MacNeal
90. *Mechanical Properties of Polymers and Composites: Second Edition, Revised and Expanded*, Lawrence E. Nielsen and Robert F. Landel
91. *Mechanical Wear Prediction and Prevention*, Raymond G. Bayer
92. *Mechanical Power Transmission Components*, edited by David W. South and Jon R. Mancuso
93. *Handbook of Turbomachinery*, edited by Earl Logan, Jr.
94. *Engineering Documentation Control Practices and Procedures*, Ray E. Monahan
95. *Refractory Linings Thermomechanical Design and Applications*, Charles A. Schacht
96. *Geometric Dimensioning and Tolerancing: Applications and Techniques for Use in Design, Manufacturing, and Inspection*, James D. Meadows
97. *An Introduction to the Design and Behavior of Bolted Joints: Third Edition, Revised and Expanded*, John H. Bickford
98. *Shaft Alignment Handbook: Second Edition, Revised and Expanded*, John Piotrowski
99. *Computer-Aided Design of Polymer-Matrix Composite Structures*, edited by Suong Van Hoa
100. *Friction Science and Technology*, Peter J. Blau
101. *Introduction to Plastics and Composites: Mechanical Properties and Engineering Applications*, Edward Miller
102. *Practical Fracture Mechanics in Design*, Alexander Blake
103. *Pump Characteristics and Applications*, Michael W. Volk
104. *Optical Principles and Technology for Engineers*, James E. Stewart
105. *Optimizing the Shape of Mechanical Elements and Structures*, A. A. Seireg and Jorge Rodriguez
106. *Kinematics and Dynamics of Machinery*, Vladimír Stejskal and Michael Valásek
107. *Shaft Seals for Dynamic Applications*, Les Horve
108. *Reliability-Based Mechanical Design*, edited by Thomas A. Cruse
109. *Mechanical Fastening, Joining, and Assembly*, James A. Speck
110. *Turbomachinery Fluid Dynamics and Heat Transfer*, edited by Chunill Hah
111. *High-Vacuum Technology: A Practical Guide, Second Edition, Revised and Expanded*, Marsbed H. Hablanian
112. *Geometric Dimensioning and Tolerancing: Workbook and Answerbook*, James D. Meadows

113. *Handbook of Materials Selection for Engineering Applications*, edited by G. T. Murray
114. *Handbook of Thermoplastic Piping System Design*, Thomas Sixsmith and Reinhard Hanselka
115. *Practical Guide to Finite Elements: A Solid Mechanics* Approach, Steven M. Lepi
116. *Applied Computational Fluid Dynamics*, edited by Vijay K. Garg
117. *Fluid Sealing Technology*, Heinz K. Muller and Bernard S. Nau
118. *Friction and Lubrication in Mechanical Design*, A. A. Seireg
119. *Influence Functions and Matrices*, Yuri A. Melnikov
120. *Mechanical Analysis of Electronic Packaging Systems*, Stephen A. McKeown
121. *Couplings and Joints: Design, Selection, and Application, Second Edition, Revised and Expanded*, Jon R. Mancuso
122. *Thermodynamics: Processes and Applications*, Earl Logan, Jr.
123. *Gear Noise and Vibration*, J. Derek Smith
124. *Practical Fluid Mechanics for Engineering Applications*, John J. Bloomer
125. *Handbook of Hydraulic Fluid Technology*, edited by George E. Totten
126. *Heat Exchanger Design Handbook*, T. Kuppan
127. *Designing for Product Sound Quality*, Richard H. Lyon
128. *Probability Applications in Mechanical Design*, Franklin E. Fisher and Joy R. Fisher
129. *Nickel Alloys*, edited by Ulrich Heubner
130. *Rotating Machinery Vibration: Problem Analysis and Troubleshooting*, Maurice L. Adams, Jr.
131. *Formulas for Dynamic Analysis*, Ronald L. Huston and C. Q. Liu
132. *Handbook of Machinery Dynamics*, Lynn L. Faulkner and Earl Logan, Jr.
133. *Rapid Prototyping Technology: Selection and Application*, Kenneth G. Cooper
134. *Reciprocating Machinery Dynamics: Design and Analysis*, Abdulla S. Rangwala
135. *Maintenance Excellence: Optimizing Equipment Life-Cycle Decisions*, edited by John D. Campbell and Andrew K. S. Jardine
136. *Practical Guide to Industrial Boiler Systems*, Ralph L. Vandagriff
137. *Lubrication Fundamentals: Second Edition, Revised and Expanded*, D. M. Pirro and A. A. Wessol
138. *Mechanical Life Cycle Handbook: Good Environmental Design and Manufacturing*, edited by Mahendra S. Hundal
139. *Micromachining of Engineering Materials*, edited by Joseph McGeough
140. *Control Strategies for Dynamic Systems: Design and Implementation*, John H. Lumkes, Jr.
141. *Practical Guide to Pressure Vessel Manufacturing*, Sunil Pullarcot
142. *Nondestructive Evaluation: Theory, Techniques, and Applications*, edited by Peter J. Shull
143. *Diesel Engine Engineering: Thermodynamics, Dynamics, Design, and Control*, Andrei Makartchouk
144. *Handbook of Machine Tool Analysis*, Ioan D. Marinescu, Constantin Ispas, and Dan Boboc

145. *Implementing Concurrent Engineering in Small Companies*, Susan Carlson Skalak
146. *Practical Guide to the Packaging of Electronics: Thermal and Mechanical Design and Analysis*, Ali Jamnia
147. *Bearing Design in Machinery: Engineering Tribology and Lubrication*, Avraham Harnoy
148. *Mechanical Reliability Improvement: Probability and Statistics for Experimental Testing*, R. E. Little
149. *Industrial Boilers and Heat Recovery Steam Generators: Design, Applications, and Calculations*, V. Ganapathy
150. *The CAD Guidebook: A Basic Manual for Understanding and Improving Computer-Aided Design*, Stephen J. Schoonmaker
151. *Industrial Noise Control and Acoustics*, Randall F. Barron
152. *Mechanical Properties of Engineered Materials*, Wolé Soboyejo
153. *Reliability Verification, Testing, and Analysis in Engineering Design*, Gary S. Wasserman
154. *Fundamental Mechanics of Fluids: Third Edition*, I. G. Currie
155. *Intermediate Heat Transfer*, Kau-Fui Vincent Wong
156. *HVAC Water Chillers and Cooling Towers: Fundamentals, Application, and Operation*, Herbert W. Stanford III
157. *Gear Noise and Vibration: Second Edition, Revised and Expanded*, J. Derek Smith
158. *Handbook of Turbomachinery: Second Edition, Revised and Expanded*, edited by Earl Logan, Jr. and Ramendra Roy
159. *Piping and Pipeline Engineering: Design, Construction, Maintenance, Integrity, and Repair*, George A. Antaki
160. *Turbomachinery: Design and Theory*, Rama S. R. Gorla and Aijaz Ahmed Khan
161. *Target Costing: Market-Driven Product Design*, M. Bradford Clifton, Henry M. B. Bird, Robert E. Albano, and Wesley P. Townsend
162. *Fluidized Bed Combustion*, Simeon N. Oka
163. *Theory of Dimensioning: An Introduction to Parameterizing Geometric Models*, Vijay Srinivasan
164. *Handbook of Mechanical Alloy Design*, edited by George E. Totten, Lin Xie, and Kiyoshi Funatani
165. *Structural Analysis of Polymeric Composite Materials*, Mark E. Tuttle
166. *Modeling and Simulation for Material Selection and Mechanical Design*, edited by George E. Totten, Lin Xie, and Kiyoshi Funatani
167. *Handbook of Pneumatic Conveying Engineering*, David Mills, Mark G. Jones, and Vijay K. Agarwal
168. *Clutches and Brakes: Design and Selection, Second Edition*, William C. Orthwein
169. *Fundamentals of Fluid Film Lubrication: Second Edition*, Bernard J. Hamrock, Steven R. Schmid, and Bo O. Jacobson
170. *Handbook of Lead-Free Solder Technology for Microelectronic Assemblies*, edited by Karl J. Puttlitz and Kathleen A. Stalter
171. *Vehicle Stability*, Dean Karnopp
172. *Mechanical Wear Fundamentals and Testing: Second Edition, Revised and Expanded*, Raymond G. Bayer
173. *Liquid Pipeline Hydraulics*, E. Shashi Menon

174. *Solid Fuels Combustion and Gasification*, Marcio L. de Souza-Santos
175. *Mechanical Tolerance Stackup and Analysis*, Bryan R. Fischer
176. *Engineering Design for Wear,* Raymond G. Bayer
177. *Vibrations of Shells and Plates: Third Edition, Revised and Expanded*, Werner Soedel
178. *Refractories Handbook*, edited by Charles A. Schacht
179. *Practical Engineering Failure Analysis*, Hani M. Tawancy, Anwar Ul-Hamid, and Nureddin M. Abbas
180. *Mechanical Alloying and Milling*, C. Suryanarayana
181. *Mechanical Vibration: Analysis, Uncertainties, and Control, Second Edition, Revised and Expanded*, Haym Benaroya
182. *Design of Automatic Machinery*, Stephen J. Derby
183. *Practical Fracture Mechanics in Design: Second Edition, Revised and Expanded*, Arun Shukla
184. *Practical Guide to Designed Experiments*, Paul D. Funkenbusch
185. *Gigacycle Fatigue in Mechanical Practive*, Claude Bathias and Paul C. Paris
186. *Selection of Engineering Materials and Adhesives*, Lawrence W. Fisher
187. *Boundary Methods: Elements, Contours, and Nodes*, Subrata Mukherjee and Yu Xie Mukherjee
188. *Rotordynamics*, Agnieszka (Agnes) Muszńyska
189. *Pump Characteristics and Applications: Second Edition*, Michael W. Volk
190. *Reliability Engineering: Probability Models and Maintenance Methods*, Joel A. Nachlas
191. *Industrial Heating: Principles, Techniques, Materials, Applications, and Design*, Yeshvant V. Deshmukh
192. *Micro Electro Mechanical System Design*, James J. Allen
193. *Probability Models in Engineering and Science*, Haym Benaroya and Seon Han
194. *Damage Mechanics*, George Z. Voyiadjis and Peter I. Kattan
195. *Standard Handbook of Chains: Chains for Power Transmission and Material Handling, Second Edition*, American Chain Association and John L. Wright, Technical Consultant
196. *Standards for Engineering Design and Manufacturing*, Wasim Ahmed Khan and Abdul Raouf S.I.
197. *Maintenance, Replacement, and Reliability: Theory and Applications*, Andrew K. S. Jardine and Albert H. C. Tsang
198. *Finite Element Method: Applications in Solids, Structures, and Heat Transfer*, Michael R. Gosz
199. *Microengineering, MEMS, and Interfacing: A Practical Guide*, Danny Banks

Microengineering, MEMS, and Interfacing

A Practical Guide

Danny Banks
Monisys Ltd.
Birmingham, England

CRC Press is an imprint of the
Taylor & Francis Group, an **informa** business
A TAYLOR & FRANCIS BOOK

CRC Press
Taylor & Francis Group
6000 Broken Sound Parkway NW, Suite 300
Boca Raton, FL 33487-2742

First issued in paperback 2019

ISBN-13: 978-0-8247-2305-7 (hbk)
ISBN-13: 978-0-367-39102-7 (pbk)

Library of Congress Cataloging-in-Publication Data

Catalog record is available from the Library of Congress

Visit the Taylor & Francis Web site at
http://www.taylorandfrancis.com

and the CRC Press Web site at
http://www.crcpress.com

Dedication

To Amanda Lamb

Acknowledgments

I would like to thank everyone who has contributed material and assistance. Material contributions should be acknowledged in the text, and I can only apologize if any of these have been accidentally omitted. To you, and everyone else, many thanks.

The Author

Danny Banks first studied electronic engineering at Leicester Polytechnic (now DeMontfort University), U.K., graduating in 1990 with a B.Eng. (Hons). He then joined the University of Surrey, U.K., as a Ph.D. student. His research involved modeling and experimental investigation of micromachined microelectrodes for recording neural signals from peripheral nerve trunks. He was awarded his Ph.D. in 1995. Subsequently, he was employed as a postdoctoral research fellow in the biomedical engineering group and was able to spend a further three years on this research. From 1997 to 1999, he was employed as a postdoctoral fellow at the European Molecular Biology Laboratory in Heidelberg, Germany. His work involved the investigation of microfabricated devices for biochemical analysis of single cells. He was also involved in the promotion of artificial microstructures for applications in molecular biology.

Since 1999 Dr. Banks has been employed at Monisys, a small company specializing in embedded systems, sensors, and instrumentation R&D, located in Birmingham, U.K. He is presently technical director.

Dr. Banks is a member of the Institute of Electrical Engineers (IEE), the Society for Experimental Biology of the Institute of Electrical and Electronics Engineers (IEEE) and Euroscience.

Table of Contents

Part 1
Micromachining .... _1_

I.1 Introduction 1
  I.1.1 What Is Microengineering? 1
  I.1.2 Why Is Microengineering Important? 3
  I.1.3 How Can I Make Money out of Microengineering? 5
References 7

Chapter 1 Photolithography 9

1.1 Introduction 9
1.2 UV Photolithography 10
  1.2.1 UV Exposure Systems 11
    1.2.1.1 Mask Aligners 12
    1.2.1.2 UV Light Sources 15
    1.2.1.3 Optical Systems 15
      1.2.1.3.1 Contact and Proximity Printing 16
      1.2.1.3.2 Projection Printing 17
      1.2.1.3.3 Projection and Contact Printing Compared ... 18
    1.2.1.4 Optical Oddities 19
      1.2.1.4.1 The Difference between Negative and Positive Resists 19
      1.2.1.4.2 Optical Aberrations and Distortions 19
      1.2.1.4.3 Optical Proximity Effects 20
      1.2.1.4.4 Reflection from the Substrate 20
  1.2.2 Shadow Masks 21
  1.2.3 Photoresists and Resist Processing 21
    1.2.3.1 Photoresists 22
    1.2.3.2 Photoresist Processing 24
      1.2.3.2.1 Cleaning the Substrate 25
      1.2.3.2.2 Applying Photoresists 27
      1.2.3.2.3 Postexposure Processing 28
1.3 X-Ray Lithography 28
  1.3.1 Masks for X-Ray Lithography 29
1.4 Direct-Write (E-Beam) Lithography 30
1.5 Low-Cost Photolithography 32
1.6 Photolithography — Key Points 34
References 35

Chapter 2 Silicon Micromachining 37

2.1 Introduction 37
2.2 Silicon 37
2.2.1 Miller Indices 39
2.3 Crystal Growth 39
2.4 Doping 40
2.4.1 Thermal Diffusion 41
2.4.2 Ion Implantation 41
2.5 Wafer Specifications 42
2.6 Thin Films 45
2.6.1 Materials and Deposition 45
2.6.1.1 Depositing Thin Films 47
2.6.1.1.1 Thermal Oxidation 47
2.6.1.1.2 Chemical Vapor Deposition 47
2.6.1.1.3 Sputter Deposition 49
2.6.1.1.4 Evaporation 50
2.6.1.1.5 Spinning 50
2.6.1.1.6 Summary 50
2.6.2 Wet Etching 52
2.6.3 Dry Etching 56
2.6.3.1 Relative Ion Etching 56
2.6.3.2 Ion-Beam Milling 57
2.6.4 Liftoff 58
2.7 Structures in Silicon 59
2.7.1 Bulk Silicon Micromachining 59
2.7.1.1 Pits, Mesas, Bridges, Beams, and Membranes with KOH 59
2.7.1.2 Fine Points through Wet and Dry Etching 63
2.7.1.3 RIE Pattern Transfer 64
2.7.1.4 Reflow 64
2.7.2 Surface Micromachining 64
2.7.3 Electrochemical Etching of Silicon 67
2.7.4 Porous Silicon 67
2.7.5 Wafer Bonding 67
2.8 Wafer Dicing 68
2.8.1 The Dicing Saw 68
2.8.2 Diamond and Laser Scribe 69
2.8.3 Releasing Structures by KOH Etching 70
References 72

Chapter 3 Nonsilicon Processes 73

3.1 Introduction 73
3.2 Chemical–Mechanical Polishing 73
3.3 LIGA and Electroplating 74

3.4 Photochemical Machining 75
3.5 Laser Machining 75
3.5.1 IR Lasers 76
3.5.2 Excimer Laser Micromachining 77
3.6 Polymer Microforming 79
3.6.1 Polyimides 80
3.6.2 Photoformable Epoxies (SU-8) 80
3.6.3 Parylene and PTFE 81
3.6.4 Dry Film Resists 81
3.6.5 Embossing 82
3.6.6 PDMS Casting 83
3.6.7 Microcontact Printing 86
3.6.8 Microstereolithography 87
3.7 Electrical Discharge Machining 89
3.8 Photostructurable Glasses 90
3.9 Precision Engineering 91
3.9.1 Roughness Measurements 92
3.10 Other Processes 93
References 94

Chapter 4 Mask Design 95

4.1 Introduction 95
4.2 Minimum Feature Size 95
4.3 Layout Software 95
4.3.1 File Formats 97
4.3.1.1 Technology Files 98
4.3.1.1.1 Units 99
4.3.1.2 Further Caveats 100
4.3.2 Graphics 100
4.3.3 Grid 101
4.3.4 Text 101
4.3.5 Other Features 102
4.3.6 Manhattan Geometry 102
4.4 Design 103
4.4.1 The Frame and Alignment Marks 104
4.4.1.1 Scribe Lane 104
4.4.1.2 Alignment Marks 105
4.4.1.3 Test Structures 107
4.4.1.4 Layer and Mask Set Identification Marks 108
4.4.1.5 Putting It All Together 108
4.4.1.6 Another Way to Place Alignment Marks 111
4.4.2 The Device 111
4.5 Design Rules 117
4.5.1 Developing Design Rules 120

4.6 Getting the Masks Produced 122
4.6.1 Mask Plate Details 122
4.6.2 Design File Details 123
4.6.3 Mask Set Details 123
4.6.4 Step and Repeat 124
4.6.5 Placement Requirements 124
4.7 Generating Gerber Files 124
4.8 Mask Design — Key Points 126

Part II
Microsystems *127*

II.1 Introduction 127
II.1.1 Microsystem Components 128

Chapter 5 Microsensors 131

5.1 Introduction 131
5.2 Thermal Sensors 131
5.2.1 Thermocouples 131
5.2.2 Thermoresistors 132
5.2.3 Thermal Flow-Rate Sensors 133
5.3 Radiation Sensors 134
5.3.1 Photodiodes 134
5.3.2 Phototransistors 135
5.3.3 Charge-Coupled Devices 135
5.3.4 Pyroelectric Sensors 136
5.4 Magnetic Sensors 137
5.5 Chemical Sensors and Biosensors 138
5.5.1 ISFET Sensors 138
5.5.2 Enzyme-Based Biosensors 140
5.6 Microelectrodes for Neurophysiology 141
5.7 Mechanical Sensors 143
5.7.1 Piezoresistors 143
5.7.2 Piezoelectric Sensors 144
5.7.3 Capacitive Sensors 144
5.7.4 Optical Sensors 145
5.7.5 Resonant Sensors 145
5.7.6 Accelerometers 146
5.7.7 Pressure Sensors 146

Chapter 6 Microactuators 147

6.1 Introduction 147
6.2 Electrostatic Actuators 147

6.2.1 Comb Drives 148
6.2.2 Wobble Motors 149
6.3 Magnetic Actuators 150
6.4 Piezoelectric Actuators 151
6.5 Thermal Actuators 151
6.6 Hydraulic Actuators 152
6.7 Multilayer Bonded Devices 153
6.8 Microstimulators 153

Chapter 7 Micro Total Analysis Systems 155

7.1 Introduction 155
7.2 Basic Chemistry 156
7.2.1 Inorganic Chemistry 157
7.2.1.1 Bond Formation 159
7.2.1.2 pH 161
7.2.2 Organic Chemistry 162
7.2.2.1 Polymers 164
7.2.2.2 Silicones 166
7.2.3 Biochemistry 167
7.2.3.1 Proteins 168
7.2.3.2 Nucleic Acids 170
7.2.3.3 Lipids 172
7.2.3.3.1 Fats 173
7.2.3.3.2 Phospholipids 173
7.2.3.3.3 Cholesterol 174
7.2.3.4 Carbohydrates 175
7.3 Applications of Microengineered Devices in Chemistry and Biochemistry 176
7.3.1 Chemistry 177
7.3.1.1 Synthesis 177
7.3.1.2 Process and Environmental Monitoring 177
7.3.2 Biochemistry 177
7.3.3 Biology 178
7.3.3.1 Microscopy 178
7.3.3.2 Radioactive Labeling 179
7.3.3.3 Chromatography 180
7.3.3.4 Electrophoresis 181
7.3.3.5 Mass Spectrometry 182
7.3.3.6 X-Ray Crystallography and NMR 182
7.3.3.7 Other Processes and Advantages 183
7.4 Micro Total Analysis Systems 183
7.4.1 Microfluidic Chips 183
7.4.2 Laminar Flow and Surface Tension 184
7.4.3 Electroosmotic Flow 185

7.4.4 Sample Injection 186
7.4.5 Microchannel Electrophoresis 186
7.4.6 Detection 190
7.4.6.1 Laser-Induced Fluorescence (LIF) 190
7.4.6.1.1 Derivatization 190
7.4.6.1.2 Advantages and Disadvantages of LIF Detection 190
7.4.6.2 Ultraviolet (UV) Absorbance 191
7.4.6.2.1 Advantages and Disadvantages of UV Absorption 191
7.4.6.3 Electrochemical Detection 192
7.4.6.3.1 Cyclic Voltammetry 193
7.4.6.3.2 Advantages and Disadvantages of Cyclic Voltammetry 194
7.4.6.4 Radioactive Labeling 194
7.4.6.5 Mass Spectrometry 194
7.4.6.6 Nuclear Magnetic Resonance 195
7.4.6.7 Other Sensors 195
7.5 DNA Chips 196
7.5.1 DNA Chip Fabrication 196
7.6 The Polymerase Chain Reaction (PCR) 197
7.7 Conducting Polymers and Hydrogels 197
7.7.1 Conducting Polymers 198
7.7.2 Hydrogels 198
References 199

Chapter 8 Integrated Optics 201

8.1 Introduction 201
8.2 Waveguides 201
8.2.1 Optical Fiber Waveguides 201
8.2.1.1 Fabrication of Optical Fibers 202
8.2.2 Planar Waveguides 204
8.3 Integrated Optics Components 204
8.4 Fiber Coupling 205
8.5 Other Applications 205
8.5.1 Lenses 205
8.5.2 Displays 206
8.5.3 Fiber-Optic Cross-Point Switches 206
8.5.4 Tunable Optical Cavities 206

Chapter 9 Assembly and Packaging 209

9.1 Introduction 209
9.2 Assembly 209

9.2.1 Design for Assembly 209
9.2.1.1 Auto- or Self-Alignment and Self-Assembly 210
9.2.1.2 Future Possibilities 211
9.3 Passivation 211
9.4 Prepackage Testing 212
9.5 Packaging 212
9.5.1 Conventional IC Packaging 213
9.5.2 Multichip Modules 214
9.6 Wire Bonding 214
9.6.1 Thermocompression Bonding 214
9.6.2 Ultrasonic Bonding 214
9.6.3 Flip-Chip Bonding 215
9.7 Materials for Prototype Assembly and Packaging 215

Chapter 10 Nanotechnology 217

10.1 Introduction 217
10.2 The Scanning Electron Microscope 217
10.3 Scanning Probe Microscopy 219
10.3.1 Scanning Tunneling Electron Microscope 219
10.3.2 Atomic Force Microscope 220
10.3.3 Scanning Near-Field Optical Microscope 221
10.3.4 Scanning Probe Microscope: Control of the Stage 221
10.3.5 Artifacts and Calibration 221
10.4 Nanoelectromechanical Systems 222
10.4.1 Nanolithography 222
10.4.1.1 UV Photolithography for Nanostructures 222
10.4.1.1.1 Phase-Shift Masks 223
10.4.1.2 SPM "Pens" 224
10.4.2 Silicon Micromachining and Nanostructures 224
10.4.3 Ion-Beam Milling 225
10.5 Langmuir–Blodgett Films 227
10.6 Bionanotechnology 228
10.6.1 Cell Membranes 229
10.6.2 The Cytoskeleton 230
10.6.3 Molecular Motors 230
10.6.4 DNA-Associated Molecular Machines 232
10.6.5 Protein and DNA Engineering 233
10.7 Molecular Nanotechnology 233
10.7.1 Buckminsterfullerene 234
10.7.2 Dendrimers 234
References 235

Part III
Interfacing *237*

III.1 Introduction 237
References 238

Chapter 11 Amplifiers and Filtering 239

11.1 Introduction 239
11.1.1 Quick Introduction to Electronics 239
11.1.1.1 Voltage and Current Conventions 239
11.1.1.2 The Ideal Conductor and Insulator 241
11.1.1.3 The Ideal Resistor 241
11.1.1.4 The Ideal Capacitor 242
11.1.1.5 The Ideal Inductor 242
11.1.1.6 The Ideal Voltage Source 243
11.1.1.7 The Ideal Current Source 243
11.1.1.8 Controlled Sources 243
11.1.1.9 Power Calculations 244
11.1.1.9.1 Switching Losses 244
11.1.1.10 Components in Series and Parallel 245
11.1.1.11 Kirchoff's Laws 246
11.2 Op-Amp 247
11.2.1 The Ideal Op-Amp 248
11.2.1.1 Nonideal Sources, Inverting, and Noninverting Op-Amp Configurations 251
11.2.2 Nonideal Op-Amps 253
11.2.2.1 Bandwidth Limitations and Slew Rate 254
11.2.2.2 Input Impedance and Bias Currents 255
11.2.2.3 Common-Mode Rejection Ratio and Power Supply Rejection Ratio 256
11.2.3 Noise 257
11.2.3.1 Combining White Noise Sources 257
11.2.3.2 Thermal Noise 258
11.2.4 Op-Amp Applications 258
11.2.4.1 The Unity-Gain Buffer Amplifier 258
11.2.4.2 AC-Coupled Amplifiers 260
11.2.4.3 Summing Amplifiers 261
11.2.4.4 Integrators and Differentiators 261
11.2.4.5 Other Functions 263
11.3 Instrumentation Amplifiers 263
11.4 Wheatstone Bridge 265
11.4.1 The Capacitor Bridge 266
11.5 Filtering 268
11.5.1 RC Filters 268

11.5.2 Butterworth Filters 273
11.5.2.1 Synthesizing Butterworth Active Filters 276
11.5.2.2 Approximating the Frequency Response of a Butterworth Filter 278
11.5.3 Switched-Capacitor Filters....279
References 280

Chapter 12 Computer Interfacing 281

12.1 Introduction....281
12.1.1 Number Representation....281
12.2 Driving Analog Devices from Digital Sources 282
12.2.1 Pulse-Width Modulation (PWM)....283
12.2.1.1 Estimating the PWM Frequency....284
12.2.1.2 Digital Implementation and Quantization....285
12.2.1.3 Reproducing Complex Signals with PWM....286
12.2.2 R-2R Ladder Digital-to-Analog Converter (DAC)....286
12.2.3 Current Output DAC 287
12.2.4 Reproducing Complex Signals with Voltage Output DACs....288
12.3 Analog-to-Digital Conversion....288
12.3.1 Sample Rate 289
12.3.1.1 Antialiasing Filters 290
12.3.2 Resolution....290
12.3.3 Signal Reconstruction: Sampling Rate and Resolution Effects 291
12.3.4 Other ADC Errors 292
12.3.4.1 Missing Codes 292
12.3.4.2 Full-Scale Error 292
12.3.5 Companding 292
12.4 Analog-to-Digital Converters 292
12.4.1 Sample-and-Hold Circuit 293
12.4.2 PWM Output ADCs....293
12.4.2.1 Integrating ADC 293
12.4.2.2 Conversion Time....294
12.4.3 Successive Approximation 294
12.4.4 Flash ADC....295
12.4.5 Sigma-Delta Converter....295
12.5 Converter Summary 296
References 296

Chapter 13 Output Drivers....297

13.1 Introduction....297
13.2 Controlling Currents and Voltages with Op-Amps 297
13.2.1 Op-Amp Current Control....297

13.2.1.1 Four-Electrode Configuration 298
13.2.2 Op-Amp Voltage Control 299
13.3 Transistors 300
13.3.1 The BJT 300
13.3.2 The MOSFET 303
13.4 Relays 306
13.4.1 Relay Characteristics 307
13.4.2 Relay Types 307
13.5 BJT Output Boost for Op-Amps 308
13.6 Optoisolators 309
Index 311

Part I

Micromachining

I.1. INTRODUCTION

I.1.1 WHAT IS MICROENGINEERING?

Microengineering and Microelectromechanical systems (MEMS) have very few watertight definitions regarding their subjects and technologies. Microengineering can be described as the techniques, technologies, and practices involved in the realization of structures and devices with dimensions on the order of micrometers. MEMS often refer to mechanical devices with dimensions on the order of micrometers fabricated using techniques originating in the integrated circuit (IC) industry, with emphasis on silicon-based structures and integrated microelectronic circuitry. However, the term is now used to refer to a much wider range of microengineered devices and technologies.

There are other terms in common use that cover the same subject with slightly different emphasis. Microsystems technology (MST) is a term that is commonly used in Europe. The emphasis tends towards the development of systems, and the use of different technologies to fabricate components that are then combined into a system or device is more of a feature of MST than MEMS, where the emphasis tends towards silicon technologies.

In Japan, particularly, the term micromachines is employed. There is a tendency toward miniaturization of machines, with less emphasis on the technologies or materials employed. This should not be confused with micromachining, the processes of fabricating microdevices.

The most rigorous definition available was proposed by the British government, which defined the term microengineering as working to micrometer tolerances. An analogous definition for nanotechnology was advanced. Although these definitions can be used effectively for policy setting, for example, they tend to lead to some anomalies: very large precision-engineered components that one would not normally consider to be MEMS were being classified as such. For this reason, the definition tends to be used with qualifications in technical literature.

This volume will attempt to standardize the definitions for this technology given in the glossary for microengineering and MEMS:

Microengineering: The techniques, technologies, and practices involved in the realization of structures and devices with dimensions on the order of micrometers

MEMS: Microengineered devices that convert between electrical and any other form of energy and rely principally on their three-dimensional mechanical structure for their operation

In this way, microengineering is a very broad term, as one may expect. It not only covers MEMS but also IC fabrication and more conventional microelectronics. As a rule of thumb, devices in which most of the features (gap or line width, step height, etc.) are at or below 100 μm fulfill the "dimensions in the order of micrometers" criteria.

The definition of MEMS as transducers means that the term can be used a little more generally than other definitions would allow. For instance, infrared displays that use suspended structures to thermally isolate each pixel fit nicely into this definition as their operation relies on the three-dimensional suspended structure even though there is no moving mechanical element to the device. It does, however, exclude devices such as Hall effect sensors or photodiodes, which rely principally on their electrical (or chemical) structure for their operation. It also tends to exclude semiconductor lasers for similar reasons, and components such as power MOSFET transistors that are formed by etching V grooves into the silicon substrate are also excluded as they are purely electrical devices.

Once one is happy with the term *microengineering*, one can create all the relevant subdisciplines that one requires simply by taking the conventional discipline name and adding the prefix micro to it. Thus, we have microfluidics, micromechanics, microlithography, micromachining, etc., and, of course, microelectronics. This flippant comment does not mean that these disciplines are simply the macroscale discipline with smaller numbers entered into the equations. In many cases this can be done, but in others this can cause erroneous results. It is intended to point out that there are relatively few surprises in the nomenclature.

At this point, it is worth highlighting the difference between science and engineering as it is of considerable import to the microengineer. Science aims to understand the universe and build a body of knowledge that describes how the universe operates. Engineering is the practical application of science to the benefit of humankind. The description of the universe compiled by scientists is often so complex that it is too unwieldy to be practically applied. Engineers, therefore, take more convenient chunks of this knowledge that apply to the situation with which they are concerned. Specifically, engineers employ models that are limited.

For example, when calculating the trajectory of a thrown ball, Newton's laws of motion would normally be used, and no one would bother to consider how Einstein's relativity would affect the trajectory: the ball is unlikely to be traveling at a relativistic speed where a significant effect may be expected (a substantial fraction of the speed of light).

A good engineering course teaches not only the models that the student needs to employ, and how to employ them, but also the limitations of those models. The knowledge that models are limited is of significance in microengineering because the discipline is still compiling a family of models and list of pitfalls. Despite the vast body of literature on the subject, there is still far more anecdotal knowledge available than written information. This is evidenced by the substantial traffic that MEMS mailing lists and discussion groups receive. There is only so much that can be achieved by reading and modeling, and even a little experience of the practice is of great benefit.

1.1.2 Why Is Microengineering Important?

The inspiration for nanotechnology, particularly molecular nanotechnology, is usually traced back to Richard Feynman's presentation entitled "There's Plenty of Room at the Bottom" in 1959 [1]. A few people cite this presentation as the inspiration for the field of microengineering, but it is more likely that it was the seminal paper by Kurt Petersen, "Silicon as a Mechanical Material," published in 1982 [2].

The micromachining of silicon for purposes other than the creation of electronic components was certainly being carried out at least a decade before Petersen published this work, which compiled a variety of disparate threads and technologies into something that was starting to look like a new technology. Not only was silicon micromachining in existence at this time, but many of the other techniques that will be discussed in later chapters of this volume were also being used for specialized precision engineering work. However, despite the appearance of some early devices, it was not until the end of that decade that commercial exploitation of microengineering, as evidenced by the number of patents issued [3], started to take off.

At the beginning of the 1990s, microengineering was presented as a revolutionary technology that would have as great an impact as the microchip. It promised miniaturized intelligent devices that would offer unprecedented accuracy and resolution and negligible power consumption. Batch fabrication would provide us with these devices at negligible costs: few dollars, or even just a few cents, for a silicon chip. The technology would permeate all areas of life: the more adventurous projects proposed micromachines that would enter the bloodstream and effect repairs, or examine the interior of nuclear reactors in minute detail for the telltale signs of impending failure. As with many emerging technologies, some of the early predictions were wildly optimistic. Although some of the adventurous projects proposed during this period remain inspirational for technological development, the market has tended to be dominated by a few applications — notably IT applications such as inkjet printer heads and hard disk drive read–write heads. Pressure measurement appears next on the list; some may intuitively feel that these devices, rather than inkjet printer heads, are more in tune with the spirit of microengineering.

Nonetheless, microengineered devices have significant advantages and potential advantages over other solutions. Although the road to mass production and low-cost devices is long and expensive, the destination can be reached; examine, for example, the plethora of mass-produced silicon accelerometers and pressure sensors. Beyond the direct advantages of miniaturization, integrating more intelligence into a single component brings with it improved reliability: the fewer components that need to be assembled into a system, the less chance there is that it can go wrong. One great advantage of microengineering is that new tools providing solutions to problems that have never been addressed before are still to be fully exploited. The technology is still relatively new, and innovative thinking can potentially bring some startling results.

MEMS Advantages

- Suitable for high-volume and low-cost production
- Reduced size, mass, and power consumption
- High functionality
- Improved reliability
- Novel solutions and new applications

There is, however, a reason for the aforementioned cautious historical preamble: market surveys are often conducted by groups with a particular interest in the technology or by those interested in showing the economy in a positive light. Evidence is often collected from people working in the field or companies that have invested a lot of R&D dollars into the technology. The preamble thus sets the following data in context.

It is undeniable that microengineering has had a substantial impact beyond disk drives and printers. The sensors and transducers section of any commercial electronics catalog reveals a dozen or so microengineered devices including accelerometers, air-mass-flow sensors, and pressure transducers. (Surprisingly, however, the electronics engineer may not be aware of the technological advances that have gone into these devices). The molecular biologist cannot help but be aware of the plethora of DNA chip technologies, and the material scientist cannot have missed the micromachined atomic force microscope (AFM) probe.

In the mid 1990s a number of different organizations compiled market growth projections for the following few years. These were conveniently collected and summarized by Detlefs and Pisano [3]. The European NEXUS (Network of Excellence for Multifunctional Microsystems) has been particularly active in this respect, publishing a report in 1998 [4] with a follow-up study appearing in 2002 [5]. Also, in 2002, the U.S.-based MEMS Industry Group published its own report [6]. The absolute numbers for the global market in such reports vary depending on how that market is defined. The NEXUS task force included all products with a MEMS component, whereas the other groups only considered the individual components themselves. The NEXUS 2002 report estimated the world market to have been worth approximately \$30 billions in 2000, whereas the U.S.-based MEMS Industry Group estimated it to be in the region of \$2 billions to \$5 billions. From the published summaries, it would appear that a growth of 20% per annum would be a conservative estimate for the coming few years. It should be noted, however, that many of these estimates are based on the highly volatile optical communications and IT markets, where optical MEMS in particular are expected to make a significant impact.

Detlefs and Pisano highlight microfluidics and RF MEMS, apart from optical MEMS, as having significant potential for growth. This being in contrast to the 10 to 20% growth that they ascribe to more established microengineered sensors (pressure, acceleration, etc.). This assessment is in concordance with the NEXUS 2002 findings, where IT peripherals and biomedical areas are identified as having the most significant growth potential.

Microengineering and Money

- Global market of billions of dollars
- 20% annual growth rate to 2005
- Significant areas: IT, optical and RF components, and microfluidics

1.1.3 How Can I Make Money out of Microengineering?

This is not a book that intends to give financial business or other moneymaking advice. It was inspired, in part at least, by the recognition that there is a growing market and opportunities for microengineered products, and in order to exploit these it is necessary to have some understanding of the technology. This book deals with the technologies involved in microengineering, so pithy observations about their potential exploitation are restricted to the introduction.

Firstly, nearly all the processes involved in micromachining involve a significant capital outlay in terms of clean rooms, processing equipment, and hazardous chemicals. In the past this has restricted novel developments to those that had or could afford the facilities or to those using lower-cost micromachining technologies. Multiproject processes, where designs from several different groups are fabricated on the same substrate (wafer) using the same process, are now available. This cuts the cost, but limits you to a specific fabrication sequence. One other option, if you happen to be in an area with a high density of small (R&D) clean room facilities, is to try out your designs by shipping your batch of wafers to as many laboratories as possible.

R&D, however, has not tended to be the bottleneck in commercial exploitation. The main bottleneck has been in scaling up from prototype volumes to mass production volumes. Much of the processing equipment is quite idiosyncratic and needs to be characterized and monitored to ensure that the vast majority of the devices coming off the line meet the specifications (process monitoring). Furthermore, parameters that are required for good electrical performance may result in undesirable mechanical characteristics. In short, it is highly likely that a new line will have to be set up and characterized for the product, and unlike IC foundries, it is difficult to adapt the line for the production of different devices. Additionally, if a silicon device is required with integrated electronic circuitry, the micromachining and circuit fabrication processes must be fully compatible and may be intertwined.

If you are really serious about getting your microengineered device into the market, and have the money to set up a fabrication facility (fab), one of your best options is probably to work with a company (or organization) that has its own facility and is willing to work with others (a MEMS foundry). Usually these will be companies that already produce a few microengineered products of their own, rather than companies set up for the sole purpose of providing micromachining facilities to other parties. At the time of publication, there were a few (but a growing number of) these companies that were genuinely willing to collaborate in product development. Even if you have your own small R&D facility and are serious about producing marketable devices, it would probably be a good idea to find a few of these companies at an early stage in development and align your R&D with their processes. Also, make use of their expertise — this will almost certainly save you a lot of headaches.

Exploitation Problems

- Large initial capital outlay
- Process monitoring
- Potential incompatibility with integrated microelectronics
- Dedicated foundries
- Packaging
- Is there a market for this product?

Packaging is another area that has often been neglected during device R&D. Most microengineered devices will need to interface with the outside world in a way beyond the simple electrical connections of integrated circuits. This will typically require the development of some specialized packages with appropriate tubes, ports, or lenses. The device itself will be exposed to the environment, which can contain all sorts of nasty surprises that are not found within a research laboratory. These surprises include obvious problems, such as dust, bubbles, or other contaminants in microfluidic systems, and the less obvious problems, such as air (many resonant devices are first tested in an electron microscope under vacuum — air can damp them sufficiently to prevent their working and packaging devices under vacuum can be problematical). Other unexpected problems include mechanical or other interactions with the package. Differential coefficients of thermal expansion between device and package can put transducers under strain, leading to erroneous results. Once again, resonant sensors are particularly sensitive to the mechanical properties of the package and to the mounting of dies within it.

Packaging and associated assembly stages are easily the most expensive of any fabrication process. At this stage, each die must be handled individually, as opposed to a hundred or more devices on each wafer during the earlier micromachining stages. Thus, the time spent handling individual dies should be kept to a minimum and automated as much as possible.

Incorporating Microengineering into Your Business

- Develop a novel solution to a new existing problem or gap in the market.
- Develop new products to complement your existing product line or as upgrades.
- Gain competitive advantage by incorporating new technology into your products.
- Gain competitive advantage by using the new technology in new-product development.

A thing to note is that although mass production of microengineered devices can potentially reduce their cost, the amount of R&D effort involved will probably make it necessary to sell early versions at a premium in order to recover costs. It pays, therefore, to be well aware of your market before investing in R&D.

The ideal thing to do is treat a microengineering technology as any other technology: first identify the problem and then select the most appropriate technology to solve it. Of course, identifying the most appropriate technology does assume awareness of the technologies that are available.

REFERENCES

1. Feynman, R., There's Plenty of Room at the Bottom: An Invitation to Enter a New Field of Physics, presentation given on 29 December 1959 at the annual meeting of the APS at Caltech.
2. Petersen, K., Silicon as a mechanical material, *Proc. IEEE*, 70(5), 427–457, 1982.
3. Detlefs and Pisano, US MEMS Review, 5th World Micromachine Summit, 1999.
4. NEXUS! Task Force, Market Analysis for Microsystems 1996–2002, October 1998. The document can be ordered from the NEXUS web site, www.nexus-emsto.com, and an executive summary is freely available.
5. Wechsung, R., Market Analysis for Microsystems 2000–2005 — A Report from the NEXUS Task Force, summary in MST News, April 2002, 43–44.
6. MEMS Industry Group report released at MEMS 2002, Las Vegas. A brief summary can be found at Small Times: J Fried, MEMS Market Continues to Grow, Says Industry Group's New Report, January 21, 2002. www.smalltimes.com/document_display.cfm?document_id=2949.

1 Photolithography

1.1 INTRODUCTION

The fundamental aim of microengineering — to take a design from a computer aided design (CAD) software package and manifest it in a physical manner — may be achieved through one of a number of different fabrication or micromachining technologies. Many of these technologies employ a process known generally as *photolithography*, or a variation of this process, to transfer a two-dimensional pattern from a mask into the structural material. The mask is created from the data held by the CAD package, and the structure is built up by a series of steps that involve the deposition (addition of material to the structure) and etching (removing material from the structure) of patterned layers.

The term *photolithography* refers to a process that uses light or optical techniques to transfer the pattern from the mask to the structural material. Typically, it will refer to a process that employs ultraviolet (UV) light, but it may informally be employed to refer to other lithographic processes or lithography, generally, within the context of microelectromechanical systems (MEMS) and micromachining. Other processes may employ electrons or x-rays.

The purpose of this chapter is to introduce the common forms of lithography, focusing on UV photolithography. Electron-beam (e-beam) and x-ray lithography, as well as some key design matters and processes related to photolithography, are introduced. This chapter is complemented by the matters discussed in Chapter 4 pertaining to mask design.

Features of Photolithography for MEMS

There are a number of features common in MEMS fabrication processes but that are not as common in integrated circuit (IC) fabrication; these are:

- Nonplanar substrate (i.e., relatively large three-dimensional features, such as pits)
- The use of thick resist layers (for structural purposes or for long etching times)
- Relatively high-aspect-ratio structures (in resists as well as substrates)
- Relatively large feature sizes (cf. IC processes)
- Unusual processing steps
- Unusual materials (particularly important in terms of adhesion)

1.2 UV PHOTOLITHOGRAPHY

UV photolithography is the workhorse of many micromachining processes and nearly all semiconductor IC manufacturing processes. With the continual demand for reduced transistor sizes and line widths from IC designers and manufacturers, UV lithography is being pushed to its physical limit to achieve features (line widths or gaps) with submicrometer dimensions. Generally, MEMS employ relatively large structures with dimensions ranging from a few micrometers to about 100 μm. Therefore, the techniques required to produce such small dimensions will not be mentioned here but will be touched on in Part III of this volume.

The basic principle of photolithography is illustrated in Figure 1.1. The aim is to transfer a two-dimensional pattern that is formed on a mask (aka reticle, especially when exposure systems are discussed) into a three-dimensional or two-and-a-half-dimensional pattern in a structural material. The description "two-and-a-half-dimensional" is used because, as you will see, although it is possible to produce structures with complex curves in the xy plane, many micromachining techniques only provide limited control of shapes in the vertical z dimension.

In the example in Figure 1.1, a thin film of silicon dioxide has been deposited on the surface of a silicon wafer. It is desired that this film be selectively removed

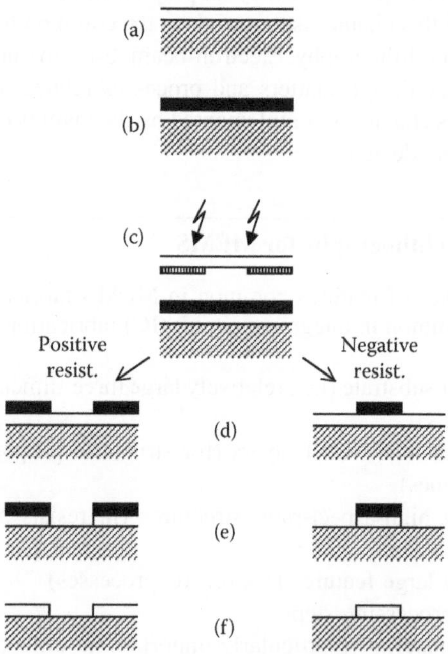

FIGURE 1.1 Basic principle of photolithography (not to scale): (a) silicon substrate with oxide coating, (b) photoresist spun on, (c) exposed to UV light through mask, (d) developed, (e) etching of underlying film, (f) photoresist stripped, leaving patterned film.

from certain areas of the wafer to expose the underlying silicon. To do this, a mask is produced. This will typically be a chromium pattern on a glass (quartz) plate, chromium being opaque to UV light and quartz being transparent. The wafer is cleaned and coated with a material that is sensitive to UV light, known as *photoresist*. The photoresist is exposed to UV light through the mask and then developed, transferring the pattern from the mask into the photoresist.

There are two basic types of photoresists: positive resists and negative resists. (These are also known, respectively, as light-field resists and dark-field resists, although this terminology can cause some confusion when several different fabrication facilities are involved in one process.) With positive resists, the chemical bonds within the resist are weakened when exposed to UV light, whereas they are strengthened in negative resists. As a result, after developing, positive resists take up a positive image of the mask (the resist remains on the mask where the chrome was) and negative resists take up a negative image, as seen in Figure 1.1. The next step involves the selective removal of the silicon dioxide film, through an etching process. A typical example would be to immerse the wafer in a bath of hydrofluoric acid. This will react with the exposed silicon dioxide, but not that protected by the photoresist, which is, as its name implies, resistant to chemical attack by the acid. Once the thin film of silicon dioxide has been etched through, the unwanted photoresist is removed with a solvent, leaving the wafer with the patterned silicon dioxide layer.

Terminology

Photoetching and photoengraving are terms that have also been used to refer to photolithographic processes, although they are not commonly used today. Although photolithography strictly refers to a process that involves light (photons), it is sometimes used in casual conversation to refer to the general sweep of lithographic processes. It would be more correct to use the terms microlithography, nanolithography, or simply lithography (or lithographic) in such cases. The term lithography itself refers to printing from a design onto a flat surface. In addition to UV photolithography, x-ray lithography and e-beam lithography will also be discussed.

1.2.1 UV Exposure Systems

The structural dimensions that can be achieved in a photolithographic process are related to the wavelength of the light employed. When light is incident upon a narrow aperture, it will be diffracted. As the dimensions of the aperture approach the wavelength of the incident light, this diffraction becomes significant. Therefore, for smaller structures, smaller-wavelength light must be used. UV light has therefore been one of the most convenient forms of illumination to employ in photolithography. It conveniently interacts with chemical bonds in various compounds, is relatively easily generated (at longer wavelengths, at least), and has a relatively small wavelength compared to visible or infrared light (from about 400 nm down to 10 nm, where it merges into the soft x-ray region of the spectrum). Also, in the upper reaches of the UV spectrum, optics can be relatively easily fabricated from quartz. UV wavelengths from 426 nm down to about 248 nm are fairly common.

1.2.1.1 Mask Aligners

Microstructures are typically built up through a series of steps in which thin films of materials are deposited and selectively etched (patterned), each photolithographic step, i.e., each pattern, requiring a different mask and each pattern having to be precisely aligned to the preceding ones. Alignment marks are placed on each layer of the design in an out-of-the-way area of the mask (i.e., somewhere where they can easily be found and can fulfill their function but will not interfere with the function of the finished device). The mask aligner is the tool used to align the marks on the mask with those existing on the substrate in order to ensure accurate registration of each layer of the design with the others, as well as to expose the photoresist through the mask to UV light. Exposure may be through a contact aligner or a step-and-repeat system.

The contact mask aligner is the system most commonly used in micromachining processes because they do not normally need the very small feature sizes that can be achieved at greater expense and complexity by step-and-repeat systems. For the contact alignment system, the mask is produced at a 1:1 scale to the finished design. This will invariably be a single large mask plate with many, usually several hundred, individual chip designs on it.

The photoresist-coated substrate (silicon wafer, glass sheet, or whatever is being micromachined) is placed in the aligner and adjusted so that the alignment marks can be located within the viewer. The mask is introduced into the machine, and the chrome-patterned face is brought into close approximation with the photoresist-coated face of the substrate, typically only micrometers apart. The alignment marks on the mask are located, and the position of the mask is adjusted so that they register with the alignment marks etched into the substrate. The mask is then brought into contact with the substrate, final alignment is checked, and the photoresist is then exposed to a pulse of UV light.

The main advantage of contact photolithography is that relatively inexpensive mask aligners and optics are required. Furthermore, the entire area of the substrate is exposed in a single exposure. One advantage of micromachining is that a number of different devices, or different versions of one device, can be placed on the same mask for fabrication on the same substrate. This is of considerable assistance, as MEMS require far more trial-and-error experiments than microelectronic circuits. Another advantage of micromachining is that the process of aligning both sides of the substrate (front and back) is a little easier; specialist double-sided alignment tools are also available. Double-sided alignment, in which micromachining is performed on both sides of a flat silicon substrate, is one feature of MEMS fabrication that is not used in conventional IC manufacture.

Contact photolithography suffers more from wear and tear of the masks than does step-and-repeat, which uses a projection system to reduce the image of the mask on the substrate. Additionally, any small damage or irregularities on the mask are reproduced in the developed photoresist structure. Although the single exposure tends to reduce the time required for photolithography, the UV intensity across the substrate may not be uniform if the system is not set up correctly. In this case, the developed image in the photoresist will not be different across the wafer, and the process yield will be affected. Finally, one does not have the option of using grayscale masks when employing contact lithographic techniques.

The wear and tear of masks can be reduced by using contact alignment's close relation, proximity alignment (or proximity printing). This proceeds in almost exactly the same manner as contact alignment, except that the mask is held at a very small distance from the photoresist. In consequence, the achievable minimum feature size is less than that possible with contact alignment methods.

A Quick Way to Calibrate the Exposure Time in Your Contact Aligner

This method is especially useful when trying out an old system for experimental purposes or trying out new resists, but not of much use if you hit problems with a calibrated setup. Work out the likely minimum and maximum exposure times. Then, subtract a bit from the one, and add a bit to the other. Apply resist to a spare wafer. Now, take a suitable mask with a slot in it (it need not be a quartz mask, but just something that will fit in the aligner). Starting at one end of the wafer (near a flat would be a good idea), put your makeshift mask in and expose it for your minimum exposure time. Now, move the strip up a bit and expose for a little longer (making sure that you note down each exposure time used and any other relevant settings). Repeat. Now, develop and examine the results under a microscope. This is not going to get you very high quality results but may be sufficient to get you started if you are just trying things out.

Contact photolithography is contrasted with the step-and-repeat process in Figure 1.2. Note that the mask face bearing the chrome pattern is the one that is brought into contact with the photoresist during contact lithography. The mask plate itself is relatively thick, typically, a few millimeters. If the chrome were not

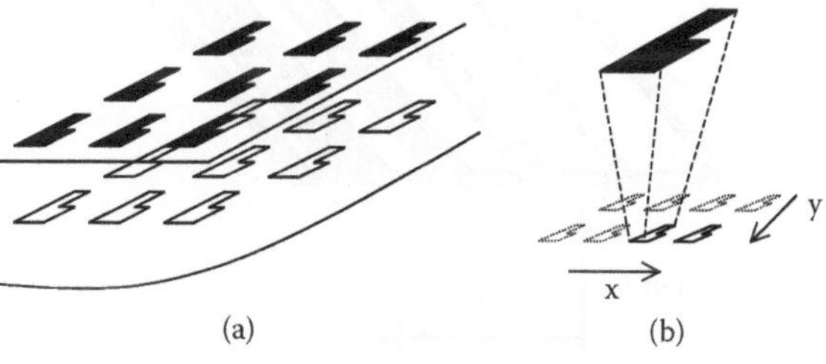

FIGURE 1.2 (a) Contact printing exposes the entire wafer at once, whereas (b) in projection printing a single mask holds the pattern for a single device. This is reduced and projected onto the coated wafer, which is stepped beneath it and receives a series of exposures.

directly in contact with the photoresist, the optical effects due to the passage of UV light through the glass plate, divergence of the source, etc., would reduce the quality of the image formed in the resist.

The step-and-repeat approach involves the use of a mask that bears a larger image of the desired pattern — usually the design for only one chip. This is placed in an optical system that reduces and projects an image of the mask onto the substrate. After each exposure, the substrate is moved (stepped) to expose the next section. Reduction will typically be a factor of about ten. In this case, note that a 1-μm blemish in the mask pattern will be reduced to a 0.1-μm blemish in the photoresist when using the step-and-repeat system but will remain as a 1-μm structure if a contact system is used. The step-and-repeat system's main strength is that it can be used to produce devices with smaller feature sizes than in the case of the contact approach, mainly due to the advantages provided by the projection system.

First, because the mask is made at a larger scale than that of the structure to be produced, it does not necessarily need to be made using a very-high-resolution technique. That is, for contact lithography with a 1-μm minimum structural feature size, the mask would have to be made using a process capable of producing 0.1-μm, or better, features in order to get a reasonable reproduction. If the same structure were to be created using a mask for 10:1 reduction in a projection system, then the minimum structural feature on the mask would be of 10-μm size. A process with better than a 1-μm minimum feature size would produce a result of the same quality as would the contact mask made using the 0.1-μm process.

Furthermore, this gives the designer a chance to control the intensity of the UV light to specific areas of the photoresist which are exposed by creating grayscale masks (Figure 1.3). These essentially incorporate meshes of small apertures in the mask design, such that when the image is reduced, the image of the aperture is beyond the resolving capacity of the photolithographic system. Thus, instead of producing a series of islands or gaps in the imaged photoresist, a reduction in the average intensity of the UV light over the area in proportion to the relative opaque area of the mask is seen. The exact implications of this and the use it can be put

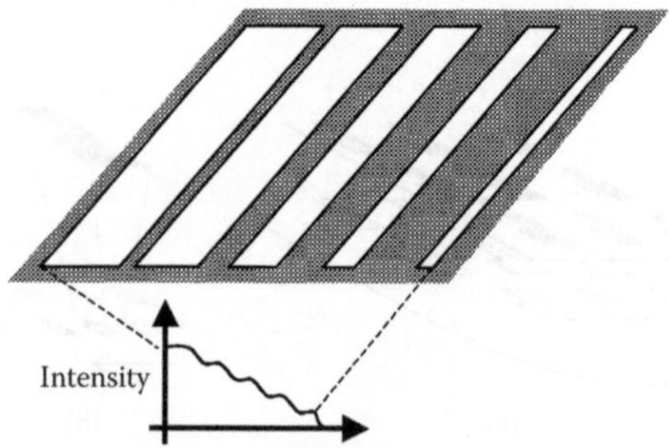

FIGURE 1.3 An example of a grayscale mask. If the openings in the mask are sufficiently small, a variation in intensity rather than distinct lines will be produced when UV light is projected onto the substrate through reducing optics.

TABLE 1.1
Advantages and Disadvantages of Contact and Projection Systems

Contact vs. Projection Lithography Systems	
Contact	**Projection**
Single exposure	More uniform light intensity
Multiple devices per wafer	Small feature sizes
Double-sided alignment	Grayscale masks
Low cost	Longer mask life

to depend on the chemistry and nature of the photoresist or the lithography system. Table 1.1 summarizes the features of contact and projection lithography systems.

1.2.1.2 UV Light Sources

For most micromachining processes involving contact lithography, the UV light source will be a broad-spectrum mercury arc lamp with a filter placed to restrict illumination to one of the spectral lines (i or g). Photoresist manufacturers supply data sheets that provide information about recommended exposure times and wavelengths. Note that whereas some resists must be exposed to light within a fairly strict spectrum in order to function correctly, others are available that also work outside the range of UV wavelengths commonly used.

For very small feature sizes, in particular, the submicrometer feature sizes typical of the most advanced IC technologies in use today, the excimer laser is used as the UV source. This is a UV laser with a torch-like beam. This means that it has to be employed in step-and-repeat processes as it cannot be used to illuminate the entire substrate at once. The excimer laser has its own place in micromachining and is discussed in more detail in Chapter 3.

Photoresists and photolithography systems are commonly referenced by the nature of the UV source: g-line, the 436-nm band of the mercury arc lamp, i line, the 365-nm band, and deep ultraviolet (DUV) at 248-nm and 193-nm wavelengths, in which excimer laser sources are preferred (Table 1.2).

1.2.1.3 Optical Systems

The resolution of an optical system is generally determined by considering its ability to distinguish between two point sources of light [1,2,3]. This work by Rayleigh in the 19th century gave rise to the Rayleigh criterion. Roughly stated, the minimum resolved distance between two peaks depends on the wavelength of light and the numerical aperture of the focusing optics:

$$d = 0.61\frac{\lambda}{NA} \tag{1.1}$$

TABLE 1.2
UV Sources and Wavelengths

Wavelength (nm)	Source	Region of Spectrum
436	Mercury arc lamp	g line
405	Mercury arc lamp	h line
365	Mercury arc lamp	i line
248	Mercury arc lamp or Excimer laser	Deep ultraviolet
193	Excimer laser	

Where λ is the wavelength of the light, and *NA* the numerical aperture of the lens. This equation was derived from optical considerations alone and based on a consideration of point light sources. In photolithography, the achievable resolution (minimum feature size) is also related to other aspects, such as the chemistry of the photoresist. Additionally, one is generally more interested in lines than point sources. Considerations for contact, proximity, and projection systems are outlined in the following subsections.

Also of interest is the depth of focus, the distance along the optical axis over which the optics produce an image of suitable quality. The Rayleigh criterion for depth of focus gives [1,2]:

$$\delta = 0.35\frac{\lambda}{NA^2} \tag{1.2}$$

As with considerations of resolution, this pure equation is not directly applicable to photolithography.

1.2.1.3.1 Contact and Proximity Printing

In contact and proximity printing, the optical limits to minimum feature sizes are due primarily to diffraction effects. In this case, the mathematics analyzes the image of a slit in a grating. This gives rise to a resolution related to the wavelength of light and the separation, *s*, between the mask plate and the substrate [2,4]:

$$d = 1.4\sqrt{\lambda s} \tag{1.3}$$

In practice, because of the dependence on process parameters, this is normally written as:

$$d = k_3\sqrt{\lambda s} \tag{1.4}$$

where k_3 is empirically derived for the process and facility. Peckerar et al. give a practical value of k_3 as 1.6, whereas Reche suggests that it can be as low as 1.5.

In the case of contact printing, the distance *s* will be half the thickness of the photoresist. Note that this can be quite substantial in micromachining applications (tens of micrometers) and that raised and indented micromachined features can mean that the surface of the resist may be considerably more rippled or featured than one normally finds. In the case of proximity printing, one may assume that the distance between mask and substrate is significantly greater than the thickness of the resist, so *s* will take this value, and the thickness of the resist may be neglected. Once again, beware of assumptions that may be invalidated by the unusual nature of MEMS processing.

As mentioned previously, one of the advantages of contact or proximity printing is that the entire area of the substrate can be exposed in a single-process step. Unfortunately UV sources such as the mercury arc lamp appear somewhat point-like. These, therefore, require special optics to expand and homogenize (make the intensity uniform across the area of the substrate that is being exposed) the beam. Somewhat unintuitively, the best results are not provided by collimated light; a divergence of a few degrees will smooth out peaks that appear in the intensity towards the edge of the pattern [3]. The optics for a contact aligner are shown schematically in Figure 1.4.

1.2.1.3.2 Projection Printing

The key parameters for projection printing are derived from the Rayleigh criteria for resolution and depth of focus (Equation 1.1 and Equation 1.2, respectively; [1,2,4]):

$$d = k_1 \frac{\lambda}{NA} \tag{1.5}$$

$$\delta = k_2 \frac{\lambda}{NA^2} \tag{1.6}$$

Once again, k_1 and k_2 are empirically derived for the process in question. In practice, k_1 will be between 0.5 and 1, typically, about 0.7 [1,4], and k_2 will be somewhere about 0.5 ([4]; Peckerar et al. suggest that is closer to 1). Reche also

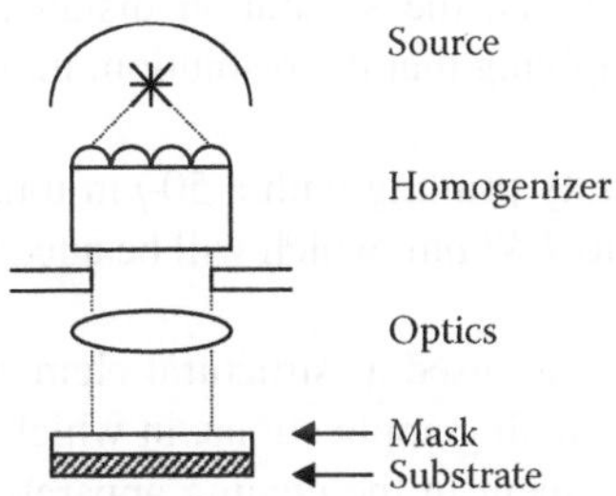

FIGURE 1.4 Contact aligner exposure optics schematic. Alignment is usually performed through a binocular microscope system, not shown, which focuses at two points near the center of the wafer. The relative position of the mask and wafer are adjusted, and the optical components of the aligner are moved out of the way during exposure.

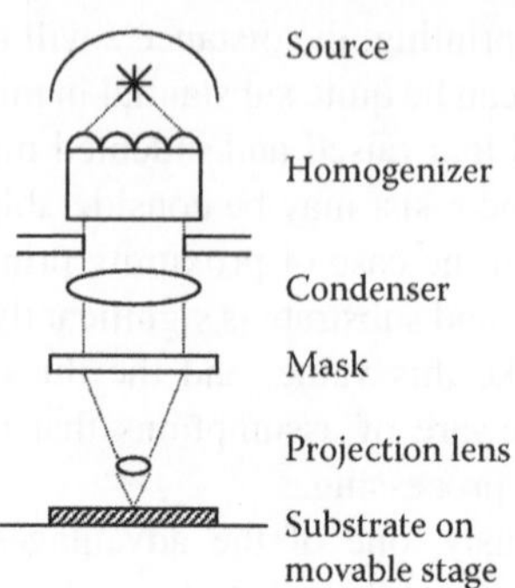

FIGURE 1.5 Schematic outline of a projection printing system.

gives the economically practical value of a numerical aperture as being no more than 0.5 for one-to-one projection printing. With reduction optics, it may be increased to 0.6 [1], although economically this would amount to using a production line stepper around the clock. The optics of a projection system are shown in outline in Figure 1.5.

1.2.1.3.3 Projection and Contact Printing Compared

Working with Equation 1.4, Equation 1.5, and Equation 1.6 and taking values of 0.7, 1, and 1.6 for k_1, k_2, and k_3, respectively, we find, with g-line (436 nm) exposure for a 1:1 projection system with a numerical aperture of 0.5, the achievable resolution will be approximately 0.61 μm with a depth of field of 1.7 μm. This would be adequate for many applications, but consider the situation in which a 10-μm thick resist is required. A trade-off between depth of field and resolution can be seen by examining Equation 1.5 and Equation 1.6. For a 10-μm depth of field (greater, preferably, to accommodate positioning and other errors), the resolution goes up to about 1.53 μm. Note that projection printing would typically be used for high-resolution printing on thin films of resist.

Using the same numbers, contact printing would give a 3.34-μm resolution with the 10-μm resist. In this case, we have considered the entire thickness of the resist film as the separation distance, which will give a worst-case estimate of resolution. For thin resists, the separation distance can be set to half the thickness of the resist (implying that the resolution, in this case, is unlikely to be better than 2.36 μm).

If we consider proximity printing with a 50-μm total separation, our achievable resolution increases to 7.47 μm, which will be adequate for many microengineering applications.

Typically, thick resists are used as structural elements in MEMS. They may also be desirable in deep-etching applications, in which a thick resist is required to withstand long periods spent in the etching apparatus. In the latter situation, high resolutions can be achieved by the use of a *hard mask*. A thin layer of resist can be used to pattern an underlying layer of more resilient material for the etching of the next process stem: a metal film, for instance. This is the hard mask; the pattern in this would then be transferred to the underlying material during a long etch process before the hard mask (etch mask) is stripped.

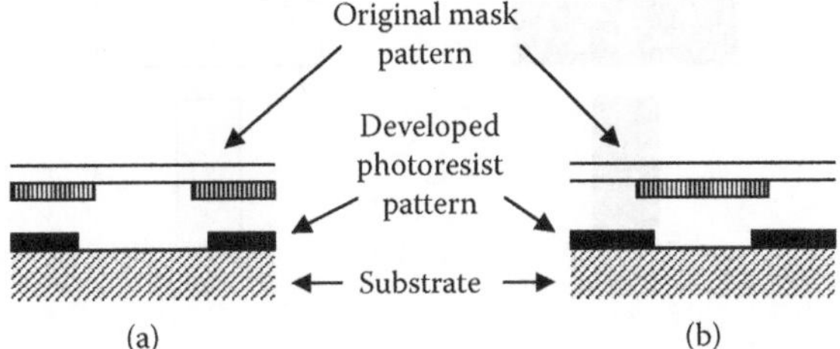

FIGURE 1.6 (a) Positive resists tend to develop with slightly wider than desired openings; (b) negative resists tend to develop with slightly smaller openings than mask features.

1.2.1.4 Optical Oddities

Optical systems cannot be made completely free of aberrations or distortions, and further problems may be introduced by the nature of mask or substrate. A few of these are discussed in the following paragraphs, and some are covered in greater detail in Part III.

1.2.1.4.1 The Difference between Negative and Positive Resists

Light will be scattered when it enters the resist layer. As illustrated in Figure 1.6, when the resist is overexposed, this leads to gaps in the developed resist that are larger than the mask features for positive resists and smaller than the mask features for negative resists. Because many etching procedures undercut the resist, particularly many wet etches, this has resulted in a preference to the use of negative resists in order to more closely reproduce the features in the mask.

1.2.1.4.2 Optical Aberrations and Distortions

The results of any photolithographic process would be limited by the quality of the optical system. Typically, these will be more severe further from the optical axis.

Astigmatism, arising from asymmetry in the optics for instance, will typically result in slightly poorer resolution in one horizontal direction than in others. It may also have knock-on consequences in terms of optical proximity effects, etc., mentioned later.

Chromatic aberrations are particularly problematic with lens-based systems, as opposed to reflective focusing systems. Although lens-based optical systems normally achieve higher numerical apertures than reflective systems, the refractive index of the material employed is dependent on the wavelength of the light being transmitted. Some photoresists are sensitive to a specific wavelength of light, whereas other broadband resists are sensitive to a broad spectral range. In the latter case, projection printing results, in particular, will suffer because of chromatic aberrations unless a filter is employed.

Distortions can sometimes be introduced because the resist is capable of reproducing very-high-resolution features. In some forms of 1:1 projection and contact printing, for instance, the fly-eye homogenizer employed can introduce patterns in the resist.

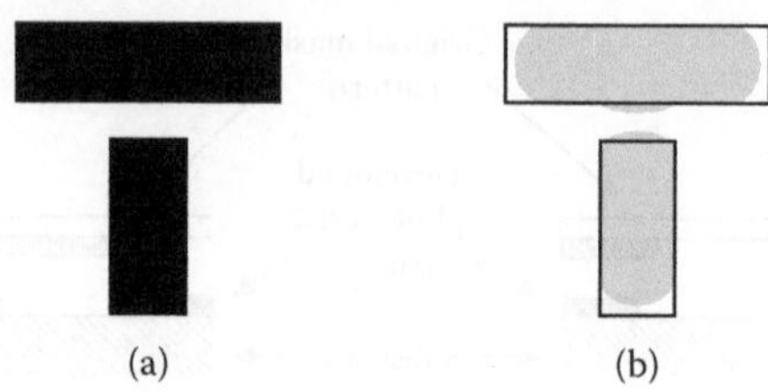

FIGURE 1.7 Optical proximity effects (exaggerated): (a) the original mask pattern, (b) the pattern reproduced (shaded area), the lines are foreshortened, corners rounded, and the small gap partially filled.

1.2.1.4.3 Optical Proximity Effects

Optical proximity effects are another aspect of photolithography that are felt most acutely with modern high-resolution projection systems. They are exemplified by the situation illustrated in Figure 1.7. Diffraction effects in the gap have led to partial exposure of the resist there and poor reproduction of the mask pattern.

Similar effects can also be seen in the rounding of corners and poor dimensional reproduction illustrated in Figure 1.7b. Note that corners in particular represent very-high-resolution objects, and thus it can be difficult to achieve good reproduction of sharp corners.

These effects can be compensated for by mask design, but with the resolutions typically used in microengineering and MEMS, they do not normally represent significant problems. High-resolution nanolithography is dealt with in Part III of this book.

1.2.1.4.4 Reflection from the Substrate

The classic example of an effect caused by reflection from the substrate is the striated or wavy patterns that appear in otherwise vertical resist sidewalls (Figure 1.8). These are a result of standing waves set up between incident light and that reflected from the substrate below the resist.

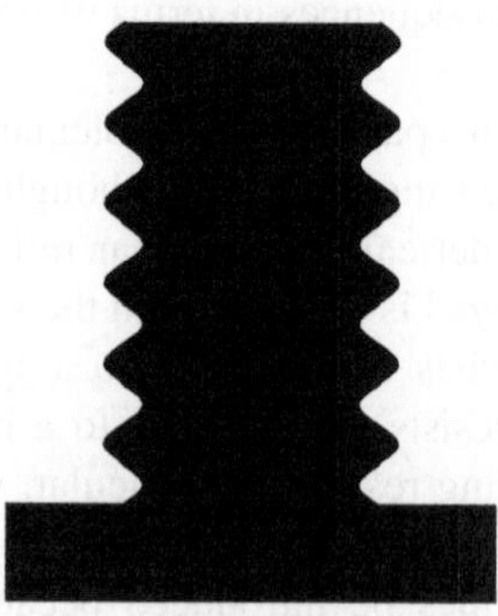

FIGURE 1.8 Schematic illustration of standing wave effects on resist (cross section).

Another problem that may be experienced is that of reflective proximity effects, in which a slope in the substrate reflects incident light horizontally into the resist. This can interfere with exposure of the resist in an adjacent area leading to overexposure.

Antireflective coatings are available from suppliers of photolithographic chemicals, and these are the solution of first resort in cases in which reflected light causes a problem.

1.2.2 Shadow Masks

An alternative to chrome on quartz masks is the use of stencils. Commonly, these would be laser-cut stainless steel stencils that are used in printed circuit board (PCB) manufacture. These are termed *shadow masks* and have two applications. The first is in certain thin-film deposition processes, notably sputtering and evaporation, in which the mask is clamped over the face of the substrate. The deposition process covers the entire surface so that when the mask is removed, unwanted material goes with it, leaving a stenciled pattern on the substrate. The second use, obviously, is in photolithography.

Shadow masks cannot be used to achieve very high precision or small feature sizes. An additional problem with the use of stainless steel, particularly in deposition processes that develop heat, is that its dimensions change because of thermal expansion, giving rise to blurred edges. Cutting masks from alloys such as invar can reduce this problem. The main advantage of using shadow masks, however, is their low cost.

Tolerance Examples for Laser-Cut Stencils

- Stainless steel, 0.1- to 0.2-mm thick
- Design resolution, 0.5 μm
- Precision, ±10 μm
- Pitch (spacing between pads), 0.3 mm

Although a very small design resolution is quoted, the ±10-μm precision limits the design minimum feature size. The pitch will be given for component pads on a PCB. Note that because this is a stencil, holes can be smaller than the spacing between them, and the designer has to consider mechanical support and stability across the design. Sub-100-μm holes may be achievable, but larger spaces (at least 100 μm) should be left between them.

1.2.3 Photoresists and Resist Processing

The aim of the exercise is to produce controlled and repeatable profiles in the developed photoresist. The ideal profile has vertical sidewalls as shown in Figure 1.9a. For some applications, it may be desirable to employ different resist profiles; one of the most useful of these is the undercut profile for liftoff processing (see

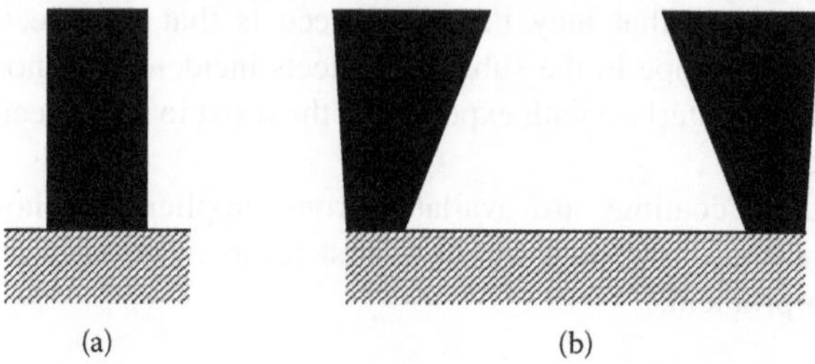

FIGURE 1.9 Photoresist profiles: (a) the ideal with near-vertical walls, this profile would be slightly narrower at the top than the bottom; (b) undercut profile required for the lift-off process.

section 2.6.4), as in Figure 1.9b. The optics, resist chemistry, and resist processing steps combine to produce the desired sidewall profile.

1.2.3.1 Photoresists

A photoresist is normally supplied in liquid form. Most resists consist of two chemical components in an organic solvent. The first component is sensitive to light. The chemical products resulting from exposure of this component to light drive a polymerization in the other resist component. The developing process then removes the unpolymerized resist in the case of negative resists. In the case of positive resists, the result of exposure is usually that the second component in the exposed areas becomes more soluble in the developer. Generally, polymerization and development in negative resists are accompanied by dimensional changes that limit the resolution of the process more than for positive resists. However, there are now a variety of specialized chemistries available for both positive and negative resists, providing the engineer with a range of different options. Table 1.3 lists some photoresists that are popular for microengineering and MEMS. Table 1.4 lists some photoresist suppliers.

Desirable Properties of Photoresists for Microengineering

1. Good resolution
2. Good adhesion to the substrate
3. Resistance to etching processes
4. Resistance to other micromachining processes (e.g., electroplating)
5. Ability to coat nonplanar topographies
6. Ability to apply and process thick coatings (5 μm to more than 100 μm)
7. Mechanically resilient

Items 1 to 3 are general for any photoresist. Items 4 to 7 are more specific to micromachining, with items 5 and 6 being of particular interest when the resist is to be used as a structural component of the design.

TABLE 1.3
Popular Photoresists for Microengineering

Resist	Source[a]	+/–	Features
SU-8	MCC	–	Epoxy-based resist, 2–200 μm thickness, very resilient, can be difficult to remove, excellent structural resist, adhesion promoters not normally required, image reversal possible, near-UV 350–400 nm
SJR5740	S	+	High-aspect-ratio positive resist up to >20 μm thickness, broadband resist, good for electroplating
S1800	S	+	Good general-purpose positive resists, 0.5–3 μm
AZ4562	AZ	+	Thick positive resist
AZ9260	AZ	+	Thick positive resist
AZ5214	AZ	+	Image-reversible positive resist

[a] +/– Signifies a positive or negative resist

Source: MCC: MicroChem Corp., Newton, MA (www.microchem.com); S: Shipley (Rohm & Haas), Marlborough, MA (electronicmaterials.rohmhaas.com); AZ: Clariant Corp., Somerville, NJ (www.azresist.com).

TABLE 1.4
Some Photoresist Suppliers

MicroChem Corp., 1254 Chestnut Street, Newton, MA 02464, USA.

Shipley: Rohm and Haas Electronic Materials, 455 Forest Street, Marlborough, MA 01752, USA.
electronicmaterials.rohmhaas.com

Clariant Corp. AZ Electronic Materials, 70 Meister Avenue, PO Box 3700, Somerville, NJ 08876, USA.
www.azresist.com

Wacker-Chemie GmbH, Hanns-Seidl-Platz 4, 81737 Munich, Germany.
www.wacker.com

GELEST, 11 East Steel Road, Morrisville, PA 19067, USA.
www.gelest.com (for PDMS)

Dow Corning, Midland, MI, USA.
www.dowcorning.com

SHE: Shin Etsu, 6-1, Ohtmachi 2-chome, Chiyoda-ku, Tokyo 100-0004, Japan.
www.shinetsu.co.jp

Futurrex Inc., 12 Cork Hill Road, Franklin, NJ 07416, USA.
www.futurrex.com

Eastman Kodak Company – PCB Products, 343 State Street, Rochester, NY 14650-0505, USA.
www.kodak.com

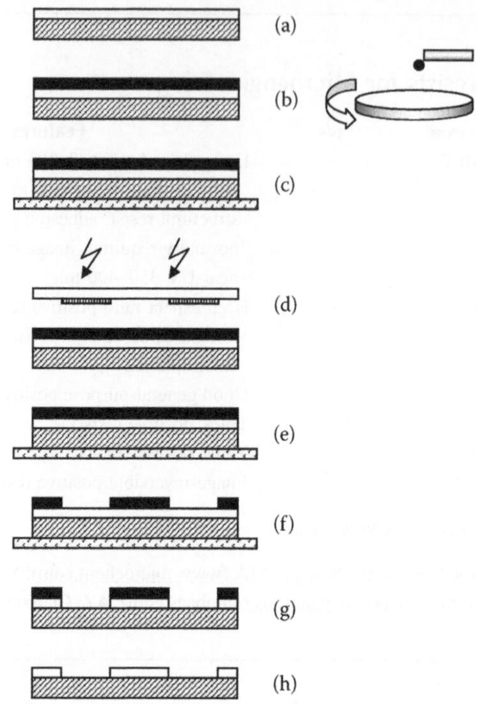

FIGURE 1.10 Steps in photoresist processing (not to scale): (a) clean substrate with film to be patterned, (b) spin-coat with resist, (c) soft-bake on a hot plate, (d) expose, (e) postexposure-bake, (f) develop, rinse, and hard-bake, (g) pattern substrate, (h) resist-strip.

1.2.3.2 Photoresist Processing

Presented in Figure 1.10 is a generic outline for photoresist processing, which proceeds as follows:

Clean
Apply primer or adhesion promoter*
Coat
Soft-bake
Expose
Postexposure-bake*
Develop
Rinse*
Hard-bake*
Pattern substrate*
Strip*

The outline has been based on the slightly more complicated SU-8 guidelines produced by MicroChem Corp [5,6], and some of the steps, marked *, are optional or not required for some resists. For example, SU-8 does not normally require adhesion promoters, positive resists do not normally require a postexposure-bake, the rinse step may be included in recommendations for developing, and the hard-bake may be replaced by a drying step (the hard-bake would ensure completion of polymerization in negative resists and would particularly be employed if the resist were to be a structural component in the design). If the resist is to be a structural component, the normal pattern substrate and strip steps would be omitted.

The Yellow Room

Photoresist processing is normally carried out in a special clean room known as the *yellow room* or *yellow area* because of the yellow lighting that is employed. This is to reduce the chance that the external lighting would age or affect the photoresist.

1.2.3.2.1 Cleaning the Substrate

Within the clean room, there are a number of common facilities and procedures for cleaning substrates prior to and during processing. These are normally wet processes, involving liquid solvents. An outline of these are:

- Organic solvents
  - Acetone
  - Isopropyl alcohol (aka IPA, propanol, propan-1-ol)
- Deionized (DI) water (filtered to remove particles)
- Corrosive cleaning processes
  - Piranha etch/clean
  - RCA clean
- Drying processes
  - Nitrogen
  - Oven (may be supplied with vacuum)
  - Hot plate
- Ultrasonic bath

Acetone is a very common organic solvent and can be used to clean wafers of a variety of contaminants, including the less stubborn photoresists. The substrate is easily dried off with a jet of nitrogen gas. Acetone will leave residual marks on the wafer, which can be rinsed off in IPA, another volatile organic that can be dried off with nitrogen. Acetone and IPA may well be adequate for the final cleaning, but during processing it is undesirable to have organic contamination on the surface of the wafer. Therefore, cleaning steps normally end with a rinse in deionized (DI) water. This has to be dried off on a hot plate or in an oven. Note that intersolvent drying is not required when transferring from one liquid solvent to the next.

One of the most convenient DI water facilities to have is a continuous flow through a tank, with water spilling from one reservoir to the next. This is normally monitored for pH or conductivity. In this way, it is possible to determine when acid (or alkaline) etchants have been fully rinsed from the substrate.

Another common facility is the ultrasonic cleaning bath (also referred to as megasonic when high-frequency ultrasound, above about 800 kHz, is employed). This provides a mechanically assisted cleaning process. The ultrasound induces cavitation, small bubbles, in the solvent. These collapse, releasing mechanical energy. This can facilitate the access of solvents to deep narrow holes (it is also used to enhance etching of such structures), as well as assisting mechanically in the removal of stubborn contaminants (e.g., burnt-on resist). Beware, however, because this is a mechanical process, it can cause damage to delicate microstructures.

Finally, when it comes to mechanical assistance, do not forget the existence of automatic (magnetic) stirrers or the simple expedient of tipping the petri dish containing solvent and substrate.

There are two more extreme processes used in the preparation and cleaning of substrates. Both of these are intended to remove organic and metallic contaminants. Note that the chemicals and procedures involved are very hazardous; if you carry them out ensure that you follow the local clean room or laboratory guidelines.

The piranha clean refers to more or less any combination of sulphuric acid (H_2SO_4) and hydrogen peroxide (H_2O_2) and is carried out at high temperatures, 80 to 120°C. Composition (H_2SO_4: H_2O_2) varies from 50:1 to about 3:1. At the higher hydrogen peroxide concentrations, the mixture is self-heating. The mixture has a short shelf life because of the hydrogen peroxide, and so needs to be used shortly after mixing. It is an oxidizing solution, and may enhance the native oxide film over bare silicon. Cleaning times should be short (on the order of 10 min or so). It is a corrosive mixture, so it may remove more than expected (metals will be attacked, for example).

The RCA clean was named after the Radio Corporation of America (RCA), where it was developed. It is composed of three steps, interspersed with DI rinses and subsequent drying. The first step is intended to remove organic contaminants and is performed between 70°C and about 100°C with a solution of ammonium hydroxide, hydrogen peroxide, and water (NH_4OH : H_2O_2 : DI, 1:1:5) for about 10 min. The second step is a dip etch in dilute hydrofluoric acid, HF (50:1 HF:DI). This removes a thin layer of oxide in which metal ions may have accumulated. Finally, ionic and metal contaminants are removed by a 10-min etch in hydrochloric acid and hydrogen peroxide solution (HCl : H_2O_2 : DI, 1:1:6), again at an elevated temperature. Note that this will also have an adverse effect on non-noble metals (e.g., aluminum, which is a very commonly used metal for conductors on ICs). Also, be aware that the same material deposited by two different methods may be affected differently (see Chapter 2).

1.2.3.2.2 Applying Photoresists

The goal is to achieve a flat even film of a specific, controlled thickness.

Photoresists are normally applied by spinning. The substrate is mounted on the chuck of a spinner, a measured volume of resist is dispensed onto the center of the substrate, and the spinner is accelerated to a predetermined speed. The substrate is then removed and heated on a hot plate or in an oven to drive off the solvent from the resist. The data sheet for the particular resist should provide process details, including spin speeds, and profiles for specific film thickness.

The first complication that may be added to this process is the requirement for priming or application of an adhesion promoter (commonly HMDS — hexamethyldisilizane). Second, it may be desirable to remove the bead of resist that can form around the edge of the substrate. This is normally performed by applying solvent (edge bead remover) to the edge of the substrate while it is still on the spinner.

A further complication comes in applying thick films that may be required either to cover large features (pits, mesas, etc.) on the substrate or as structural elements themselves. The particular problems relate to achieving an even coverage with high-viscosity resists and drying and developing the resist film without cracking or peeling and without trapped gases. A number of the resists listed in Table 1.3 were specifically formulated for this particular problem, and information can be obtained from the manufacturers' data sheets.

Thick-Resist Processing

This has been subject to some discussion on the MEMS e-mail list [7]. The following have been extracted as ideas to be tried out if attempting to apply thick resist films (over difficult topographies):

- Close down the lid of the spinner, if available, to prevent premature evaporation of solvent.
- Flood the substrate with thinner prior to applying the resist.
- Use a hot plate for baking; alternatively, use a vacuum oven (but contact the manufacturer for ideas first — most are very helpful).
- Use an alternative method for applying resists.

There are several alternative methods by which photoresists may be applied to a substrate. The first of these is spraying (e.g., in Reference [8], one source for the equipment required is EV Group, E Thallner GmbH, Schärding, Austria. www.evgroup.com). This can cover quite awkward substrates with thick resist coats.

The next alternative is to apply several coats of resist. Normally the approach would be to apply, image, and postexposure-bake one layer, then repeat for the next layer. Final development of different layers then takes place all in one go. The process does need to be tailored to some extent; it would be undesirable for the solvent in one layer to interfere with the previously imaged layer.

Alternative resist technologies are also available. The PCB industry employs dry-film solid resists than can be used as structural materials in some MEMS applications, and a recent development is the electrodeposited resist. These can be electroplated onto the substrate.

1.2.3.2.3 Postexposure Processing

The postexposure-bake is normally only required for negative resists; its purpose is to drive the cross-linking reactions that harden the resist to the developer. The resist will be developed in an appropriate solution, either in a bath or a spray system. Many developers contain alkali metal ions (such as potassium, K^+, and sodium, Na^+). These can cause problems with subsequent processing and also with the performance of electronic circuits, so the substrate should be cleaned of these contaminants at the earliest possible stage.

A subsequent hard-bake step may be useful for some negative resists, particularly epoxy-based chemistries, to complete the cross-linking reactions following development. This can make the resist very resilient to subsequent processing or suitable as a structural material. It can also make it very difficult to remove when desired. If all else fails, plasma ashing (etching in an oxygen plasma) is normally the last resort (see the section on dry etching in Chapter 2).

1.3 X-RAY LITHOGRAPHY

One of the goals in micromachining and MEMS microfabrication has been the search for high-aspect-ratio microstructures, i.e., structures with a large ratio of height (or depth) to width. With UV exposure systems, this has been limited by the optics of the system and penetration of the UV photons into the resist (SU-8 is a bit of an exception and has found considerable application in microengineering).

The x-ray region of the spectrum begins at wavelengths of greater than 1 nm and extends to wavelengths beyond (less than) 0.01 nm. X-rays are produced by electron bombardment of materials, or electron deceleration, in contrast to gamma rays, which are produced by events within the nucleus of the atom. X-rays with wavelengths of above about 0.1 nm are referred to as soft x-rays, whereas those with shorter wavelengths are referred to as hard x-rays.

X-rays have two advantages when considering high-aspect-ratio structures in thick resists. First, their exceedingly short wavelength means that they are unlikely to be affected by diffraction effects through the mask, unless they are being used for very fine lithography. Second, high-energy x-rays can penetrate into very thick layers of resist, with relatively little attenuation; thus, the resist is evenly exposed through its thickness. The main limit to resolution, in terms of micrometer feature sizes, is that several x-ray sources appear as blurred point sources.

Handy Equations

1. Relationship between frequency f, wavelength λ, and velocity of light c:
   $c = f\lambda$
2. Relationship between photon energy and frequency, where h is Planck's constant:
   $E = hf$
3. Approximate values for:
   - c : 3×10^{8} ms^{-1}
   - h : 6.63×10^{-34} Js

Energies are often given in thousand electron volts, keV. One electron volt is approximately equal to 1.6×10^{-19} J and represents the energy imparted to an electron when accelerated across a potential difference of 1 V in a vacuum.

Although attenuation of x-radiation of specific energy, or wavelength, can be related to the atomic number and density of the material in question, the relationship is not simple and, in practice, discontinuities are found. In addition to materials reference books (e.g., The CRC Handbook [9]), NIST maintains a database of x-ray attenuation coefficients for different materials [10]. The fraction of an incident beam of unit intensity that penetrates through a layer of material of thickness x can be estimated as:

$$I = e^{-\alpha x} \tag{1.7}$$

where α is the mass attenuation coefficient determined for the material employed and the energy of the incident photons. X-ray exposure of resist is effected by electrons that are liberated into the resist when x-ray photons are absorbed. Thus, e-beam resists can commonly be employed as x-ray resists; polymethylmethacrylate (PMMA)-based resists are the most common.

1.3.1 Masks for X-Ray Lithography

The aim is to provide a mask with sufficient contrast. To this end, patterns of highly attenuating material are formed on thin films of highly transmitting material; microengineering techniques are often used to create such masks (Figure 1.11). Gold deposited on a silicon nitride or polyimide film is commonly used as the attenuating material.

A problem with x-ray exposure is that the mask, subjected to very high levels of x-radiation, heats up. Corresponding thermal expansion can then limit resolution and performance of the process, particularly with masks that incorporate polymer membranes. Because x-ray sources are relatively small, many exposure systems incorporate scanning of the mask and substrate across the source, which allows a little time for cooling.

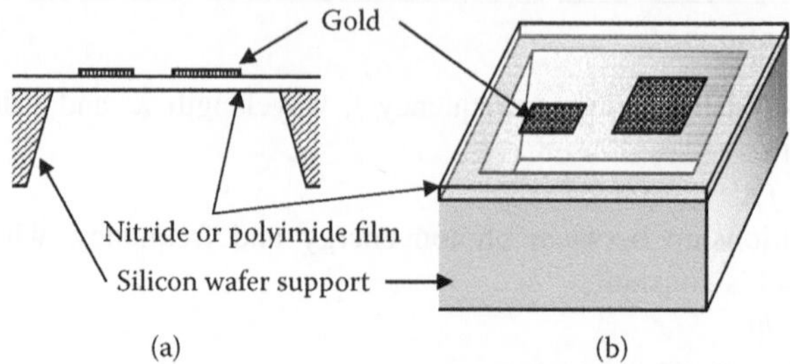

FIGURE 1.11 Example of how a micromachined silicon wafer can be used to form a mask for x-ray lithography (schematic not to scale): (a) cross section, (b) overview.

Overall, x-ray lithography is a relatively expensive process, particularly when considering high-energy x-rays in which a synchrotron x-ray source is required. It tends to be reserved for specialized applications.

1.4 DIRECT-WRITE (E-BEAM) LITHOGRAPHY

Throughout the previous sections of this chapter, masks have remained as rather enigmatic objects; it has been implied that such things exist, but not how they may be produced. A hint was provided with the discussion of projection systems: reduction optics allow for the production of relatively large masks that do not require high design tolerances, which can then be reduced through a projection system to produce the final design. This approach has been used, usually, with two intermediate reduction steps. However, the most common practice in microengineering is to write the design directly onto the mask plate. This cuts out the need to produce expensive intermediate masks that are only going to be used once.

Both UV and e-beam direct-write systems are available, although the latter are more commonly used; UV systems are reserved for situations in which resolution is not critical but flexibility is an important requirement (MEMS development is one such scenario). The basic principle is the same for both, but the mechanisms differ. This section primarily deals with e-beam systems, and the reader is invited to make inferences regarding UV systems.

The de Broglie hypothesis gives the equivalent wavelength of a particle as:

$$\lambda = \frac{h}{p} \tag{1.8}$$

where h is Planck's constant, and p is the momentum of the particle. The momentum can be derived from the voltage used to accelerate the electrons, giving a de Broglie wavelength related to accelerating voltage:

$$\lambda \approx \frac{1.23 \times 10^{-9}}{\sqrt{V}} \tag{1.9}$$

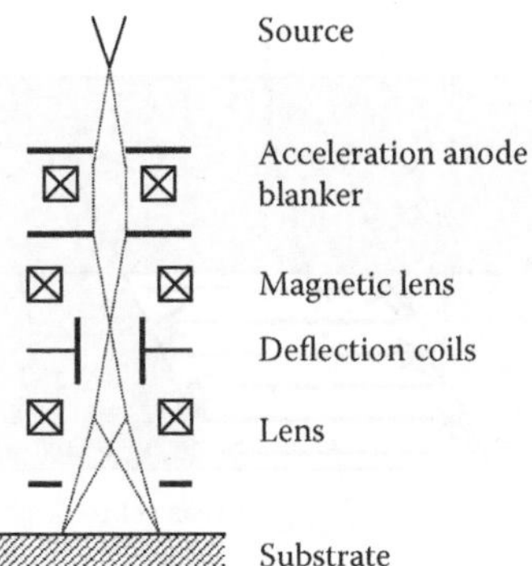

FIGURE 1.12 Elements of a direct-write e-beam system. Details vary with manufacturer. Magnetic lenses and coils are preferred to electrostatic ones because electrostatic lenses and deflection plates have to be placed closer to the beam line.

Typical direct-write e-beam systems employ acceleration voltages of 10 kV to 50 kV (giving corresponding electron energies of 10 keV to 50 keV). From this it can be seen that the wavelength of the electron is much smaller (picometers) than that of UV light (nanometers).

A direct-write e-beam system is outlined in Figure 1.12; note that the system is under vacuum. The electron source has commonly been a heated tungsten wire (a thermionic source), although field emission sources (a needle, again commonly tungsten, with an applied potential difference across it) are now popular as they provide bright point sources; the two approaches are often combined. The emitted electrons are accelerated and focused into a bright spot on the substrate. The position of the spot is controlled by scanning coils, and blanking plates turn the spot on and off by diverting the beam electrostatically. The system is very reminiscent of a scanning electron microscope (SEM), and SEMs can easily be turned into direct-write systems, although the source, optics, and control are not optimized for the job. Similarly, confocal microscopes can be turned into optical direct-write systems, although the choice of resist is limited to broadband resists that are at least partially sensitive to the wavelength of the laser employed (if it is a reddish laser, you will probably be out of luck).

There are two obvious problems immediately apparent to this approach. One is the size of the spot, and the other is the scanning mechanism employed. A third, perhaps less obvious problem, is the high energies employed.

Consider the spot size. Four-inch-diameter wafers are commonly employed in MEMS fabrication, and 6-inch wafers will probably become increasingly common in the near future. To write a mask design to a 4-inch mask plate with a very small spot size will take a very long time. For this reason, it is necessary to specify the design's minimum feature size when mask plates are fabricated; the largest reasonable spot size can be used, reducing the time taken and reducing the cost of the plate. E-beam systems can be focused to spots from about 10 nm to 1000 nm or so.

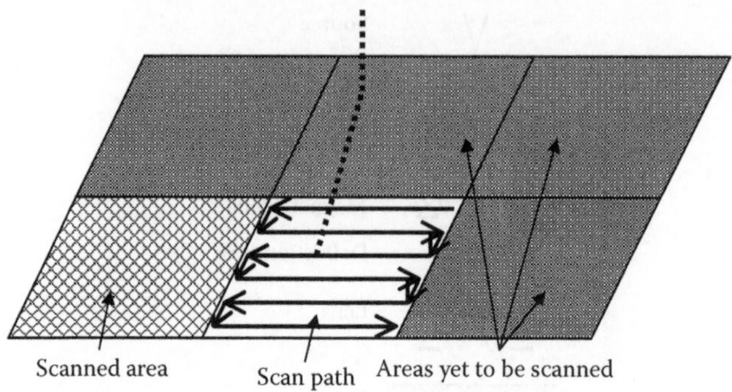

FIGURE 1.13 Raster scanning with e-beam. The beam is scanned over a small area in a zigzag manner and turned on and off to expose different parts of the resist. The substrate is then translated beneath the beam to an unexposed area, and the procedure is repeated until the entire substrate has been written.

The problem of writing to a large area is compounded by the scanning mechanism. The beam is only perpendicular to the photoresist at one point: along the axis of the system. As it is scanned across, it enters the resist at a different angle, giving rise to distortion. This limits the overall area (field) that can be covered before the mask plate being written has to be translated (moved) in a horizontal or vertical direction. The field area will depend on the system, beam diameter and shape, and, eventually, on the design's minimum feature size. However, it is unlikely that it will be on the order of more than a few millimeters.

Figure 1.13 therefore illustrates the writing of a mask plate using a raster-scan mode. A small area is written by scanning the beam back and forth, and the mask plate is translated horizontally and vertically beneath the beam until the entire area has been covered. The system may be modified somewhat by directing the beam specifically to the areas to be exposed and filling these in (vector-scan approach). This is most efficacious if only a relatively small area requires exposure.

One further problem, arising from the use of high-energy electrons, is proximity effects. Electrons have a tendency to scatter at a wide variety of angles (including 180°) when entering a solid. This makes it very important to pay attention to proximity effects when setting up e-beam direct-write systems. This is not a feature of x-ray systems discussed in the previous section, because the electrons liberated by x-ray irradiation generally have energies orders of magnitude lower than those employed in e-beam systems.

1.5 LOW-COST PHOTOLITHOGRAPHY

Because the cost of a mask is set relatively high, particularly, when producing only a few devices for research, several different approaches have been tried to reduce costs. Generally speaking, these are only useful for designs that incorporate one or two masks and have relatively large feature sizes; channels for microfluidics would be a good example. These approaches are:

- Laser printer
- High-resolution printer
- Shadow mask (laser-cut stencil)
- PCB artwork
- Scrounged and modified SEM or confocal microscope

These are best restricted to in-house development work. If a design is to be fabricated by a MEMS foundry, then a conventional approach should be used; discuss your requirements with the foundry or service provider in question.

The first two options are variations on a theme. Transparencies can be produced on a desktop laser printer and used to create masks. These tend to produce designs with ragged edges and large features (100 μm or more). The ragged edges are not so much of a problem if an anisotropic etchant is to be used to create the final structure, notably potassium hydroxide etching of silicon, which is limited by crystal planes and will therefore produce well-defined edges from ragged masks. The preferred route at present, however, is to have the transparency printed by a commercial print shop on a high-resolution printer. Resolutions of several thousand dots per inch can be achieved, with good even coverage, allowing the production of features down to a few tens of micrometers.

The resulting transparency can be taped to a glass plate for use in an aligner. Note, however, that many polymer films tend to be relatively opaque to UV light, and the resulting mask will not be very robust in any case. It may well be preferable to produce a mask plate from the film. Again, several options present themselves. Chromium-coated mask blanks are available from several suppliers; these may be quartz, but given that dimensions are not going to be critical in this case, lower-cost glass alternatives may be used. Probably the most convenient approach is to purchase a presensitized mask blank, i.e., one already coated with photoresist. Mask blanks with other metal film coatings are also available, although perhaps not so readily. These may present an advantage in that a metal etchant may be more conveniently found, depending on the laboratory setup, although other metals may not be as durable as chrome.

Exposure will, again, depend on the particular facilities available. The film will need to be clipped to the mask blank or sandwiched between the blank and a glass plate. UV exposure will not only depend on the photoresist used but also on the absorption of the UV light by the film and other components of the assembly; it may take a long time.

Following development of the resist, it will be necessary to etch the underlying metal layer. A variety of premixed wet etches are commercially available. Alternatively, refer to Chapter V-1 on chemical etching in Vossen & Kern's book *Thin Film Processes* (1978) [11]. This lists several chromium etches from which an appropriate formula can be selected.

The use of printers should enable the production of a mask plate for less than \$100, given prices at the time of writing. One of the drawbacks, as with most of the other approaches listed, is that conventional mask-design software packages cannot be employed because a PostScript output file is normally required. In the past, it has been possible to find freeware or shareware CIF to PostScript converters.

Laser-cut stencils, used in the PCB industry, have already been touched on as possible masks for photolithography. The Gerber file format is commonly used for data interchange within the PCB industry, so a PCB-design software package will probably be required to generate these. PCB design packages can be downloaded from the Internet, but are usually in cut-down trial versions.

A further alternative is to approach a company that produces artwork for the PCB industry. These films are usually fairly robust and capable of fairly high resolution but are relatively expensive.

The last approach, which has also been touched on previously, is to beg, steal, borrow, scrounge, or hijack an SEM or a confocal microscope! This then needs to be adapted to control the scanning and optics from a design file. Both have been done, despite the problems in matching the resist to the available optical spectrum in the confocal, but the approach requires one to be technically confident. Nonetheless, the learning involved in adapting these systems does bring on a considerable understanding of the processes involved and the capabilities of the tool, which can then be used for unusual applications. One example of this would be performing lithography on nonplanar substrates, such as fine capillaries.

1.6 PHOTOLITHOGRAPHY — KEY POINTS

A few important facts and precautions in relation to the process of photolithography are:

1. Mask making is extremely expensive; make sure that the design is correct before embarking on this process. Furthermore,
   a. Ensure that everyone involved in fabrication has reviewed the design files.
   b. Get someone else to review them, too.
   c. Spend adequate time on the process and device design; it will save a lot of money and grief later.
2. If involved in the photolithography process, note the following:
   a. Read the instructions.
   b. Especially, read the instructions when dealing with corrosive or hazardous materials.
   c. The process may involve either a negative or positive resist.
   d. Positive resists are generally capable of higher resolutions than the negative.
   e. Resists are usually applied in liquid form by spinning, then soft-baked to drive off the solvent; an adhesion promoter or primer may be required for some resists.
   f. The process then proceeds with exposure (normally, with a proximity or contact printer), development, hard-bake, etching, and resist-strip.

   g. UV light is employed in the exposure step; short wavelength implies higher resolution.
   h. Thick resist films may require special treatment.
3. Masks are normally produced by direct-write e-beam systems. Masks are normally chrome on quartz or low-expansion glass.
4. X-ray lithography can achieve high resolution and high-aspect-ratio structures, but it is difficult to produce masks.

REFERENCES

1. Levinson, H.J. and Arnold, W.H., Optical lithography, in *Handbook of Microlithography, Micromachining, and Microfabrication — Volume 1: Microlithography*, Rai-Choudry, P., Ed., SPIE Optical Engineering Press, Bellingham, WA, 1997, chap. 1.
2. Peckerar, M.C., Perkins, F.K., Dobisz, E.A., and Glembocki, O.J., Issues in nanolithography for quantum effect device manufacture, in *Handbook of Microlithography, Micromachining, and Microfabrication — Volume 1: Microlithography*, Rai-Choudry, P., Ed., SPIE Optical Engineering Press, Bellingham, WA 1997, chap. 8.
3. Watts, R.K., Lithography, in *VLSI Technology*, 2nd ed., Sze, S.M., Ed., McGraw-Hill, New York, 1988, chap. 4.
4. Reche, J., Photolithography for thin film MCMs, Süss MucroTec Application Note, www.suss.com.
5. MicroChem Corp, NANO SU-8 Negative Tone Photoresists Formulations 2–25, data sheet from MicroChem Corp, Newton, MA.
6. MicroChem Corp, NANO SU-8 Negative Tone Photoresists Formulations 50 and 100, data sheet from MicroChem Corp, Newton, MA.
7. MEMS Clearinghouse: http://www.memsnet.org/.
8. Pham, N.P., Sarro, P.M., and Burghartz, J.N., Spray coating of AZ4562 photoresist for MEMS applications, *Proc. SAFE 2001*, November 28–29, 2001, Veldhoven, The Netherlands.
9. Lide, D.R. Ed., *CRC Handbook of Chemistry and Physics*, 85th ed., CRC Press, Boca Raton, FL, 2004.
10. Hubbell, J.H. and Seltzer, S.M., Tables of x-ray mass attenuation coefficients and mass energy-absorption coefficients, version 1.03, 1997, [Online]. Available: http://physics.nist.gov/xaamdi/. Originally published as NISTIR 5632. National Institute of Standards and Technology, Gaithersburg, MD, 1995.
11 Vossen, J.L. and Kern, W., Eds., *Thin Film Processes*, Academic Press, New York, 1978.

2 Silicon Micromachining

2.1 INTRODUCTION

The concepts of microengineering and MEMS originally grew from the integrated circuit (IC) industry, and the seminal paper commonly referred to by Petersen, "Silicon as a Mechanical Material" [1]. Silicon micromachining remains one of the best developed microengineering techniques and the mainstay of the industry. It possesses the additional advantage that signal processing circuitry can be integrated with MEMS on the same silicon chip (substrate), giving rise to the mass production of cheap, intelligent microsensors and actuators.

There are a large number of different micromachining techniques. Different foundries or laboratories use different sets and combine them in different ways to produce different devices. This chapter, and the one that follows, contrasts with Chapter 1 in that it will only briefly introduce each micromachining technique. When read in conjunction with Part II, it will become apparent how these techniques may be combined. The task for the microengineer is, then, to determine the desired end product and then simultaneously work out a suitable fabrication sequence and find an organization or organizations that can implement it. In the case of a multiuser multiproject process provider, the available fabrication sequence is restricted. An in-house facility may restrict the processes available, whereas if the substrates can be shipped between different service providers, this is not such a problem (instead, one has logistic, yield, and process compatibility problems to overcome).

This chapter, then, introduces silicon and silicon micromachining processes.

2.2 SILICON

Silicon (Si), is an elemental semiconductor that is found in group IV of the periodic table. In its monocrystalline form, it is lighter and harder than steel. Its electrical properties can be altered by introducing impurities, i.e., from elements of groups III and V of the periodic table. Silicon occurs naturally in abundance and also forms hard insulating oxides (glass) and nitrides.

In MEMS, silicon is found in both *monocrystalline* and *polycrystalline* forms. As these names suggest, in the former, all the atoms are aligned in the same crystal lattice arrangement. In the latter case, the material is made up of many smaller crystals of silicon. The silicon wafers used as a substrate for micromachining are all of the monocrystalline form.

Silicon has a face-centered cubic (FCC) unit cell, which is shown in Figure 2.1. The atoms within this cell are, however, arranged in a diamond configuration.

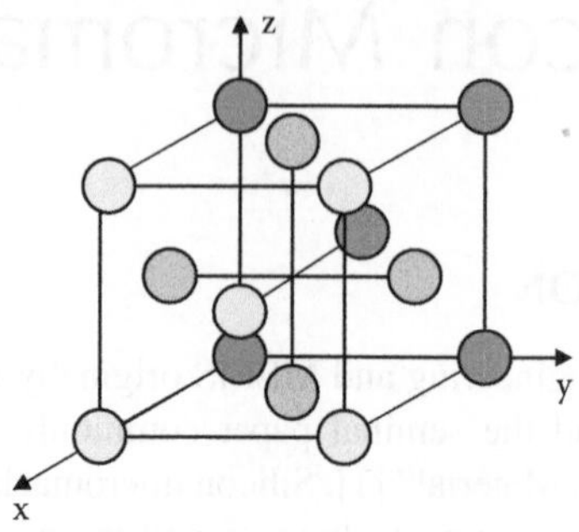

FIGURE 2.1 FCC cell.

This is shown in Figure 2.2a, and Figure 2.2b shows how this arrangement fits into the FCC unit cell. Different crystal planes of the silicon lattice have different densities of chemical bonds between silicon atoms. This is made use of in some micromachining processes in which the etchant attacks different crystal planes at different rates.

Crystal planes and orientations (directions) are identified using the Miller Indices, as described in the following section.

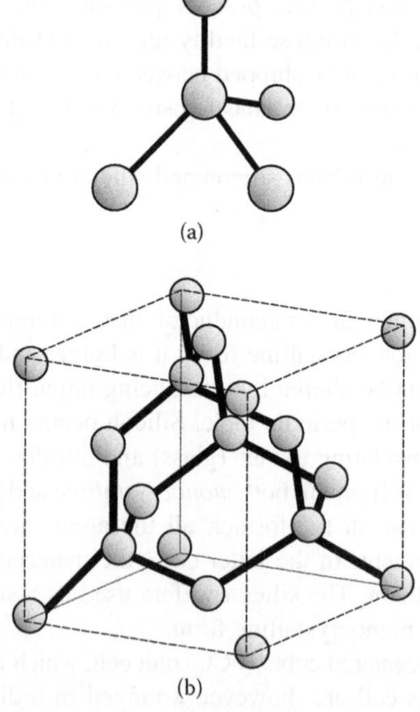

FIGURE 2.2 (a) Tetrahedral arrangement of atoms, (b) Si unit cell.

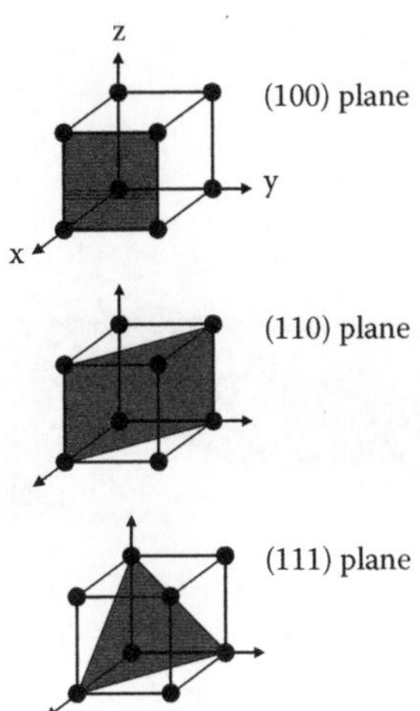

FIGURE 2.3 (100), (110), and (111) planes.

2.2.1 Miller Indices

Descriptions of crystal planes and directions are based on the concept of the unit cell; that is, the smallest repeating unit of the crystal structure. Dimensions are normalized to the length of the edge of a unit cell — thus, a unit cell is a 1 unit ∞ 1 unit ∞ 1 unit cube.

To describe a unit plane, the intercept of the plane on each axis is determined, the reciprocal of the each intercept is found (with the reciprocal of infinity being zero), and this is written in parentheses. Directions are perpendicular to the plane and are written with angular brackets.

Planes: (1/x-intercept 1/y-intercept 1/z-intercept)
Directions: < 1/x-intercept 1/y-intercept 1/z-intercept >

Thus, the <111> direction is perpendicular to the (111) plane. Figure 2.3 shows the most commonly referred to planes, (100), (110), and (111).

Indices are determined on the unit cell, thus higher-order planes are (211) or (411), with the x intercept one half or one quarter of the way along the axis, respectively.

2.3 CRYSTAL GROWTH

To produce silicon of sufficient purity, a process known as *zone refining* is used to purify polycrystalline silicon. The material is passed through a heating coil in a manner that causes a molten zone to pass along the length of the solid bar, taking impurities with it. Monocrystalline silicon ingots are then produced using the *Czochralski process*. Purified silicon is melted in a crucible, and a seed crystal

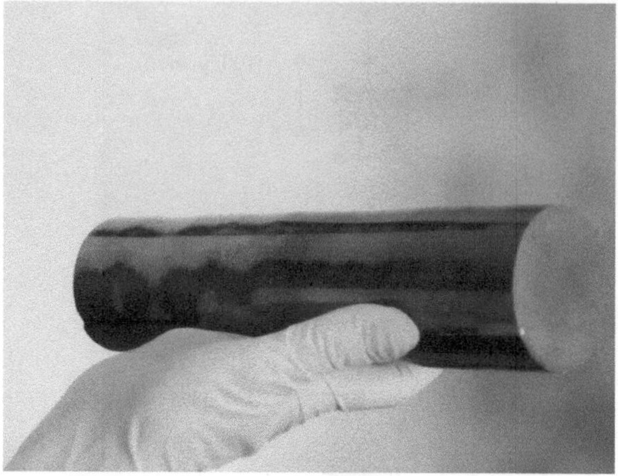

FIGURE 2.4 Silicon bar. (Image courtesy of Compart Technology Ltd, Peterborough, U.K., www.compart-tech.co.uk and Forest Software, Peterborough, U.K., www.forestsoftware.co.uk.)

in a rotating clamp is brought up to the surface of the melt. As the seed crystal is slowly withdrawn from the crucible, it draws out the cooling silicon with it. As this solidifies, it takes on the same crystal structure as that of the seed crystal.

The result is a cylindrical bar or ingot (Figure 2.4) of up to 12-in. (300-mm) diameter. (Note: wafer diameters are often specified in imperial units).

2.4 DOPING

Impurities are normally introduced into the silicon melt to dope it as either a *p-type* or *n-type* semiconductor. In the case of *p-type* semiconductors, a group III element, boron (B), is introduced. Group V elements, phosphorous (P) or arsenic (As), are used to form *n-type* silicon. The introduction of a small proportion of B impurities into the silicon reduces the number of electrons available from carrying current, whereas n-type dopants such as P or As increase the number of available electrons. The physical effects induced by this processing form the basis of electronic components such as diodes and transistors.

Dopant Levels

Silicon is referred to as p-type, p$^+$, or p^{++} (also n-type, n, or n) silicon, depending on the degree of doping. Silicon wafers will be either n- or p-type. Electronic devices will usually be constructed of n-type and p-type layers, with more heavily doped n or p$^+$ regions being used to connect to conductive interconnects. Heavily doped n and p^{++} silicon is highly conductive and not normally used in devices, except as short-conducting tracks.

In the fabrication of both electronic circuits and MEMS, it is desirable to introduce controlled levels of impurities into the silicon substrate in specific areas. The engineer has two basic options for achieving this: thermal diffusion and ion implantation.

2.4.1 Thermal Diffusion

This is normally carried out in a furnace at temperatures in excess of 1000°C. As such, it must be one of the earliest processes engaged in or the temperatures will damage (melt) subsequent parts of the structure. The process is fairly straightforward in concept and consists of the following:

- A layer of high-quality silicon dioxide (thermal oxide, or densified chemical vapor deposition [CVD] oxide) is deposited and patterned to form a mask (the photoresist being stripped).
- The wafer is brought into contact with ceramic tiles rich in the appropriate impurity (a diffusion source). A doped spin-on glass can also be used if deep diffusion of high impurity concentration is not required.
- The wafer and diffusion source are introduced into a furnace heated to sufficient temperature for an appropriate time. (For example, at temperatures of 1175°C an 8-h diffusion will result in > 8-μm-thick structures released by concentration-dependent etching. It will be necessary to use an oxide mask of at least 1-μm thickness in such cases.)
- Following diffusion, the mask needs to be stripped. This is normally a difficult process as the material will have been affected by the diffusion; a wet etch may not suffice, and dry-etching processes (discussed later in this chapter) may have to be employed.

Note that diffusion is an anisotropic process — the impurities diffuse laterally under the mask as well as vertically into the substrate. Diffusion profiles are, therefore, somewhat rounded.

2.4.2 Ion Implantation

In contrast to thermal diffusion, ion implantation is a very precise and isotropic process. Charged ions of the chosen impurity are accelerated towards the substrate. They will reach a depth that can be determined by the momentum that the ion gains through acceleration. Note that ions with a stronger charge can be accelerated more rapidly towards the target and, thus, implanted to greater depths. Nonetheless, ion implantation normally targets only the top 1 μm of the substrate. It is possible to implant ions to a depth of 4 μm into the surface of the substrate, but this can leave it mildly radioactive (it will have to be subjected to a cooling period).

Ion implantation has several advantages over thermal diffusion, such as the following:

- It is carried out at room temperature (but the substrate will get hot unless mounted on a cooled chuck).

- It is isotropic.
- It can be used with far more elements than thermal diffusion (oxygen can be implanted, for instance, to form an insulating layer beneath the surface of the wafer).

One application of ion implantation in electronics is to create self-aligned gates on transistors. This effectively utilizes part of the electrical structure of the transistor as the mask (see Figure 2.7). Ion implantation is not, however, a magical process by which the impurities simply appear in their targeted locations. When passing through the wafer, there is a chance that ions will cause damage to the crystal lattice on the way.

2.5 WAFER SPECIFICATIONS

The first thing to consider when ordering the wafer is the dopant and the degree of doping required. This will normally be done by specifying the resistivity of the material — for example, p-type (boron), 10 to 30 Ωcm.

Silicon ingots are sawn into individual wafers. In addition to the diameter of the resulting wafer, the thickness, crystal orientation, and flats should be specified. Wafers are supplied with diameters of 2, 4, 6, 8, or even 12 in. (50, 100, 150, 200, or 300 mm). Wafers of 2-in. diameters and the equipment required to process them are becoming rare, except for wafers composed enpotic materials. At the time of writing, 12-in. wafers are only available in the most advanced IC fabs, and most MEMS work is performed on 4- and 6-in. wafers.

Wafer suppliers will normally supply wafers with what are regarded as optimal thickness — around 525 μm for 4- or 6-in. wafers. Thicker wafers waste silicon, and thinner wafers stand a greater chance of breaking during processing. It is possible, however, to specify wafers from several millimeters thick to about 10 μm thick (Figure 2.5). Very thin wafers are flexible. During processing these will need to be bonded to a thicker supporting wafer. Very thin wafers are produced by grinding and chemical–mechanical polishing of thicker wafers. Note that the thickness will vary across the wafer because of the natural variations in the mechanical machining process.

Crystal orientation is specified as the plane along which the ingot is cut, and tolerance may also be specified. It is normal to grind flats on the sides of the wafer (see Figure 2.5a). These are specified by the purchaser (e.g., 4-in.-diameter (100) wafer with one (110) flat), although there is a loose convention that p-type wafers have one flat ground and n-type wafers have two. These flats are useful for aligning the wafer for anisotropic etching, but it should be recalled that they are mechanically produced and will not align exactly to the crystal plane.

Finally, single- or double-sided polishing, as appropriate, should be specified. If photolithography is to be performed on both sides of the wafer, then it is necessary to specify double-sided polishing.

Further refinements may be added to the specification. Silicon-on-insulator (SOI) wafers are becoming increasingly popular for MEMS applications. These

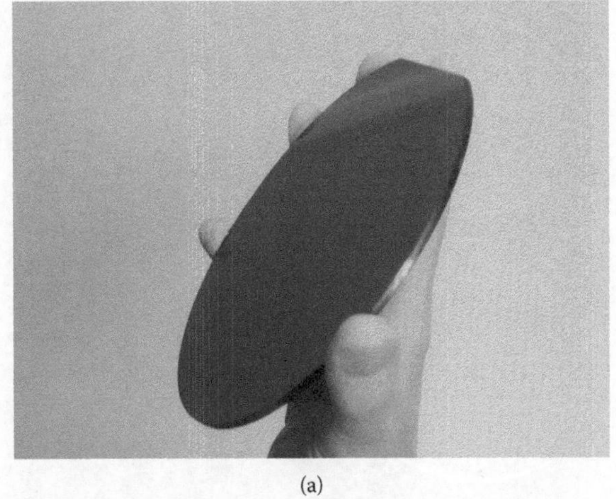

(a)

(b)

FIGURE 2.5 (a) 3-mm thick 6-in. diameter wafer, (b) 10-μm-thick 4-in.-diameter wafer. (Images courtesy of Compart Technology Ltd., Peterborough, U.K., www.compart-tech.co.uk and Forest Software, Peterborough, U.K., www.forestsoftware.co.uk.)

normally consist of a standard-thickness wafer with a thick (user specified — usually about 1 μm) insulating silicon dioxide layer on the surface. A second wafer is bonded to this and etched back to provide a sandwich structure of a few microns of silicon on top of a 1-μm oxide layer on a standard silicon wafer. This is referred to as a bonded and etched-back SOI (BESOI). Other methods of producing SOI wafers include ion implantation (a very thin oxide layer being produced within a micron or so of the surface) and silicon-on-sapphire wafers.

Most suppliers will offer to deposit various thin films over the surface of the wafer. The foremost of which is the growth of an epitaxial silicon layer (epi layer),

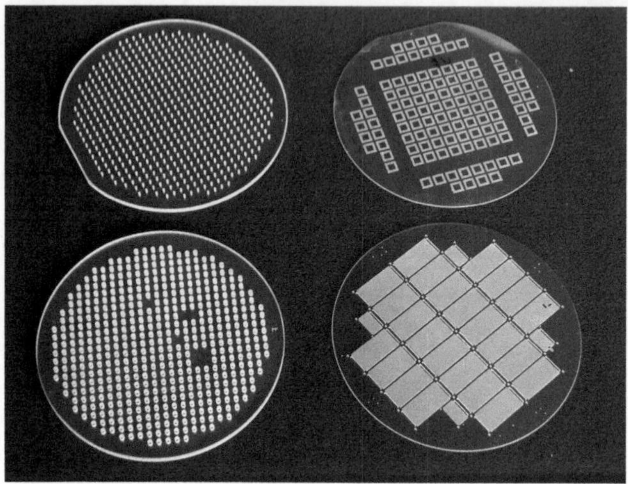

FIGURE 2.6 Machined glass wafers. (Image courtesy of Compart Technology Ltd., Peterborough, U.K., www.compart-tech.co.uk and Forest Software, Peterborough, U.K., www.forestsoftware.co.uk.)

which provides a device-quality surface for circuit fabrication. Silicon dioxide, silicon nitride, and aluminum films are also commonly provided on request.

Wafer suppliers are also normally able to supply glass wafers (Figure 2.6), which are also commonly used for MEMS devices, III–V (three–five) semiconductor wafers (i.e., gallium arsenide, or GaAs, which is used for RF, optical, and high-frequency electronic circuits and, less frequently, for MEMS), sapphire wafers, and other unusual materials. Some can also supply other shapes in silicon, such as cylinders, etc.

Specifying the Wafer

The specifications for a wafer are as follows:

- Dopant — impurity and resistivity
- Diameter, thickness
- Orientation, flats
- Polishing
- Special requirements (e.g., SOI, thin-film deposition)

Remember to specify tolerances to critical parameters.

One final point to recall is that a very thin *native* oxide layer forms on silicon when exposed to air. This can be stripped by dipping the wafers in a wet oxide etch prior to processing, but for critical processes, it may be necessary to perform a sequence of processing steps in an evacuated chamber without breaking the vacuum, for which special equipment is required.

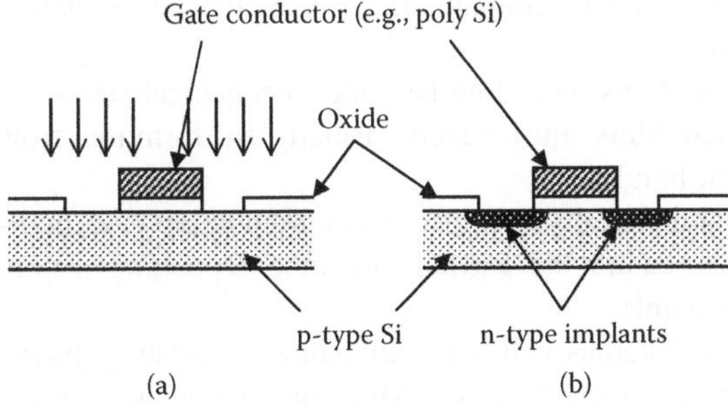

FIGURE 2.7 Self-aligned MOSFET gate (enhancement-type, n-channel): (a) the gate structure acts as a mask for ion implantation, (b) the resulting precisely aligned implants; note that the oxide outside the gate structure would normally be protected by additional resist (not to scale).

2.6 THIN FILMS

There are a number of different techniques that facilitate the deposition or formation of very thin films (on the order of micrometers or less) of different materials on a silicon wafer or other suitable substrate. These films can then be patterned using photolithographic techniques and suitable etching techniques. Common materials include silicon dioxide (oxide), silicon nitride (nitride), polycrystalline silicon (polysilicon or poly), and aluminum.

A number of other materials can be deposited as thin films, including noble metals such as gold. Noble metals will contaminate microelectronic circuitry, causing it to fail, so silicon wafers with noble metals on them have to be processed using equipment specially set aside for the purpose. Noble-metal films are often patterned by a method known as *lift off*, rather than wet or dry etching. Other thin-film materials that are gaining in popularity are very hardwearing such as silicon carbide, diamond-like carbon, and diamond (deposited as a polycrystalline thin film).

Often, a photoresist is not tough enough to withstand the etching required. In such cases, a thin film of a tougher material (e.g., oxide or nitride) is deposited and patterned using photolithography. The oxide or nitride then acts as an etch mask during etching of the underlying material. When the underlying material has been fully etched, the masking layer is stripped away.

2.6.1 Materials and Deposition

There are a variety of different methods that can be used to deposit films of different materials; each method has its strengths and weaknesses, and each method is suited to different materials. It is possible to make the following general points as an introduction:

- Different deposition techniques produce different qualities of films.
- Films may contain impurities.
- Films may contain pin holes.

- Equipment may be contaminated if used for the deposition of different materials.
- Deposited films will often be under mechanical stress.
- Deposited films must adhere (usually by forming strong [covalent] chemical bonds).
- Different materials have different melting points (i.e., high-temperature processes cannot be carried out after depositing a material of low melting point).
- Different materials will have different coefficients of thermal expansion (this may cause cracking, wrinkling, or delamination during fabrication).
- Some deposition processes may coat all exposed surfaces (i.e., be conformal); others may not coat vertical sidewalls at all (this being described as the degree of "step coverage").

Because the properties of the deposited material are so dependent on the deposition process, it is common to use both the name of the process and the name of the material together; thus: LTO, meaning low-temperature oxide, aka LPCVD oxide, is a film of silicon dioxide deposited by the low-pressure chemical vapor deposition (CVD) technique. Additional comments on thin-film materials will, therefore, be left until after discussion of the deposition processes. Table 2.1 introduces the most common thin-film materials that can be found in silicon fabs.

TABLE 2.1
Common Thin-Film Materials

Chemical Symbol	Full Name	Abbreviated Name	Comments
SiO_2	Silicon dioxide	Oxide	An electrical insulator
Si_3N_4	Silicon nitride	Nitride	An electrical insulator
	Polycrystalline silicon	Poly or polysilicon	Silicon film that is made up of multiple crystalline regions at different orientations to each other (cf. the monocrystalline silicon wafer — all atoms aligned in a single lattice); this is a poor electrical conductor and is usually doped to improve its conductance
Al	Aluminum		
	Noble metals		Gold (Au), platinum (Pb)
	Other metals		Tantalum (Ta), tungsten (W), chrome (Cr), titanium (Ti); Ta and W are sometimes used as conductors, more often to form conductive metal-silicides (more conductive poly); Ti is used as a conductor, but also with Cr as an adhesion layer or barrier layer for noble-metal films

2.6.1.1 Depositing Thin Films

Common deposition processes are shown in Table 2.2 along with some comments. Thermal diffusion has also been included for comparison.

In most of the processes described, the thickness of the film is mainly determined by the time taken in depositing it (deposition time).

2.6.1.1.1 Thermal Oxidation

Thermal oxidation can only be applied to exposed silicon. The substrate is immersed in a furnace at a temperature of above 1000°C in an oxygen-rich atmosphere. Steam may also be introduced (wet thermal oxidation). A chemical reaction takes place at the surface of the wafer, whereby silicon is converted to silicon dioxide. This produces a very-high-quality conformal film, but because the oxygen molecules have to diffuse through a thickening layer of silicon dioxide before they can react with the silicon, the process is very slow. The thickness of the resulting film can be controlled down to 10 nm or so, but films in excess of a few 100 nm are unusual because of the high temperatures and slow growth rate. Notice that the film is not deposited on the surface of the silicon; as it forms (grows), then the underlying silicon is converted into the film itself. Thermal oxide films used as sacrificial layers can produce very small structures.

2.6.1.1.2 Chemical Vapor Deposition

CVD in its various forms produces a film by reacting with precursor gases in a chamber. The product of this reaction is deposited on the substrate as a thin film. There are two common derivative forms of CVD: low-pressure CVD (LPCVD) and plasma-enhanced CVD (PECVD), which achieve the results through slightly different approaches. The kinetics, chemistry, and different reaction systems are not dealt with here, and the reader is referred to more detailed texts [2].

All three forms are capable of depositing the basic insulators: silicon dioxide (SiO_2, or oxide) and silicon nitride (Si_3N_4, or nitride). Polycrystalline silicon (polysilicon, or poly) can be deposited by CVD at medium to high temperatures, although LPCVD processes are commonly used, and it is also possible to deposit epitaxial silicon layers by CVD. PECVD can be used to produce a form of polysilicon contaminated with hydrogen; this has found application in solar cells and similar devices. Generally speaking, the higher-temperature processes produce higher-quality films. LPCVD oxide can be enhanced by densification — heating to high temperatures in a furnace in an oxygen or wet oxygen atmosphere. PECVD films (oxide, nitride, and poly) are normally contaminated with considerable amounts of hydrogen. This reduces their qualities as electrical insulators and makes them etch faster. On the other hand, PECVD films normally grow faster than LPCVD, which, in turn grow faster than CVD, which is faster than thermal oxidation. So, PECVD can normally be used to deposit relatively thick films (microns).

A further factor limiting film thickness and the structures that can be created is mechanical stress in the deposited films. Too much stress will lead to the structure buckling or the films' wrinkling or cracking. High-temperature nitride films have a

TABLE 2.2
Deposition Processes

Process	Film Quality	Temperature	Step Coverage	Comments
Thermal oxidation	Very good	High	Conformal	Oxide only, slow, usually submicron thickness
Chemical vapor deposition (CVD)	Good	Medium–high	Medium (material and process dependent)	Normally inorganic films only
TEOS-based CVD	Good	Medium	Conformal	For oxide films only
Low-pressure CVD (LPCVD)	Medium	Low	Medium (material and process dependent)	Used for inorganic, sometimes organic films; can produce thicker films (microns)
Plasma-enhanced CVD (PECVD)	Low	Low to room temperature	Medium (process and material dependent)	May be used for organic and inorganic films; inorganic materials often contaminated with hydrogen; can produce very thick films
Sputtering	Good (depends on material)	Low	Medium to poor	May be used for a variety of materials
Evaporation	Good	Low to room temperature	Poor	Usually used for metals and e-beam and thermal varieties
Spinning	Usually good but uneven	Material dependent	Usually good	Usually produces relatively thick films; used to apply photoresist and a variety of materials
Thermal diffusion		High		See Subsection 2.4.1

Note: Temperature: high — 1000°C, medium — 500–600°C, low — 200°C; in VLSI processing, 500–600°C are low temperatures; step coverage medium: can cover step heights of greater but comparable depth to the film thickness.

particular problem: they exhibit high tensile stresses within the film and cannot be deposited directly onto silicon. A stress-relieving layer of oxide is required. It is not advisable to attempt PECVD nitride deposition directly onto silicon, unless this particular process has been very well characterized for MEMS applications. Note that processes that produce the best electronic devices do not necessarily produce the best mechanical devices. It is possible to control the stress in films by altering the deposition parameters or the composition of the resulting film (using PECVD to deposit a hydrogen-contaminated silicon oxynitride layer, for instance). The mechanical, electrical, and chemical (etching) properties of the film will all be affected by the deposition parameters used, so it is necessary to carefully characterize and monitor each process (a demanding and time-consuming job) or seek out a foundry that has experience with the processes required for the device under development.

Oxide vs. Nitride

Oxide films are excellent electrical insulators, easily deposited, and easily etched (commonly wet-etched in buffered HF; see Subsection 2.6.2). Nitride films have higher stress but are mechanically harder and chemically more resilient (to attach to and with respect to diffusion of ions or moisture). PECVD nitride films are extensively used as protective coatings.

One oxide CVD process, TEOS has become quite popular. This is a LPCVD process based on tetraethoxysilane (i.e., TEOS) and produces high-quality conformal oxide films.

CVD processes are quite versatile. LPCVD is often used to deposit other inorganic films, such as silicon carbide, tungsten, and metal silicides. Carbon films that range from polycrystalline diamond (CVD processes) to diamond-like carbon films (PECVD) can be deposited. PECVD, in particular, is being exploited to deposit polymeric films (for example, parylene). A further form of CVD, metalorganic CVD (MOCVD), is used to deposit III–V semiconductors (these will not be dealt with here as they are not commonly used in MEMS as yet).

Signs of Stress

Stress in films may cause one of the following several problems:

- Cracking or wrinkling of the film
- Strings peeling off from sharp corners
- Twisting or buckling of structures (particularly, cantilever beams)
- Buckling of the silicon wafer (in extreme cases)

2.6.1.1.3 Sputter Deposition

The sputter deposition process is performed in a chamber at low pressures and temperatures. A target consisting of the material that is to be deposited is placed above the substrate, and a plasma of inert gas (argon) is formed in the chamber.

The ions of the plasma are accelerated towards the target, where they knock atoms from its surface. Some of these displaced atoms make their way to the substrate where they settle, forming a thin film with a chemical composition and structure which approximates that of the target. Deposition rates are controllable and can be relatively high. Sputtering is commonly used to deposit metal films and, less commonly, to deposit simple inorganic compounds.

2.6.1.1.4 Evaporation

The substrate is placed in a chamber opposite a source (target) of the material that is to be deposited. The chamber is evacuated, and the material is heated to form a vapor in the chamber, which condenses on the substrate (and the walls of the chamber, etc.). The heating is normally effected by a filament (or sometimes inductive heating of a crucible) or an electron beam, giving rise to thermal or e-beam evaporation processes. As a result, the materials involved are usually limited; evaporation is commonly used to deposit elemental metals, particularly the noble metals. It is important to select an appropriate combination of source, filament, crucible, etc., to avoid contamination problems (facility manuals or the literature, e.g., Vossen and Kern [2], should be referred to if in doubt). This is a line-of-sight process, so step coverage is usually very poor, but it is at a low temperature and (depending on the material) does not usually result in high-stress films.

High-purity targets and sources for sputtering or evaporation are readily available from specialist suppliers.

2.6.1.1.5 Spinning

The section on photoresist processing in Chapter 1 introduced spin-coating as a method for applying films of photoresist prior to processing. By varying the solvent content and viscosity of the film, and the spin speed and profile, it is possible to apply films from less than 1 μm thick to 100 or 200 μm thick in a single step. It is even possible, with some trial and error, to form a reasonable coating on fairly rough (micromachined) substrates, although spray-coating is generally preferred.

The spin-coating process can be adapted to apply to a variety of different materials. Of particular interest in micromachining work is the application of solgels. These are suspensions of very fine (nanometer dimension) ceramic in a liquid. A film is applied by spinning, and then the substrate and applied film are heated in an oven or furnace to drive off the liquid and to melt or fuse the film into a continuous ceramic layer. This approach is used, particularly, for the application of low-melting-point glasses (these formulations may be referred to as spin-on glass [SOG]) and piezoelectric materials (e.g., PZT). Coatings applied in this manner will typically be a few microns thick, and uniformity and consistency of coating can be a problem.

2.6.1.1.6 Summary

Table 2.3 summarizes the deposition methods commonly used for various common thin-film materials (note that not all deposition methods are listed for each material).

Annealing

One process often referred to is annealing. Once a film has been deposited, the workpiece is heated in a furnace to a particular temperature for a particular time, usually in an inert atmosphere, and then cooled (sometimes, in a controlled manner). This changes the properties of the film; the atoms rearrange, usually, into a more ordered structure, and some more volatile compounds (e.g., hydrogen in PECVD films) get driven off to some extent. The resulting film is usually of a better quality and may have fewer pinholes, and the stress characteristics may have changed. Annealing may involve high temperatures, and therefore it cannot always be employed.

TABLE 2.3
Summary — Materials and Deposition

Oxide	Thermal oxidation LPCVD PECVD	Insulating, masking, and sacrificial material.
Nitride	PECVD LPCVD	Insulating and masking material, often exhibits high internal stress
Silicon carbide	CVD	Very hard material; not yet common in MEMS
Silicon	LPCVD MBE	Polysilicon used as a structural and conducting material (normally phosphorous doped to enhance conductivity); MBE is molecular beam epitaxy, used to create high-quality thin layers of Si with the same crystal structure as the underlying wafer (for electronic devices)
Refractory metal silicides	LPCVD PECVD	Effectively, highly conductive metal-doped polysilicon; Ta, W, and Ti common; may require annealing
Al	Evaporation sputter	Most commonly used conducting layer; CVD techniques also used
Noble metals	Evaporation sputter	Gold and platinum common; these do not adhere well to oxide, and chrome or Ti adhesion layer is commonly used; Au, in particular, will form an eutectic with silicon (an Au-Si alloy)
Other metals	Sputter evaporation	Common metals: Ti, W, Ta, Cr; CVD may also be used
Carbon	CVD PECVD	Polycrystalline diamond can be deposited by CVD techniques; PECVD is used to form an amorphous diamond-like-carbon; diamond MEMS are uncommon

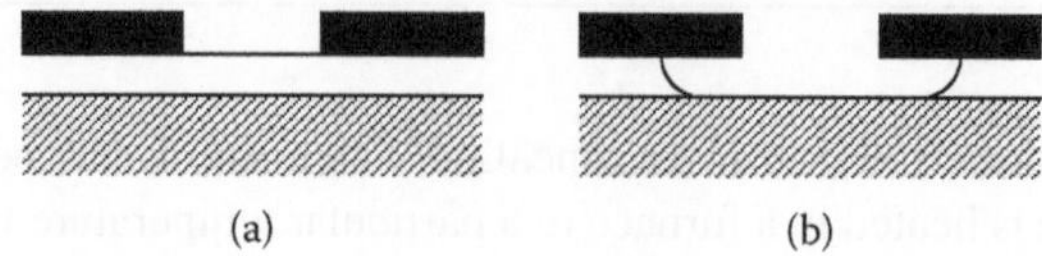

FIGURE 2.8 (a) Oxide film on silicon wafer with developed photoresist mask, (b) after extended wet etching; the etch has progressed beneath the mask (not to scale).

2.6.2 Wet Etching

Wet etching is a general term for the removal of material by immersing the wafer in a liquid bath of the chemical etchant. Wet etchants fall into two broad categories: isotropic and anisotropic etchants.

Isotropic etchants attack the material being etched at the same rate in all directions. Anisotropic etchants attack the (material) silicon wafer at different rates in different directions, and therefore there is more control of the shapes produced. Some etchants attack silicon at different rates depending on the concentration of the impurities in the silicon (concentration-dependent etching).

Isotropic etchants are available for oxide, nitride, aluminum, polysilicon, gold, and silicon. Because isotropic etchants attack the material at the same rate in all directions, they remove material horizontally under the etch mask (undercutting) at the same rate as they etch through the material. This is illustrated for a thin film of oxide on a silicon wafer in Figure 2.8; here, an etchant that etches the oxide faster than the underlying silicon (e.g., hydrofluoric acid) is used. The characteristics of some of the more common wet etchants are listed in Table 2.4. The interested reader can find a number of comprehensive reviews of the topic, such as that by Williams and Muller [3] and Kern and Deckert [4].

TABLE 2.4
Common Etch Rates

	Si	LPCVD Oxide	PECVD Oxide	LPCVD Nitride	PECVD Nitride	Al
10:1 HF		Fast	Very fast	Medium	High	Fast
BHF		Fast	Very fast	Very slow	Slow	Fast
Phosphoric acid (hot)			Very slow	Medium	Fast	Fast
HF-HNO_3–CH_3COOH	Fast	Medium	Fast	Very slow	Very slow	Fast

Note: Observe relation between speed and use for masking.

Hydrofluoric acid (HF) is commonly used to etch oxides. The stronger concentrations are used to rapidly strip oxides or as a dip etch, whereas when buffered with ammonium fluoride it is used to pattern oxides. The latter is known as buffered HF (BHF) or buffered oxide etch (BOE); its etch rate is more controlled because of the buffering, and it does not peel photoresist as does more concentrated HF. Phosphoric acid in various combinations with other compounds is used to etch either nitride or aluminum; oxide is commonly used as the mask (phosphoric acid attacks photoresist). Isotropic silicon etchants are based on hydrofluoric and nitric acids in combination with either methanol or water. These can be formulated for etch rates that vary from polishing to wafer-thinning applications.

Note that most etches will strip aluminum from a wafer, including the Piranha etch, which was introduced in Chapter 1 for precleaning substrates. Gold is commonly etched with an iodine-based solution, but noble metals are also etched by *aqua regia*, a mixture of hydrochloric and nitric acids (3:1).

Most wet etches are purchased premixed directly from specialist dealers because handling them is very dangerous. Notice also that some wet-etching processes have to be performed at high temperatures or under reflux conditions (i.e., the etching solution is boiled in an apparatus fitted with a condenser so that no vapor is lost to the environment). Concentrations are normally given in ratio by volume of standard (as supplied) components. Percentage values are normally given by weight. This 10:1 HF is in fact ten parts of HF to one part water, the HF being supplied as 49% by weight HF (the remainder being water). Water in a clean room will normally be deionized (DI) and filtered.

Postetch Rinsing

Immediately upon the completion of wet etching, the wafers are rinsed in DI water. This is normally performed in a series of three basins connected by small waterfalls. DI water is continually supplied so that the overspill from the first flows into the second, and the second flows into the third. The output is monitored by either a conductivity or pH meter, and flows into an acid drain. The etched wafers are placed first in the basin nearest to the inlet and then moved forward.

Anisotropic etchants that etch different crystal planes in silicon at different rates are available. The most popular anisotropic etchants are potassium hydroxide (KOH) and tetramethyl ammonium hydroxide (TMAH). Common anisotropic etchants are compared in Table 2.5.

The simplest structures that can be formed using KOH to etch a silicon wafer with the most common crystal orientation (100) are shown in Figure 2.9. These are V-shaped grooves or pits with right-angled corners and sloping sidewalls. Using wafers with different crystal orientations can produce grooves

TABLE 2.5
Anisotropic Etchants Compared

KOH	Potassium hydroxide solution. 200:400:1 (100):(110):(111) etch ratios. Attacks aluminum, etches oxide (especially PECVD) slowly, nitride (LPCVD) preferred for masking. Carried out at 80C. Etch rate drops by 3 orders of magnitude for a boron concentration above 1.5×10^{20} atoms/cm^3.
TMAH	Tetramethyl ammonium hydroxide. 20:1 (100):(111), depending on dopant. Alkali ion free (potassium and sodium ions can damage the performance of electronic components, although many photoresist developers involve them). Al not attacked. Oxide may be used as an etch mask.
EDP	Ethylene diamine, Pyrocatechol and water. 50:30:3 (100):(110):(111). No alkali ion contamination. Attacks Al slowly. Oxide may be used as etch mask. Performed under reflux. Etch stop at boron concentrations > 7×10^{19} atoms/cm^3. Very dangerous.
Hydrazine	25:11 (100):(111). No alkali ion contamination. No attack on Al. Very dangerous.

Note: "Oxide" usually means good quality, densified LPCVD oxide, not PECVD. Nitride can also be used.

or pits with vertical walls. Both oxide and nitride etch slowly in KOH. Oxide can be used as an etch mask for short periods (i.e., for shallow grooves and pits). For longer periods, nitride is a better etch mask as it etches more slowly in the KOH.

High levels of boron in silicon will reduce the rate at which it is etched in KOH by several orders of magnitude, effectively stopping the etching of the

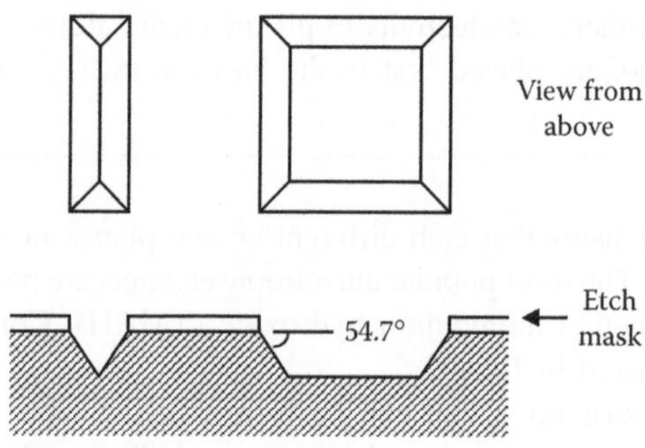

FIGURE 2.9 V grooves and pits etched by KOH (not to scale).

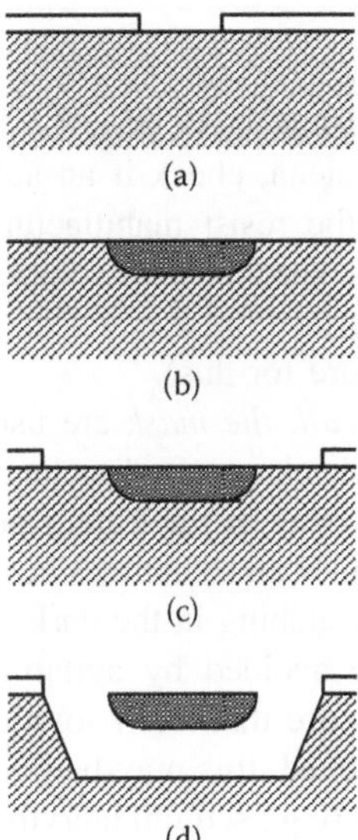

FIGURE 2.10 Concentration-dependent etch process: (a) mask for boron diffusion, (b) oxide mask stripped following diffusion, (c) mask for KOH etch, (d) boron-doped structure released by KOH etch (cross section, not to scale).

boron-rich silicon. This is termed *concentration-dependent etching*. The boron impurities are usually introduced into the silicon by diffusion. A thick oxide mask is formed over the silicon wafer and patterned to expose the surface of the silicon wafer onto which the boron is to be introduced (Figure 2.10a). The wafer is then placed in a furnace in contact with a boron-diffusion source. Over a period of time, boron atoms migrate into the silicon wafer. Once the boron diffusion is completed, the oxide mask is stripped off (Figure 2.10b).

A second mask may then be deposited and patterned (Figure 2.10c) before the wafer is immersed in the KOH etch bath. The KOH etches the silicon that is not protected by the mask and etches around the boron-doped silicon (Figure 2.10d).

Boron can be driven into the silicon as far as 20 μm over periods of 15 to 20 h; however, it is desirable to keep the time in the furnace as short as possible. With complex designs, etching the wafer from the front in KOH may cause problems when slow-etching crystal planes prevent it from etching beneath the boron-doped silicon. In such cases, the wafer can be etched from the back; however, this is not without disadvantages (longer etching times, more expensive wafers, etc.). The high concentration of boron required means that the microelectronic circuitry cannot be fabricated directly on the boron-doped structure.

Improving Results

Ragged lines are usually a symptom of poor masking or poor mask adhesion. Make sure the substrate is clean, check if an adhesion promoter is required for the photoresist, check the resist manufacturer's guidelines, check with shorter etch times, and make sure the etch solution is not contaminated. With isotropic silicon etchants, surface finish can usually be improved by adding to the recipe (check the literature for this).

Cavities appearing beneath the mask are usually due to pinholes in the mask. If using photoresist, check as stated earlier. If a thin-film mask is being used, try and improve the quality of the film or perform multiple depositions. Try an alternative film (chrome can sometimes be used). With some anisotropic etchants (KOH or EDP), try etching in the dark.

Uneven etching can be avoided by agitating, stirring, or bubbling air through the etch bath to ensure that fresh solution circulates. If small holes are not being completely etched, this may be due to poor solution access — surfactants added to the etch (e.g., sodium lauryth sulfate) or use of ultrasound (particularly if the process develops gas).

2.6.3 Dry Etching

There are various etching processes that are carried out in a chamber at low pressure, using either inert or reactive gasses. The two principle advantages for MEMS processing are the following:

- Higher selectivity of target material over masking material
- No problems with surface tension causing microstructures to bend and adhere

There are two main classes of dry etching: reactive ion etching (RIE), which involves chemical processes, and ion beam milling, which involves purely physical processes.

2.6.3.1 Relative Ion Etching

This is the most common form of dry etching for micromachining applications (a summary can be found in Williams and Muller [3], and the processes are described in detail in Vossen and Kern [2]). A plasma of reactive ions is created in a chamber, and these react chemically with the material to be etched. In its most basic form (embodied as the barrel etcher), RIE is an isotropic etch; however, it is more often used as an anisotropic etch. In this form, the reactive ions are accelerated toward the material to be etched, and etching is enhanced in the direction of the travel. Deep trenches and pits (up to several microns) of arbitrary shape and with vertical walls can be etched in a variety of materials including silicon, oxide, and nitride. Unlike anisotropic wet etching, RIE is not limited by the crystal planes in the silicon.

There has been considerable development of deep RIE (DRIE) processes for MEMS, aimed at creating structures with vertical sidewalls and high aspect ratios (the height-to-width ratio). The most successful of these has been the Bosch process [5]. This involves repeatedly changing the system over from RIE to CVD functions. Following a period of etching, a layer of polymer is deposited to protect the sidewalls of the structure from further etching leading to an extremely anisotropic etch process capable of creating structures of several tens of microns deep in silicon with very vertical sidewalls.

Incomplete Etching

As with wet etching, the gases of the RIE process need to gain access to the material to be etched. As a result, very-narrow-diameter holes may etch more slowly than larger holes. For this reason, it is desirable to over-etch to ensure that all structures have been etched completely. The point at which a plasma etcher finally etches through one layer to the one beneath can be identified as the plasma changes color.

Oxygen RIE is increasingly used in MEMS fabrication. In the first instance, it is used as an isotropic etch to strip polymer films — the ashing process. Polymer films used as sacrificial layers can be readily removed by plasma ashing. Its second application is to modify the surface of problem materials to improve adhesion, a brief exposure to oxygen plasma can help considerably with films such as polytetrafluoro ethylene (PTFE). Note, however, that exposure to oxygen plasma will also have the effect of oxidizing other exposed surfaces, such as silicon. Argon plasma may also be used to roughen surfaces without chemically changing them.

Doped Oxides

Phosphosilicate glass (PSG) and borosilicate glass (BSG) have different etching characteristics to (pure) silicon dioxide when etched by reactive means (both wet and dry). PSG tends to etch a lot more rapidly than plain oxide, and hence can be useful as a sacrificial material.

2.6.3.2 Ion-Beam Milling

This process uses inert ions accelerated from a source to physically remove material. There are two forms, showered-ion-beam milling (SIBM) and focused-ion-beam milling (FIBM). The former showers the entire substrate with energetic ions, whereas in the latter ions are focused to a spot that is directed to a particular part of the workpiece.

SIBM can be used as much as RIE, although it is generally slower and more controlled. FIBM can be used to trim structures to dimensions of approximately 10 nm.

Dry Etching and Sputtering

One question that could be asked is where does all the material go that is removed through physical processes? If the dry-etching process is not tuned, then the material can be sputtered onto adjacent areas, resulting in the growth of *grass*. As a further aside, note that the sputter-deposition process can also be reversed to achieve a form of dry etching. Also, note that the grass in the area being etched may be caused by dirt on the surface of the substrate acting as a kind of etch mask.

2.6.4 Liftoff

Liftoff is a stenciling technique often used to pattern noble-metal films. There are a number of different techniques; the one outlined here is an assisted-liftoff method.

A thin film of the assisting material (e.g., oxide) is deposited. A layer of resist is put over this and patterned, as for photolithography, to expose the oxide in the pattern desired for the metal (Figure 2.11a). The oxide is then wet-etched so as to undercut the resist (Figure 2.11b). The metal is then deposited on the wafer, typically by evaporation (Figure 2.11c). The metal pattern is effectively stenciled

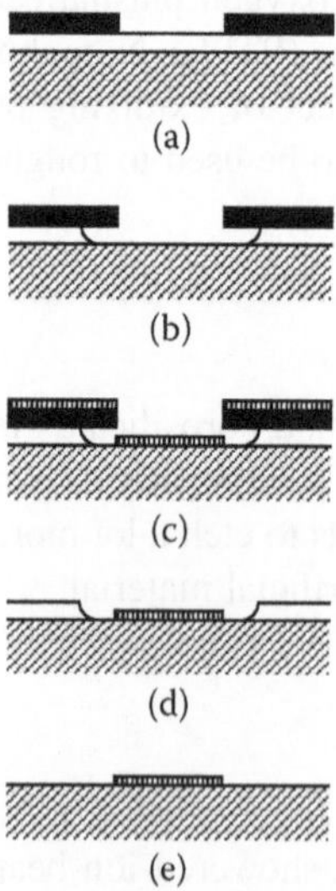

FIGURE 2.11 Oxide-assisted liftoff: (a) photoresist pattern developed, (b) oxide wet-etched to undercut the resist, (c) metal evaporated on, (d) resist strip, (e) oxide strip.

through the gaps in the resist, which is then removed, lifting off the unwanted metal with it (Figure 2.11d). The assisting layer is then stripped off too, leaving the metal pattern alone (Figure 2.11e).

In the assisted-liftoff method, an intermediate layer assists in the process to ensure a clean liftoff and well-defined metal pattern. When noble metals are used, it is desirable to deposit a thin layer of a more active metal (e.g., chrome) first to ensure good adhesion of the noble metal. There are liftoff techniques in which only a (negative) photoresist is used as the stencil, and special liftoff resists are becoming available. One further method that is used to achieve the negative (overhanging) sidewalls required for liftoff is to use an *image-reversal* process. This enables a positive photoresist (AZ5214) to be used; the exact process varies from laboratory to laboratory. Following the initial exposure step, the acid produced as part of the resist chemistry is neutralized near the surface using ammonia or is driven off in a second baking step, and the entire resist is then flood-exposed to UV and developed.

2.7 STRUCTURES IN SILICON

Silicon microstructures are formed by combining the aforementioned techniques in various ways. Most structures in silicon are fabricated using either *bulk* or *surface micromachining* processes. Bulk micromachining involves the selective removal of material from the bulk of the silicon wafer to form the desired structure. Surface micromachining involves the creation of microstructures by the successive deposition and patterning of sacrificial and structural layers on the surface of the silicon wafer. This section considers some of the simpler structures that can be created. Part II considers how these can be combined to create more complex devices.

2.7.1 Bulk Silicon Micromachining

One of the simplest possible and most obvious structures is the patterning of insulated electrical conductors. One possible application of this could be to use electric fields to manipulate individual cells.

2.7.1.1 Pits, Mesas, Bridges, Beams, and Membranes with KOH

Anisotropic etching with KOH can easily form V-shaped grooves or cut pits with tapered walls into silicon (Figure 2.9 and Figure 2.12). Notice that because KOH etching is anisotropic, arbitrary mask openings will eventually become limited by specific planes. This is illustrated in Figure 2.13.

KOH can also be used to produce mesa structures (Figure 2.14a). When etching mesa structures, the corners can become beveled (Figure 2.14b) rather than right-angled. This has to be compensated for in some way. Typically, the etch mask is designed to include additional structures on the corners.

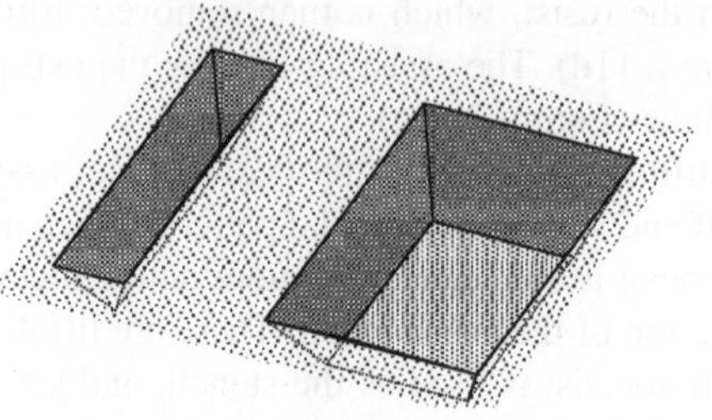

FIGURE 2.12 KOH-etched pit and groove.

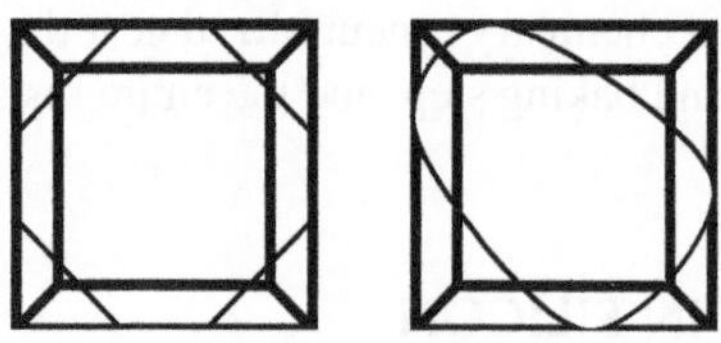

FIGURE 2.13 Illustration of how KOH etching eventually becomes limited by crystal planes, given an arbitrary mask opening (pits viewed from above).

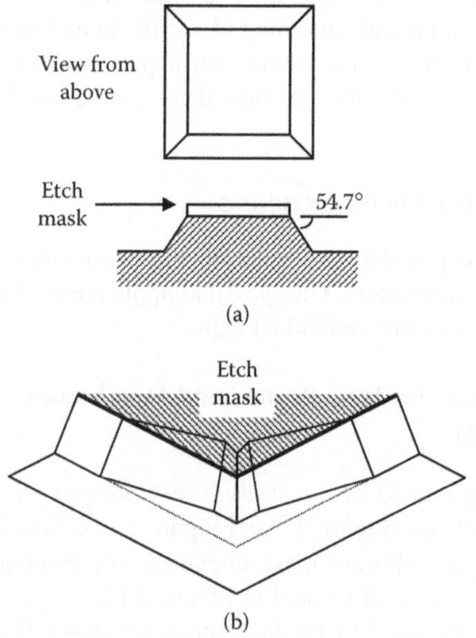

FIGURE 2.14 Mesa structures.

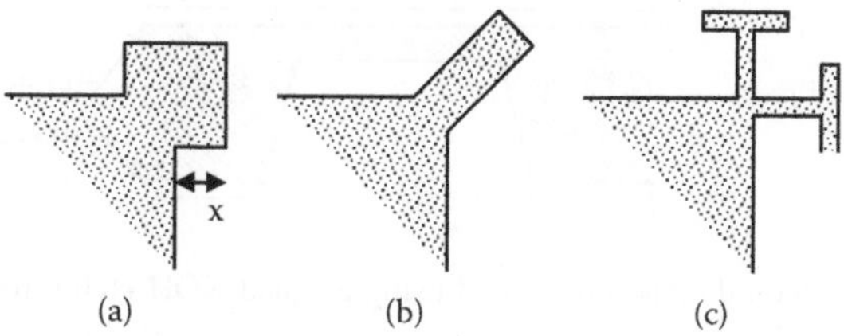

FIGURE 2.15 Corner compensation structures: (a) square, (b) <110> directed bar, (c) T shapes, which may be multiplied (not to scale).

These compensation structures are designed so that they are etched away entirely when the mesa is formed to leave 90° corners. One problem with using compensation structures to form right-angled mesa corners is that they put a limit on the minimum spacing between the mesas.

Some examples of mask designs for corner compensation structures are shown in Figure 2.15. These are aligned with specific crystal planes and are designed to be etched completely away when the desired KOH etch depth has been reached. The <110> aligned structures provide almost perfect corner compensation but tend to be quite large. The exact dimensions of each structure have to be computed, based on the etching conditions experienced in the clean room. Details are provided in papers by Puers and Sansen [6], and Sandmaier et al. [7]. The simplest structure to implement is the square corner compensation structure (Figure 2.15a). The dimensions of this (half of one side of the square) can be approximated by:

$$x = \frac{1.33h - 10}{4} \tag{2.1}$$

$$x = \frac{3.72h + 5.7}{3} \tag{2.2}$$

where h is the etching depth (height of the mesa) and the units are all in micrometers. Equation 2.1 was derived from Puers and Sansen's etch rates experienced with KOH + IPA solution (IPA is isopropyl alcohol, which some find improves the quality of KOH etches), and Equation 2.2 was derived from their results with KOH.

Alignment to Crystal Planes

Do not rely on flats ground onto wafers to provide alignment to crystal planes. These are mechanically ground on and will be aligned to within a specified tolerance error. If more precise alignment is required, then an additional masking and KOH etch step has to be introduced to etch crystal plane alignment marks into the wafer.

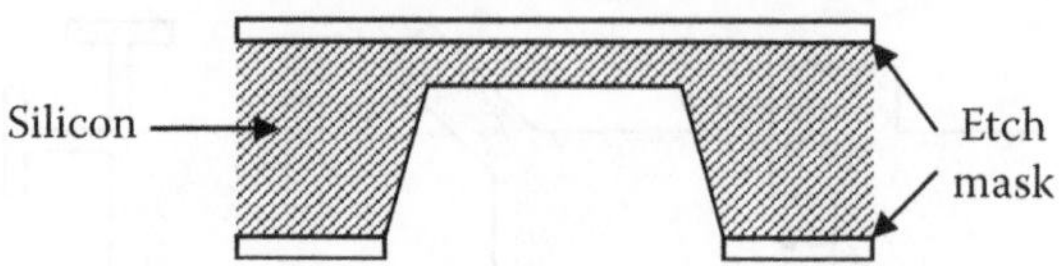

FIGURE 2.16 Silicon diaphragm created using a timed KOH etch (cross section, not to scale).

The silicon diaphragm is the basic structure used to construct pressure sensors and some accelerometers. Silicon diaphragms from about 50-μm and upward thicknesses can be made by etching through an entire wafer with KOH (Figure 2.16). The thickness is controlled by timing the etch and, therefore, is subject to errors induced by variability in the composition of the etching solution, other etching conditions, and uniformity of the wafer.

Thinner diaphragms, of up to about 20-μm thickness, can be produced using boron to stop the KOH etch (Figure 2.17) — concentration-dependent etching. The thickness of the diaphragm is dependent on the depth to which the boron is diffused into the silicon, which can be controlled more accurately than the simple, timed KOH etch.

Concentration-dependent etching can also be used to produce narrow bridges or cantilever beams. Figure 2.18a shows a bridge, defined by a boron diffusion, spanning a pit that was etched from the front of the wafer in KOH. A cantilever beam (a bridge with one end free) produced by the same method is shown in Figure 2.18b.

The bridge and beam in Figure 2.18 project across the diagonal of the pit to ensure that they will be etched free by the KOH. More complex structures are possible using this technique, but care must be taken to ensure that they will be etched free by the KOH.

If it is desired to make beams or bridges of a different orientation, the wafer can be etched through from the back in KOH (Figure 2.19). This will ensure that the structure is released from the silicon. During such etching, it is necessary to ensure that the front of the wafer is adequately protected from the long KOH etch. Another alternative could be to produce a diaphragm, and etch the desired bridge or beam shape using a reactive ion etcher (dry etching).

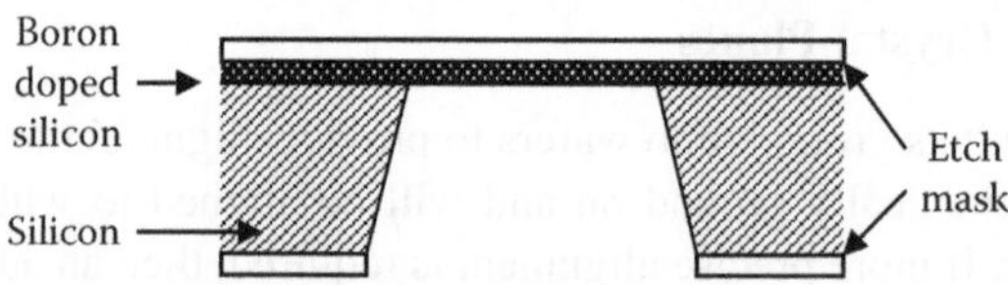

FIGURE 2.17 Thin silicon diaphragm created using KOH and boron etch stop (cross section, not to scale).

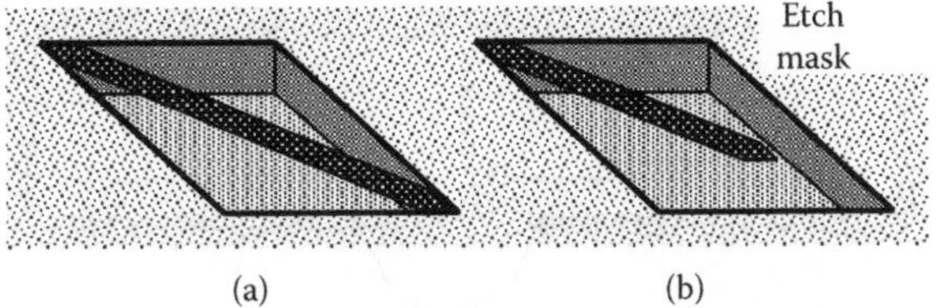

FIGURE 2.18 Formation of (a) a bridge and (b) a cantilever beam, using concentration-dependent etching.

One of the applications for these beams and bridges is as resonant sensors. The structure can be set vibrating at its fundamental frequency. Anything causing a change in the mass, length, etc., of the structure will register as a change frequency. Care has to be taken to ensure that only the quantity to be measured causes a significant change in frequency.

2.7.1.2 Fine Points through Wet and Dry Etching

A combination of dry etching and isotropic wet etching can be used to form very sharp points. First, a column with vertical sides is etched away using an RIE (Figure 2.20a). A wet etch is then used, which undercuts the etch mask, leaving a very fine point (Figure 2.20b). The etch mask is then removed.

Very fine points such as these can be fabricated at the end of the cantilever beams as probes for use in atomic force microscopy. The technique can also be used to produce sharp, small blades.

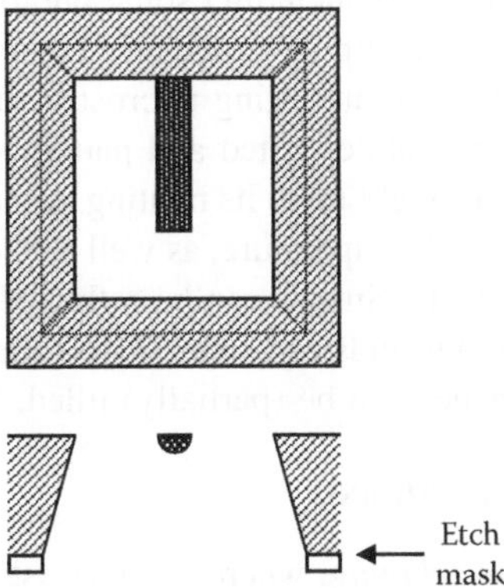

FIGURE 2.19 Etch-stop cantilever beam formed by etching through the wafer from the back.

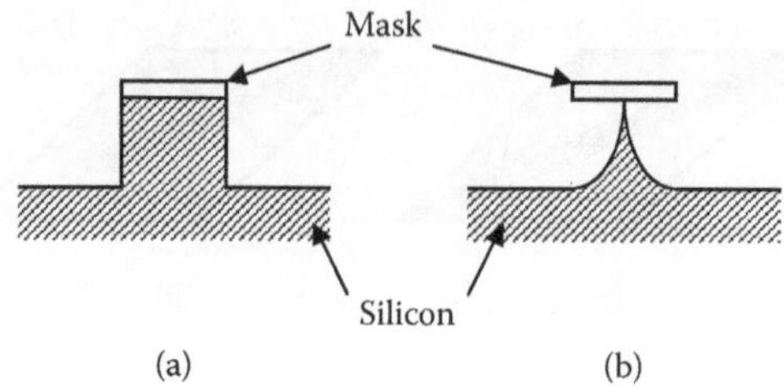

FIGURE 2.20 Formation of sharp points using a combination of RIE and wet etching: (a) RIE pillar formed; (b) wet-etched to a point (not to scale).

2.7.1.3 RIE Pattern Transfer

RIE can be quite aggressive on the mask material used. Some practitioners have, however, used this to good advantage when etching structures in silicon. If the selectivity of silicon over oxide has been well characterized for a particular process as 100:1, for example, then this provides a way by which the etch depth can be carefully controlled. An oxide mask of 200-nm thickness, e.g., is deposited and patterned on the surface of the wafer. The wafer is then etched until the oxide has been completely removed, which should be apparent from the color of the plasma. As a result, the pattern from the oxide mask will have been transferred into the silicon wafer and will now be 20 μm deep due to the differing etch rates of the two materials.

2.7.1.4 Reflow

This is not, strictly speaking, a bulk micromachining technique nor is it limited to silicon-based processes. It can be used to achieve a variety of interesting effects such as creating smooth wavelike surfaces or spheres or for filling in around tall structures and deep trenches. It is commonly performed with TEOS, although it can be used with other materials including some photoresists and other polymer films. Solder reflowing is a common procedure in many electronics processes and can be adapted to produce interesting microstructures.

The material in question is deposited and patterned. The substrate with the patterned material is then brought up to its melting point and the material *reflows*. By controlling the timing and temperature, as well as the patterning, it is possible to achieve a variety of effects. Stripes can be reflowed to form a more rounded corrugated surface, isolated islands can be made to form semispherical structures (see Figure 2.21), or trenches can be (partially) filled.

2.7.2 Surface Micromachining

In contrast to *bulk micromachining*, where, as discussed in the previous section, structures are created by removing material from the bulk of the silicon wafer or substrate, *surface micromachining* involves the gradual building up and patterning of thin films on the surface of the wafer to create the final structure. The process would typically employ films of two different materials, a structural material

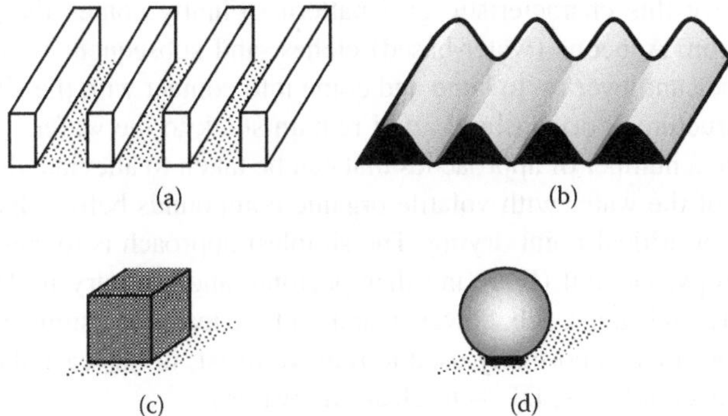

FIGURE 2.21 When reflowed different effects can be achieved such as: (a) stripes, (b) a corrugated surface, (c) a cube of solder on a small metal pad, (d) a sphere (not to scale).

(commonly polysilicon) and a sacrificial material (oxide). These are deposited and dry-etched in sequence. Finally, the sacrificial material is wet-etched away to release the structure. The more layers, the more complex the structure and the more difficult it becomes to fabricate. More recently, polymers such as photoresists have been used as the sacrificial layer, particularly in simple surface micromachining processes. These are removed by plasma-ashing, which, being a dry etch, avoids the problems caused by surface tension.

A simple surface-micromachined cantilever beam is shown in Figure 2.22. A sacrificial layer of oxide is deposited on the surface of the wafer. A layer of polysilicon is then deposited and patterned, using RIE techniques to a beam with an anchor pad (Figure 2.22a). The wafer is then wet-etched to remove the oxide layer under the beam, freeing it (Figure 2.22b). The anchor pad is under-etched; however, the wafer is removed from the etch bath before all the oxide is removed from beneath the pad, leaving the beam attached to the wafer.

This is not a very practical process. It demands that the etching rate of the oxide be well known in advance. Also, it would probably be a good idea to etch a number of holes through the beam itself to ensure that the wet etch has sufficient access to the underlying oxide. (In fact, almost all surface-micromachined

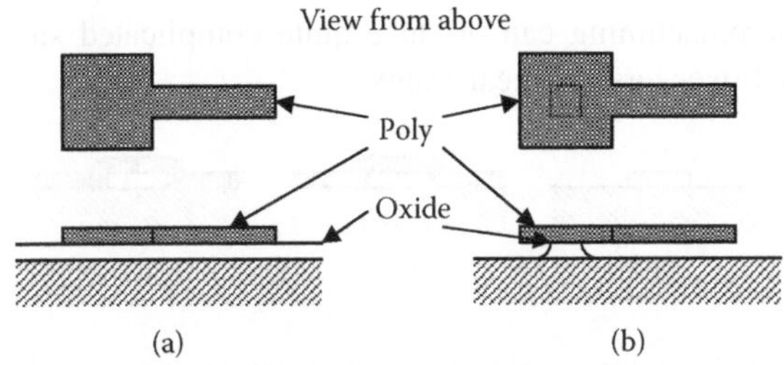

FIGURE 2.22 Simple surface-micromachined cantilever: (a) before wet etching, (b) after removal of sacrificial oxide (not to scale).

structures have this characteristic grid pattern.) Finally comes the problem of surface tension. Aqueous (water-based) etches, and subsequent washing in DI, will cause the cantilever tip to bend and come into contact with the silicon wafer. When the structure is dried, the tip will remain stuck to the wafer.

There are a number of approaches that can be taken to alleviate this problem: substitution of the water with volatile organic compounds before drying, use of dry etching, or critical-point drying. The simplest approach is to rinse the wafer first in isopropyl alcohol (IPA) and then acetone, and then dry it. This is not a particularly reliable approach, however, and acetone leaves a residue on the wafer. (Soaking in acetone is normally used to remove resist; in this case the process is reversed — acetone, IPA, DI — to clean the wafer.)

The use of anisotropic reactive-ion etching to release micromachined structures has become quite popular. A couple of points to note are that the RIE etch (release) rate will depend on the lateral diffusion of the ions, so it is not particularly suited to under-etch long distances; structures released by RIE are often more hole than structural material (i.e. lattice-like). Secondly, if long RIE times are required, it is best to be aware that some of the structural material may be removed as well, because etches are not perfectly selective.

The third option mentioned was the use of critical-point-drying equipment or carbon dioxide dryers. These can be purchased as specialist units and can dry the substrate without causing surface tension problems.

Wet Etching and Micromachined Structures

Even if surface tensions are addressed, surface-micromachined structures can be delicate to the extent that the manner in which they are withdrawn from the final etch bath can damage them or cause sticking.

A variety of different chambers can be fabricated on the surface of silicon wafers, using surface micromachining techniques. In Figure 2.23, the chamber is defined by a volume of sacrificial oxide (Figure 2.23a). A layer of polysilicon is then deposited over the surface of the wafer (Figure 2.23b). A window is dry-etched (RIE) through the polysilicon, and the wafer is then immersed in a wet etch that removes the oxide, leaving a windowed chamber (Figure 2.23c).

Surface micromachining can produce quite complicated structures such as microengineered tweezers and gear trains.

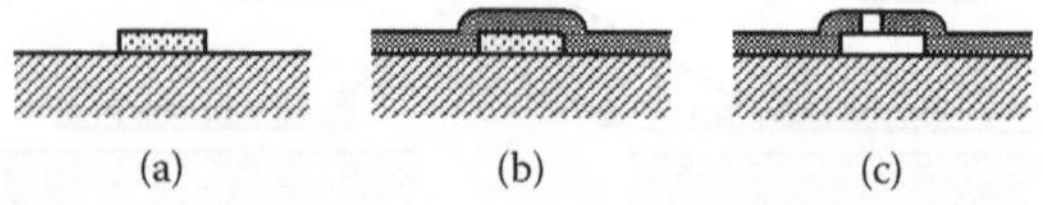

FIGURE 2.23 Forming a surface-micromachined chamber: (a) sacrificial oxide is patterned in the shape of final chamber, (b) polysilicon is deposited, (c) polysilicon patterned to open a hole or holes into the chamber, and the oxide is removed by wet etching (not to scale).

2.7.3 Electrochemical Etching of Silicon

A variety of electrochemical silicon-etching techniques are under development. One of these is the electrochemical passivation technique. A wafer with a particular impurity concentration is used, and different impurities are diffused (or implanted) into the wafer. This is done to form a diode junction at the boundary between the differently doped areas of silicon; this will delineate the structure to be produced. An electrical potential is then applied across the diode junction, and the wafer is immersed in a suitable wet etch (KOH). This is done in such a way that when the etch reaches the junction, an oxide layer (passivation layer) is formed, which protects the silicon from further etching.

This is another bulk silicon micromachining technique and is essentially similar to the boron-etch-stop technique (concentration-dependent etching). The structures that can be produced are similar to those produced by the boron etch-stop technique. The main advantage of the electrochemical method is that much lower concentrations of impurities are required, and therefore the resulting structure is more compatible with the fabrication of microelectronic circuitry. It is essential to ensure that the required electrical potential is distributed evenly across the wafer to all necessary parts, and this may involve addition of a conducting layer or other design considerations.

2.7.4 Porous Silicon

Microporous silicon is created by an electrolytic process. The silicon wafer is immersed in HF, and an electrical current is passed across the interface. Where the silicon is exposed to both the HF and the current, a very complex (submicron) porous structure is formed. The porosity can be controlled to some extent by varying the parameters used.

In addition to providing an unusual structure for use in MEMS (and electronic) devices, porous silicon has a vast surface-to-volume ratio, and consequently it is etched much more rapidly than normal silicon. By selectively anodizing different areas of the wafer (using a suitable insulating layer) and then etching away the porous silicon, it is possible to form interesting structures.

2.7.5 Wafer Bonding

There are a number of different methods available for bonding micromachined silicon wafers together, or to other substrates, to form larger, more complex devices.

A method of bonding silicon to glass that is particularly popular is anodic bonding (electrostatic bonding). The silicon wafer and glass substrate are brought together and heated to a high temperature (several hundred degrees Celsius). A large electric field is applied across the join, which causes an extremely strong bond to form between the two materials. This bond is formed by ions in the glass that migrate towards the join, and there are obvious limitations on the types of glass that can be used; in particular, the coefficient of thermal expansion has to be close to that of silicon. Both Pyrex and borosilicate glass can be used.

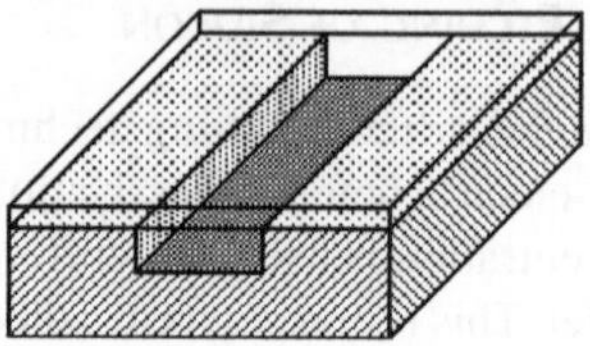

FIGURE 2.24 Forming a microchannel by bonding a glass plate over an etched silicon wafer.

It is also possible to bond silicon wafers directly together using gentle pressure under water (direct silicon bonding).

Other bonding methods include using an adhesive layer, such as a spin-on-glass or photoresist. Whereas anodic bonding and direct silicon bonding form very strong joins, they suffer from some disadvantages, including the requirement that the surfaces to be joined are very flat and clean. This can be overcome to some extent by using an adhesive layer.

Figure 2.24 shows a glass plate bonded over a channel etched into a silicon wafer (RIE), forming a pipe through which fluid can flow. Wafer-bonding techniques can potentially be combined with some of the basic micromachined structures to form the valves, pumps, etc., of a microfluid-handling system.

2.8 WAFER DICING

After wire bonding and packaging, wafer dicing is one of the more expensive processes that a device can undergo. This is because it is at this point that devices start being handled individually, so losing some of the economy inherent in batch production. Also, wafer dicing is often a mechanically stressful process. The three common techniques employed to break a silicon wafer up into individual chips or dies are as follows:

- Dicing saw
- Diamond scribe
- Laser

The wafer is normally affixed to an adhesive plastic membrane throughout the dicing process.

2.8.1 The Dicing Saw

This is probably the most common approach to wafer dicing. A diamond saw blade is used, and cuts of less than 100 μm can be made with 10-μm accuracy in position. Thus, dies with dimensions of only 1 or 2 mm per side can be produced. One of the caveats to be aware of is that because this is a process in which mechanical force is applied to the wafer, it can damage delicate micromachined structures, particularly surface-micromachined devices. Debris may also be deposited on the surface of the wafer, particularly in the vicinity of the saw blade.

It is normal to position the saw blade in such a way that it cuts completely through the wafer. It is possible, however, to position it so that it only cuts partway through the wafer (several tens of micrometers). By making many such cuts, it is possible to form trench-like or pillar-like structures, depending on the depth and spacing of the cuts (Figure 2.25). These have been combined with wet silicon etching to produce arrays of spikes (Figure 2.25c); the process involves repeated dipping and withdrawal of the structure from the etch solution in order to achieve a smooth tapered point.

2.8.2 Diamond and Laser Scribe

An alternative to using the dicing saw to break up the wafer is to break it along one of the major crystal planes. This assumes, of course, that the individual chips have been laid out aligned to an appropriate cleavage plane. A diamond-tipped scribe is then drawn over the surface of the wafer to create a scratch, and flexing the wafer causes it to break along this imperfection. Alternatively, a laser can be used to create a series of small craters along which the wafer will be broken; this is usually performed on the backside of the wafer to prevent ejected material and heat from damaging the devices on the front side. Note also that silicon is transparent to infrared light and therefore, an appropriate wavelength has to be carefully selected.

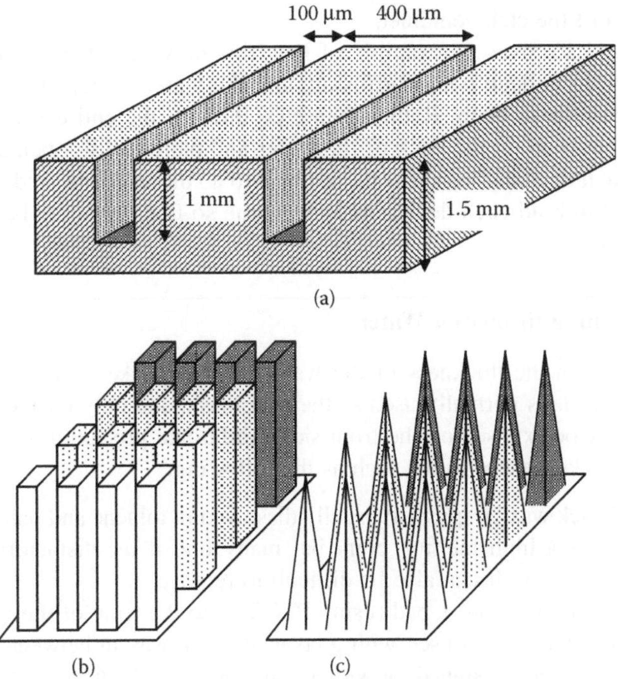

FIGURE 2.25 (a) Deep trenches can be sawn into wafers, to less than 100-μm width, which leads to the possibility of creating microstructures, (b) sawing pillars, (c) converting pillars to needles by wet etching.

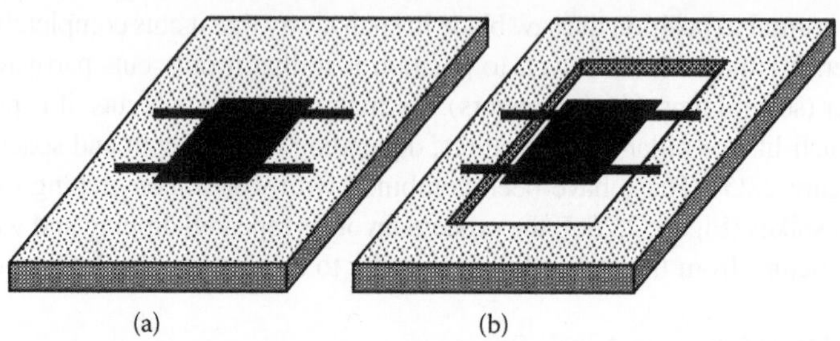

FIGURE 2.26 (a) A structure or device has been defined by boron diffusion (black) into the surface of a wafer, (b) following KOH etching of a window from the back of the wafer, the main area of the device remains attached to a silicon frame by thin bridges that can easily be broken.

2.8.3 Releasing Structures by KOH Etching

Particularly in the case of bulk micromachined structures, it is possible to consider using anisotropic etching techniques to separate the individual dies. It is usually a good idea to include some structures to hold the die into what remains of the wafer until they are required, otherwise it will be necessary to fish them out from the bottom of the etch container.

Figure 2.26 gives an example of how this may be achieved. The devices are defined by a deep boron diffusion, and the wafer is etched from the back in KOH to release them. Rather than etch through the entire wafer, a nitride mask is deposited on the back of the wafer and patterned to open up windows beneath each device. This turns the wafer into a silicon frame, and thin bridges are formed to hold each device into the frame so that they may be broken out as required.

KOH Etching through a Wafer

Depending on the thickness of the wafer, this will take some time (several hours). Nitride is normally used as the etch mask, and only nitride and noble metals may be exposed on the front side. Additional protection is often used. This may take several forms such as the following:

- "Black wax" (Apiezon) usually dissolved in toluene and painted on. It has a high melting point but many find it unsatisfactory when used alone. It can also be difficult to remove.
- A dummy wafer taped (using PTFE) onto the front of the wafer to be etched, often used with a layer of black wax in between.
- A specially designed jig with a seal that keeps the front of the wafer protected from the KOH solution. Note that the front will have to withstand short periods in KOH while the etch goes to completion.

When using this technique (or just using KOH to etch through a wafer), it is best to design the process so that the front side of the wafer does not have to be protected throughout the long etching process (see box). This means that materials unaffected by long periods on hot KOH (e.g., nitride, gold) are chosen; the bulk of the wafer should be etched through before sensitive structures are formed on the front. This can be done with SOI wafers.

Figure 2.27 illustrates this. The front of the wafer is protected by a PECVD nitride layer while the KOH etch proceeds from the back through to the buried oxide layer. The nitride layer is then stripped and material is deposited and patterned on the front side of the wafer. Finally, RIE from the front releases the individual devices or structures.

This is a very effective way to create and release microstructures. There are two points to bear in mind, however. Firstly, the cavities in the back of the wafer make it difficult for vacuum chucks to hold it in place during various processes, notably the spin-application. Secondly, regular cavities can form lines of mechanical weakness that make the wafer move susceptible to fracture during processing.

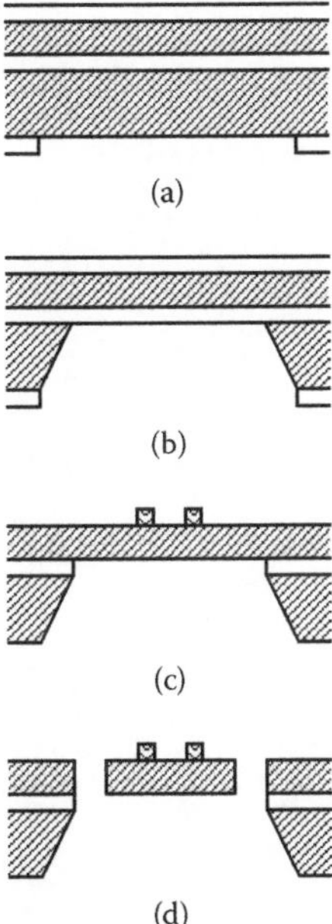

FIGURE 2.27 Use of KOH etch and SOI wafer to release free-standing microstructures without having to worry about long exposure of the front to the etch solution: (a) SOI wafer with PECVD nitride on the front, and nitride mask on the back, (b) etch through to oxide etch stop, (c) stripe front nitride and oxide, and perform any additional patterning required, (d) RIE etch to release device.

REFERENCES

1. Petersen, K.E., Silicon as a mechanical material, *Proc. IEEE*, 70(5), 420–457, 1982.
2. Vossen, J.L. and Kern, W., Eds., *Thin Film Processes II*, Academic Press, San Diego, CA, 1991.
3. Williams, K.R. and Muller, R., Etch rates for micromachining processing, *J. MEMS*, 5(4), 256–269, 1996.
4. Kern, W. and Deckert, C.A., Chemical etching, in Vossen, J.L. and Kern, W., Eds., *Thin Film Processes*, Academic Press, New York, 1978, chap. V-1.
5. Laermer, F. and Schilp, A., Method of Anisotropically Etching Silicon, U.S. patent 5,501,893, March 1996.
6. Puers, B. and Sansen, W., Compensation structures for convex corner micromachining in silicon, *Sensor. Actuators*, 1990, Vol. A21-23, 1036–1041.
7. Sandmaier, H., Offereins, H.L., Kühl, K., and Lang, W., Corner compensation techniques in anisotropic etching of (100)-silicon using aqueous KOH, *IEEE Transducers '91*, 203–214, 1991.

3 Nonsilicon Processes

3.1 INTRODUCTION

There are a vast number of processes that can be used to micromachine and microform materials other than silicon. Some of these, along with their common applications, are introduced in this chapter. Several of them have cross over applications to silicon micromachining (electroplating, for example), and the chapter begins with a technique that, for micromachining purposes, arose from silicon processing and is still extensively applied in that field.

3.2 CHEMICAL–MECHANICAL POLISHING

Chemical–mechanical polishing (CMP) is one of a range of processes that are generally used to thin and process silicon wafers (also known as *wafer lapping*). These can be applied to a wide variety of materials and can range from pure mechanical grinding on an abrasive wheel to CMP, which usually involves an abrasive powder suspended in a corrosive etching solution; in design, the mechanical component removes the bulk of the material and the chemical process provides a highly polished finish.

The process was originally developed and used to thin and polish silicon wafers sliced from the ingot. In recent years it has undergone considerable development to assist in *planarization*, a key process in producing multiple metal interconnection layers that the modern integrated circuit industry has come to rely on. To create structures with small feature sizes, a very flat surface is required for photolithography and without planarization, only a few interconnection layers (e.g., two poly and two metal) can be implemented.

The planarization process essentially involves deposition of a relatively thick insulating layer and creation of a flat planar surface using CMP while reducing the thickness of the film to a micron or so. The process can be used in micromachining, particularly where trenches or mesas cause problems with a subsequent photolithography step. In such cases, the film thickness and structural dimensions involved are often greater than those involved in IC fabrication, and the material used may well be stripped later to empty the trenches.

CMP can be applied to other materials and in other forms. It is allied to the polishing processes used to prepare geological or metallurgical samples for electron microscopy, for instance. Alternatively, an abrasive slurry can be forced through a structure to shape and polish it.

3.3 LIGA AND ELECTROPLATING

The acronym LIGA comes from the German words for the process (*Lithographie, Galvanoformung, Abformung*). LIGA uses lithography, electroplating, and molding processes to produce microstructures. It is capable of creating very finely defined microstructures up to 1000 μm high. Micron feature sizes can theoretically be implemented with aspect ratios as high as 1000:1, though achieving these in practice depends on the processing steps that follow exposure and development.

In the process originally developed, a special kind of photolithography using x-rays (x-ray lithography) is used to produce patterns in very thick layers of photoresist — usually PMMA-based resists. The x-rays from a synchrotron source are shone through a special mask onto a thick photoresist layer that covers a conductive substrate (Figure 3.1a). This resist is then developed (Figure 3.1b). The pattern formed is then electroplated with metal (Figure 3.1c). Nickel is commonly used because nickel plating is a well-developed process, and reagents and equipment are commercially available. The metal structures produced could be the final product; however, it is common to produce a metal mold (Figure 3.1d). This mold can then be filled with a suitable material, such as plastic (Figure 3.1e), to produce the finished product in that material (Figure 3.1f). This process is repeated to spread the high initial cost of masks and x-ray exposure over as many units as possible. A plastic structure can even be electroplated to form a metal structure.

As the synchrotron source makes LIGA expensive, alternatives are being developed. These include high-voltage electron beam lithography, which can be used to produce structures on the order of 100 μm high and excimer lasers capable of producing structures up to several hundred microns high.

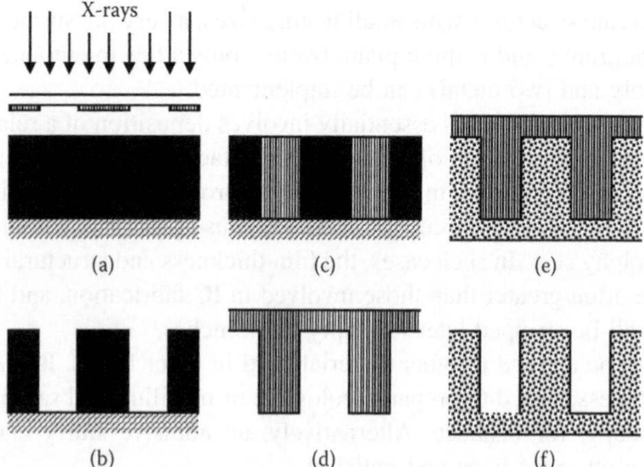

FIGURE 3.1 The LIGA process: (a) x-ray exposure, (b) resist developed, (c) electroplating to fill gaps in resist; this structure can be used as a one-off, (d) plating is continued to create a metal mold insert, (e) injection molding, (f) final structure in plastic. The steps illustrated in (e) and (f) may be repeated to mass produce structures.

Electroplating is not limited to use with the LIGA process but may be combined with other processes and more conventional photolithography to produce microstructures. One aspect of the process that is very much apparent on the microscale is its dependence on current density. The best results are obtained if current density is constant over the entire structure, but geometrical considerations mean that this is difficult to achieve. The problem is often addressed by use of pulse electroplating (rather than constant current) and overplating. Some LIGA processes involve overplating followed by mechanical milling to provide a flat surface. Because this places mechanical stress on the structure, design rules have to be employed to ensure that various parts of the structure will be sufficiently supported during milling.

3.4 PHOTOCHEMICAL MACHINING

Photochemical machining (PCM) is the equivalent of silicon photolithography and wet etching applied to metal foils. A wide variety of metal films can be machined: steel, copper, molybdenum, etc., and these can be from tens of microns to a few millimeters thick. The process is inherently isotropic and usually carried out from both sides of the substrate (especially for thicker foils) in a spray-etching system. That is, the etching solution is sprayed onto the substrate rather than being immersed in a bath. The process is usually capable of feature sizes down to 100 μm or 200 μm, although it may be pushed further.

PCM is commonly used in the mass production of the lead frames that are used in IC packaging to provide electrical connections through the package to the circuit board. If used within its normal commercial specifications and tolerances, this can be a very economical process compared to silicon micromachining. By combining it with electroplating, it is possible to produce a variety of different structures in metal, including closed channels.

3.5 LASER MACHINING

Compared to light from a conventional incandescent bulb, laser light is special: it is monochromatic (i.e., light of only one wavelength), coherent in time and space, and has a very low angle of divergence (i.e., it is collimated). The first two properties make it possible to focus the entire power contained in a laser beam onto a very small diameter spot — this results in a very high power density. In terms of micromachining, this means that structures with vertical sidewalls can be created in resist or very high aspect ratios can be created.

There are several different types of laser, but for micromachining purposes, only three common types will be considered. The first two are infrared (IR) lasers, producing IR light, and the third is the excimer laser, which produces ultraviolet (UV) light. IR lasers are commonly used in industry and are available in laser jobbing shops (companies that provide laser machining facilities on a sub-contract basis). UV lasers have developed to become more specifically microengineering orientated; one application is to produce the very-low-wavelength UV light used in DUV photolithography.

TABLE 3.1
Laser Power Outputs

	CO_2	Nd:YAG
Lasing material	Carbon dioxide, nitrogen, and helium gas	Neodymium doped yttrium aluminum garnet crystal
Wavelength	10.8 μm	1.03 μm
Output power	up to 25 kW	up to 1 kW

3.5.1 IR Lasers

There are two common IR industrial lasers — the carbon dioxide (CO_2) lasers and neodymium YAG (Nd:YAG) lasers. Table 3.1 summarizes some of their properties (note that the average power is given for Nd:YAG lasers; industrial Nd:YAG lasers are typically operated in a pulsed mode with much higher peak power outputs).

The CO_2 laser is the most commonly available of the two, but for micromachining purposes, the Nd:YAG laser is most interesting. IR lasers affect materials by localized heating. The diameter of the spot to which a laser beam can be focused is related to the wavelength of the light, so in theory, Nd:YAG lasers can be focused to much smaller diameters than CO_2 lasers. In practice, IR lasers are focused to spot diameters that range from 0.03 mm to 0.3 mm.

IR lasers can be used in any of three basic applications: drilling and cutting, welding, or heat treatment (localized hardening or annealing). Note that the material being machined should not be too transparent or reflective for the wavelength of light employed. Furthermore, note that when performing a process such as cutting or welding, the material adjacent to the cut or weld will also be affected by the heat involved in the process.

The Nd:YAG laser can produce holes of narrower diameter (down to 50-μm diameter and up to 50-mm deep) and also narrower cut (kerf) widths (down to 30 μm — cutting essentially proceeds by drilling a series of holes) than CO_2 lasers, which are typically operated with beam diameters of 100 to 300 μm. CO_2 lasers, however, are much more efficient when it comes to cutting or welding; the Nd:YAG essentially produces very-small-diameter spot welds. IR lasers will typically cut up to 10 mm in mild steel, less in aluminum and stainless steel, and nonmetallic materials up to 20 mm thick.

In order to use a laser to machine a substrate, it is necessary to register the workpiece with a reference point so that relative movement between beam and workpiece ensures that machining occurs in the appropriate location. This is commonly performed using a computer numerically controlled (CNC) system (i.e., a robotic system). These are capable of considerable accuracy, and systems developed for laser micromachining are capable of resolutions down to 2 μm over an area of 200 × 200 mm.

3.5.2 Excimer Laser Micromachining

Excimer lasers produce relatively wide beams of UV laser light. One interesting application of these lasers is their use in micromachining organic materials (plastics, polymers, etc.). This is because the excimer laser does not remove material by burning or vaporizing, unlike IR lasers, so the material adjacent to the area machined is not melted or distorted by heating effects. The process by which material is removed (ablated) is generally thought to be because of interaction between the laser light and chemical bonds within the material; individual molecules are broken up (heating involves exciting the molecules as a whole, thus they change state from solid to gas).

When machining organic materials, the laser is operated in the pulsed mode, removing material with each pulse. The amount of material removed is dependent on the material itself, the length of the pulse, and the intensity (fluence) of the laser light. Below a certain threshold fluence, depending on the material, the laser light has no effect. As the fluence is increased above the threshold, the depth of material removed per pulse is also increased. It is possible to accurately control the depth of the cut by counting the number of pulses. Considerably deep cuts (hundreds of microns) can be made using the excimer laser.

The shape of the structures produced is controlled by using chrome on quartz masks, such as the masks produced for photolithography. In the simplest system the mask is placed in contact with the material being machined, and the laser light is shone through it (Figure 3.2a). A more sophisticated and versatile method involves projecting the image of the mask onto the material (Figure 3.2b). Material is selectively removed where the laser light strikes it.

Structures with vertical sides can be created. By adjusting the optics, it is possible to produce structures with tapered sidewalls (Figure 3.3).

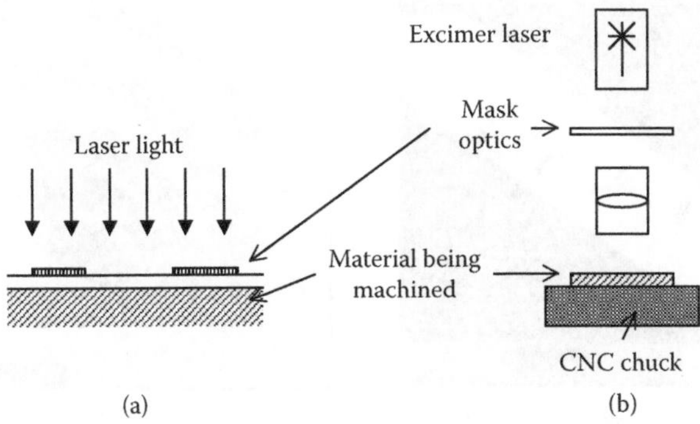

FIGURE 3.2 Schematic illustration of contact printing with (a) an excimer laser, and (b) projection system — the substrate being machined is mounted on a computer numerically controlled (CNC) chuck, which moves it relative to the mask plate; this enables structures such as those in Figure 3.4 to be created with only one mask.

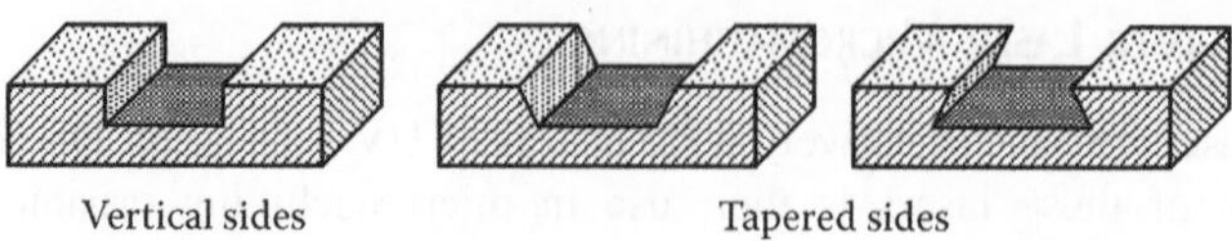

FIGURE 3.3 Different sidewall profiles can be created by adjusting excimer laser optics.

Excimer lasers have a number of applications beyond those mentioned here. One area of application is in machining the cornea of the eye to change its optical properties, thus correcting for short sight. The versatility of the excimer laser is illustrated in Figure 3.4. Notice that interesting curved and stepped

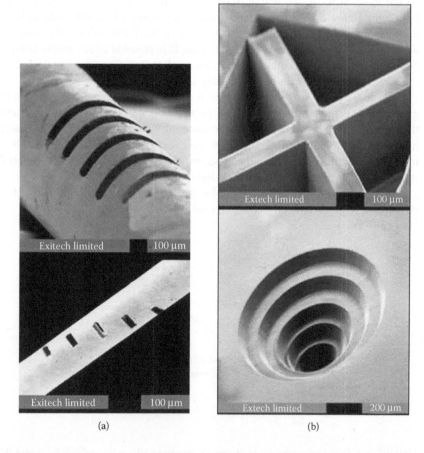

FIGURE 3.4 Examples of excimer laser machining: (a) nonplanar substrates; the upper fiber is acrylic; the lower a plastic optical fiber, (b) structures machined in polycarbonate, (c) microlenses machined in polycarbonate, (d) CVD diamond. (Images reproduced with kind permission of Exitech Ltd., U.K. www.exitech.co.uk.)

(c) (d)

FIGURE 3.4 (Continued).

structures can be created by using the CNC platform to move or drag the workpiece beneath the mask during exposure or after a set number of pulses. Additionally, because reduction optics are commonly employed to project the image of the mask onto the substrate, grayscale masks can be used (see Subsection 1.2.1, Figure 1.3) to expose different parts of the workpiece to different intensities of laser light (the depth ablated per pulse being proportional to the intensity).

3.6 POLYMER MICROFORMING

Polymers (plastics), in the macroworld, have the distinction of being incredibly inexpensive to mass produce and very versatile. Many people would like to reproduce this success with plastic microstructures, but techniques for polymer microforming lagged somewhat behind silicon and metal micromachining. The most widespread form is photolithographic processing in the form of photoresists, polyimides, and photoformable epoxies (SU-8; the latter also being a class of photoresist). At the time of writing, these were closely followed by PDMS casting, but hot embossing appears to be making its mark as a mass production technique. This section also addresses stereolithography and microcontact printing. The latter is a lithographic technique that is not restricted to polymers but can also pattern biomolecules, for instance. However it usually requires a polymer (PDMS) original.

3.6.1 Polyimides

Polyimides are UV photoformable polymers that are common in the electronics industry. These have several different trade names and different properties.

Fabrication of polyimide structures is normally performed on a polished silicon wafer to provide a convenient flat rigid substrate to which the material can be applied and which holds it flat during subsequent machining steps. Polyimides are usually spun on and patterned using conventional UV lithography techniques, usually to several microns thickness. Metal films can also be deposited, patterned, and sandwiched between layers to provide a variety of different electrode or interconnection structures. Polyimide structures are often used as part of the packaging of silicon microsystems — they are flexible and more robust than individually bonded wires. The silicon die must be bonded to the ribbon cable either by conventional wire-bonding processes or flip-chip techniques (see Chapter 9). Unfortunately many polyimides are not very resistant to ingress of water.

3.6.2 Photoformable Epoxies (SU-8)

SU-8 (from Microchem Corp., U.S.) is a photoformable epoxy (negative photoresist) that has gained a considerable following among the MEMS community. It is easy to see why — it is available in several different formulations and can be applied in films of 1 μm to 200 μm thickness in a single spin process, can be exposed using standard UV exposure equipment, and produces high-aspect-ratio structures (10:1 or better) with relatively straight sidewalls. It is also highly resilient to chemical attack. As a result, microstructures can be produced in SU-8 with a relatively low initial capital investment.

Owing to its popularity, there are several data sheets and application notes available directly from manufacturers and distributors and other data available on the Internet. Once the processes required to apply, expose, and develop SU-8 have been mastered, the main problem encountered appears to be its removal. As an epoxy, it is exceedingly stubborn to remove and, at the time of writing, it seems most appropriate to advise the users that if they need to remove the hard-baked (cured) SU-8, they are probably not going to be able to do it very well. Nonetheless, there are three possible options that have been suggested: plasma ashing, laser ablation, and use of release layer.

A release layer is particularly useful if the SU-8 is to be used with electroplating to create metal microstructures (as in the LIGA process). The release layer is basically a thin coat of photoresist that is applied beneath the SU-8 film.

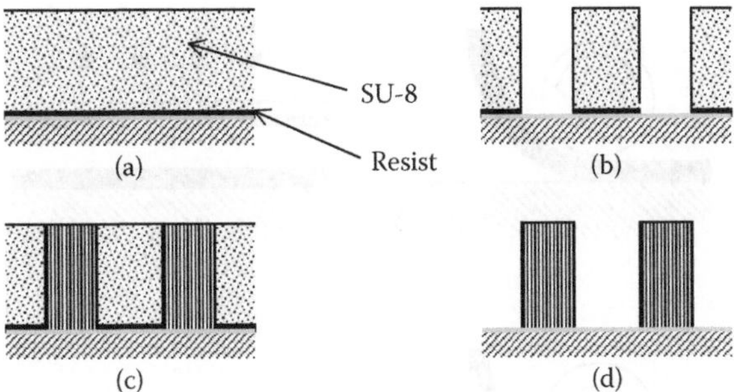

FIGURE 3.5 LIGA using SU-8: (a) a substrate (coated with a nickel seed layer for electroplating) is coated first with a thin layer of photoresist and then SU-8, (b) the SU-8 and resist are patterned, (c) electroplating is used to form a nickel structure, (d) the photoresist layer is then stripped off, taking the SU-8 with it and leaving the metal structure. (A detailed description can be found on the *OmniCoat*™ data sheet from MicroChem Corp, Newton, MA, U.S. www.microchem.com.)

When the SU-8 is to be removed, the release layer is simply stripped away, taking the SU-8 with it (see Figure 3.5).

One additional caveat: although the cured SU-8 may be particularly stubborn, this does not mean that it can sustain prolonged attack in KOH or EDP; it may suffer adhesion problems.

3.6.3 Parylene and PTFE

Two other polymers that can often be found in MEMS laboratories are parylene and polytetrafluoroethylene (PTFE). Parylene is usually deposited by CVD. It is a particularly stubborn material and difficult to pattern. It is also difficult to achieve a good conformal coating without pinholes or defects.

PTFE is normally available in spin-on form. Again, it is difficult to pattern and usually only used if absolutely necessary (such as encapsulating devices for implantation in the body). It is very difficult to get anything to adhere to PTFE, and it usually requires some sort of surface treatment if additional films are to be deposited and patterned on it. The most common of these is treatment with oxygen plasma to roughen and chemically activate the surface to some degree.

3.6.4 Dry Film Resists

Developed for printed circuit board (PCB) processing, dry film resists are not commonly used for micromachining. They can, however, be used to create various microstructures.

The resists are normally available as films of different thicknesses ranging from 50 to 100 μm. They are laminated onto the substrate (or a proceeding patterned and developed resist layer) by a roller laminator at an elevated temperature. The material can then be patterned and developed to create various microstructures,

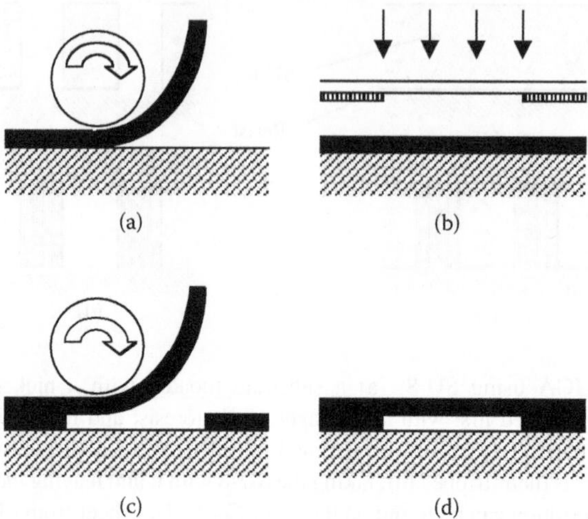

FIGURE 3.6 Creating a chamber using dry film resist: (a) the resist is applied to the substrate using a roller, (b) it is exposed through a mask to UV light, (c) the first layer is developed and a second layer is applied, (d) the second layer can then be patterned and developed as required.

although there are limitations — aspect ratios and the angle of channel walls are somewhat limited, as is the resolution achievable. By applying the resist to conventional PCB substrate material (FR4), it is possible to make use of the expertise and relatively low costs available for PCB production. Figure 3.6 illustrates construction of a simple chamber using dry film resist. Closed channels for microfluidic applications can also be produced in this manner.

3.6.5 Embossing

One of the most promising methods for mass production of microstructures is the hot-embossing process. In theory, the process is relatively simple. A mold insert is created by one of a number of micromachining processes, usually bulk silicon micromachining or nickel electroplating in a LIGA-related process. The mold and target material are heated until the chosen polymer becomes plastic, and the mold is then pressed into the plastic so that it takes up the impression of the structure. The mold is removed and the plastic sets in the desired shape (Figure 3.7).

In practice, the process is not quite so simple. The mold insert will probably have a different coefficient of thermal expansion than the polymer, so in the best of circumstances the final dimensions of the plastic structure will not be the same as those of the mold. Furthermore, the process has to be controlled to ensure a clean release, and parameters will have to be adjusted for each different material to be used.

Fortunately, however, there are two commercially available processes. The first (Figure 3.8) is based on macroscale mass production techniques and can be run continuously. The insert has to be flexible enough to be fitted around a roller,

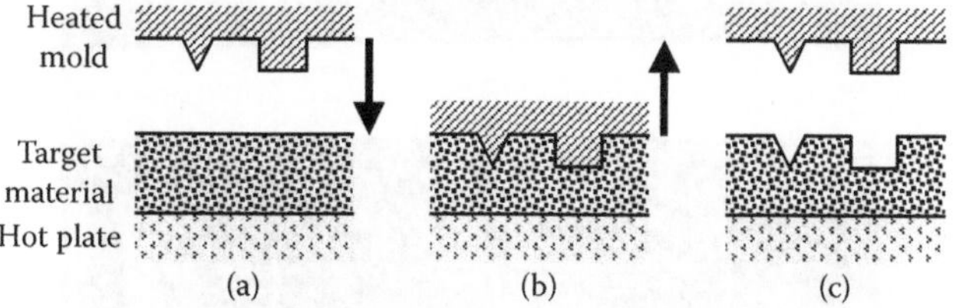

FIGURE 3.7 Principle of hot embossing: (a) the mold and target polymer are heated, (b) the heated mold is pressed into the polymer, (c) the mold is removed, leaving an impression of the structures on the mold in the polymer surface. (Not to scale.)

and this is impressed upon a continuously moving web. Although very well suited to mass production, this does introduce some additional problems for the designer: the final structures are distorted because the original is fitted onto a roller, and they tend to be stretched slightly in the direction of travel of the web.

An alternative approach that has proven to be popular in smaller laboratories is shown in Figure 3.7, in which the embossing tool is repeatedly pressed into a succession of polymer blanks. Figure 3.9 shows a typical hot-embossing machine, and Figure 3.10 shows a nickel-plated mold insert.

Different polymers can be microstructured in this way with very good results (Figure 3.11). These include PMMA (Figure 3.11a), polycarbonate, polypropylene, and others (Figure 3.11b).

3.6.6 PDMS Casting

Another material that has become popular for constructing microfluidic channels for various applications is cast polydimethylsiloxane (PDMS). This is a silicone elastomer, and is flexible and deformable, making it an interesting material for the construction of microstructures.

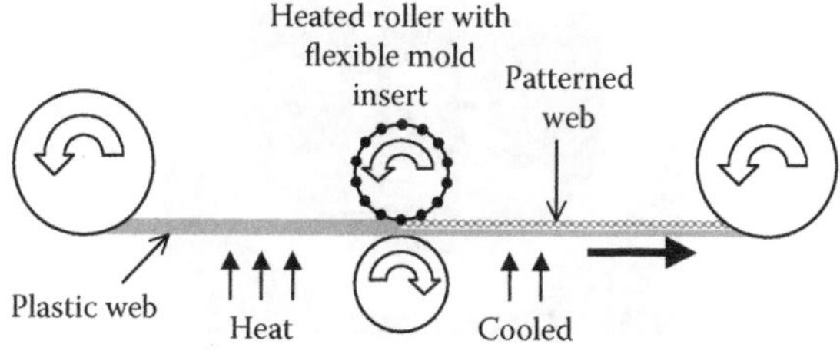

FIGURE 3.8 Principle of reel-to-reel roller polymer embossing. A web of plastic material is wound off one reel and passed between two heated rollers that are pressed together. A flexible mold insert wrapped around one of these rollers imprints the surface of the web with microstructures. The web is then taken up on a second roller.

FIGURE 3.9 Jenoptik hot-embossing machine with microstructured polymer wafer. Reproduced courtesy of Application Center for Microtechnology (AMT), Jena, Germany (www.amt-jena.de).

FIGURE 3.10 Nickel mold insert for hot embossing, 100-mm diameter. (Courtesy of Application Center for Microtechnology [AMT], Jena, Germany. www.amt-jena.de.)

(a)

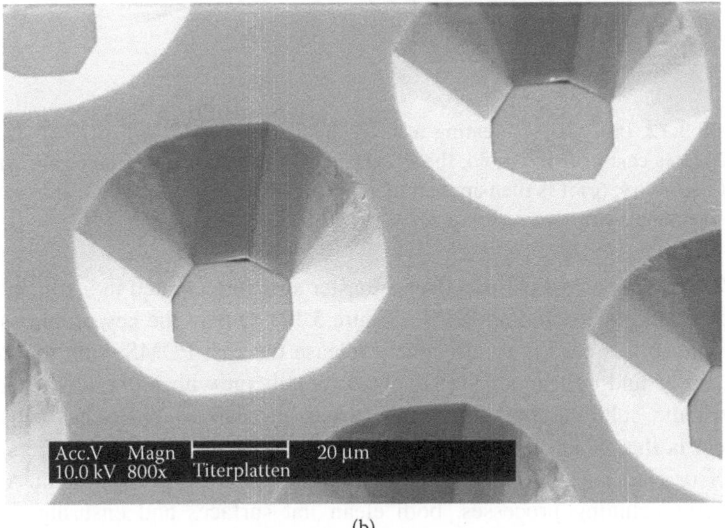

(b)

FIGURE 3.11 Polymer microstructures formed by hot embossing: (a) channels in a cyclic olefin copolymer substrate, (b) PMMA microtitre plate. (Courtesy of Application Center for Microtechnology [AMT], Jena, Germany. www.amt-jena.de.)

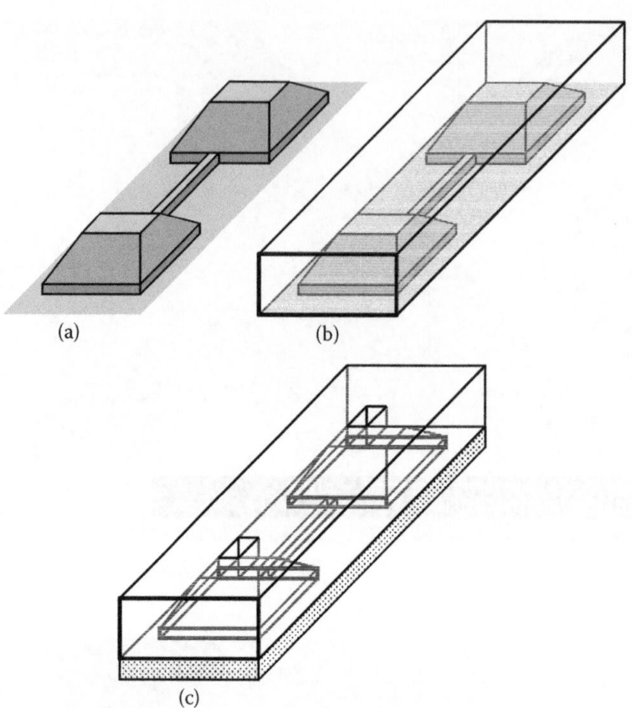

FIGURE 3.12 Process for creating a microchannel in PDMS: (a) original structure, (b) PDMS is cast over the mold, the PDMS is then peeled off, and holes are punched in it if necessary, (c) it is then applied to a flat substrate, such as a glass slide, creating a microchannel with two reservoirs.

PDMS can be cast or spun over a master structure created in a silicon wafer (or other material, including SU-8) (Figure 3.12). One of the key problems to be overcome in this case is to effectively release the cast PDMS from the master. Effenhauser and colleagues [1] silanized the silicon with a dimethyloctadecyl-chlorosilane solution, but others have tried alternative approaches. The next problem is that of ensuring good adhesion between the PDMS and the silicon, PDMS, or glass substrate that has been used to close the channel system. As with all micromachining processes, both clean flat surfaces and ensuring that the structure has not been exposed to the environment for a long period after casing, appear to help. Effenhauser and colleagues found that they could place the PDMS down and peel it off again should the channels become clogged. After cleaning the PDMS they could replace it and continue to use the device.

Others are seeking more permanent bonds for their devices or attempting to make more complicated three-dimensional structures. Various approaches have been tried with various degrees of success. These include treating both surfaces with PDMS prior to bonding and use of HDMS primer. Ensuring that the substrate is very dry (drive off water on a hot plate) also seems to be a key to achieving a good bond.

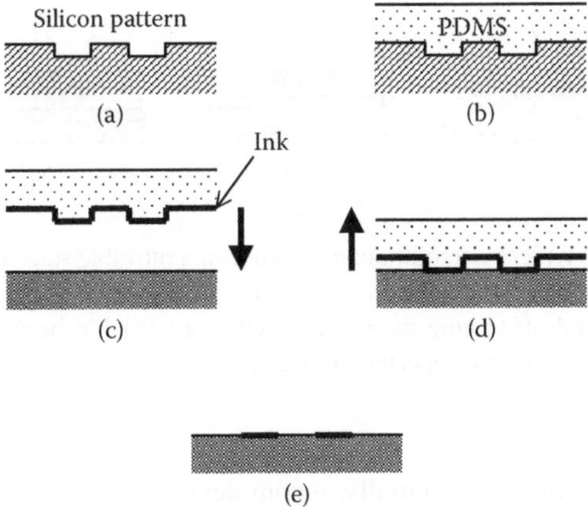

FIGURE 3.13 Outline of microcontact printing process: (a) a silicon pattern is used to create a PDMS stamp, (b) a PDMS stamp, (c) this is inked, (d) this is applied to an appropriately prepared (e.g., gold coated) substrate, (e) the chemical ink remains on the substrate at points of contact; the PDMS stamp can be wrapped around a roller and used in a continuous process.

3.6.7 Microcontact Printing

This is an interesting approach that makes use of PDMS structures to create quite complex structures, usually in metal. A variety of related techniques have been developed by Whitesides' group [2,3]. The basic approach involves the use of PDMS masters to print long-chain molecules onto an appropriate substrate. This has been used to print biomolecules onto various substrates, but the approach favored by Whitesides' group involves chemistry related to self-assembled monolayers; long-chain molecules that spontaneously self-organize when printed onto the appropriate substrate — in particular, alkanethiols on silver and gold substrates (again, this chemistry has also be used in the creation of biosensors).

The process is shown in outline in Figure 3.13. One of the advantages of this approach is that it is not limited to flat substrates; by applying it imaginatively to capillaries, Jackman and colleagues [2] have produced a variety of interesting structures.

3.6.8 Microstereolithography

Stereolithography is a well-developed process that is employed to produce three-dimensional prototype structures for macroscale engineering. The overall process is illustrated in Figure 3.14.

A stage is immersed just under the surface in a bath of UV-curable polymer. UV light is then focused onto the surface of the liquid, causing the polymer to solidify where the illumination is most intense. One layer of the structure is formed by scanning the spot over the surface of the polymer, turning the beam on and off as required. The stage is then lowered deeper into the liquid and

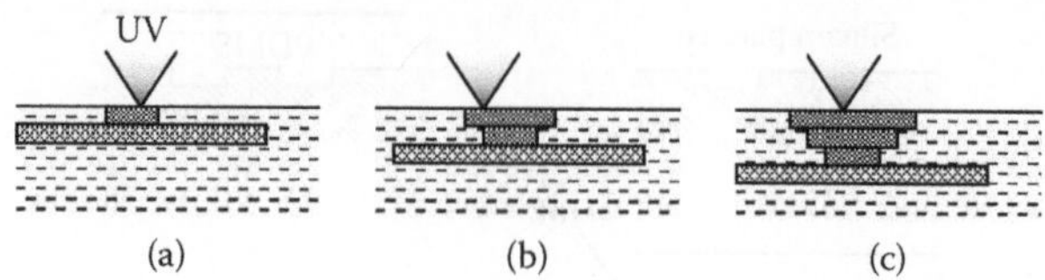

FIGURE 3.14 In a typical stereolithography process, a movable stage is immersed in a UV curable polymer. UV light is focused on the surface of the bath and scanned across it. The light is turned on and off to cure the polymer. Once a layer has been formed, the stage is moved down and the process is repeated. A three-dimensional structure is built layer by layer.

another layer is built up. Eventually, a complete and complex three-dimensional structure is produced. The main limit on the structure being produced is that it has to be supported by the stage (this prevents the construction of, for instance, caged balls without incorporating some sort of sacrificial layer or structure into the design).

Conventional stereolithography cannot be used to fabricate microstructures as it stands, mainly owing to surface tension and viscosity of the polymer deforming the structures and limiting the resolution of the process. Several groups are working on optimizing conventional stereolithographic processes for MEMS applications.

Ikuta et al. [4], however, have presented an alternative to conventional stereolithography that makes use of features (the viscosity of the fluid) that may otherwise be a problem on a microscale. Instead of focusing the UV light on the surface of the liquid, it is focused into the volume of the fluid and scanned in three dimensions (Figure 3.15). Because of the high viscosity of the fluid, the small parts, and the speed with which microstructures can be created, the solidified polymer remains floating within the volume of the fluid as the structure is constructed. This enables the construction of various unusual microstructures such as freely rotating cog wheels that would otherwise have to be anchored to the platform if a more conventional approach were used. The structures created

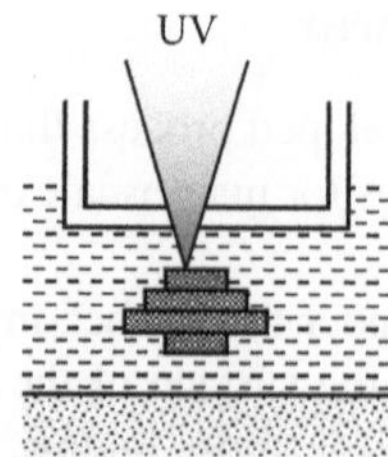

FIGURE 3.15 In the super IH micro stereolithography process described by Ikuta et al., [4] structures can be created without any supporting columns.

by the "Super IH" process appear a little more organic and less precise than those created by more conventional processes, but this can probably be improved upon.

3.7 ELECTRICAL DISCHARGE MACHINING

Electrical discharge milling (EDM), also commonly known as "spark erosion," is another precision macroscale machining process, capable of working to micrometer tolerances, that is being used or adapted for micromachining of metals.

EDM can only be used with conducting materials and is usually employed in the precision machining of very hard metal alloys. As one may anticipate, the process involves creating a series of sparks between the workpiece (substrate) and an electrode (mandrel), which is maintained at a positive voltage with respect to the workpiece. Each spark takes with it some small quantity of the material being machined. A dielectric liquid is employed to control the spark discharge process and cool the workpiece.

EDM is normally deployed in one of three modes (Figure 3.16): hole boring, shaped working electrode, and wire EDM. The first two rely on the fact that EDM is a noncontact process (so no mechanical forces are applied to the working electrode) in order to use a soft and easily shaped material to machine a much harder material. EDM hole boring is capable of creating holes with micrometer dimensions. Note, however, that the working electrode will degrade with use and may even have to be reshaped (by reversing the EDM) or replaced during the process. For this reason, wire EDM was developed (Figure 3.16c). Here, the working electrode is a wire that is continually drawn past the workpiece. Thus, there is always a new part of the electrode available for machining.

Micro-EDM systems have been developed for microengineering applications. These typically employ three-axis positioning systems with micrometer XY accuracy and a smaller working electrode. This, by itself, poses a problem because

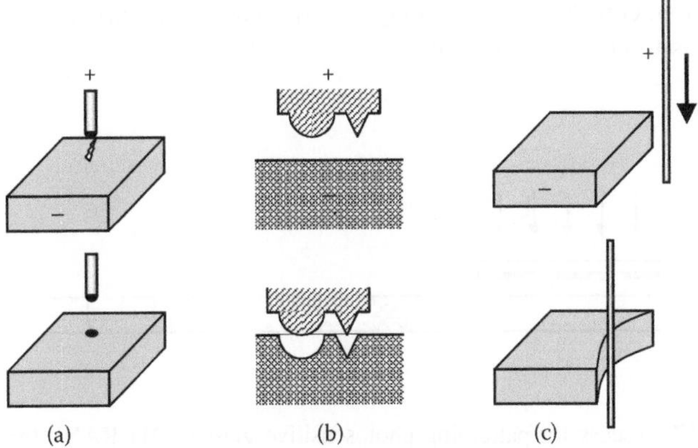

FIGURE 3.16 EDM modes: (a) hole boring, (b) shaped working electrode, (c) wire EDM.

the smaller electrode degrades very rapidly during machining. For this reason, different machining procedures and electrode shapes are under development to extend the electrode life.

3.8 PHOTOSTRUCTURABLE GLASSES

Micromachined glass is a popular alternative to silicon in many applications, especially biological, where glass is relatively inert and transparent, allowing biological processes to be viewed directly under an optical microscope. The normal process for patterning glass is to apply a photoresist, expose, and pattern it using either BHF or RIE, the latter being used where anisotropic etching is required. To be able to pattern and anisotropically etch glass plates without the use of a photoresist or RIE system would considerably simplify the process of producing glass components and result in cheaper components. To this end, various photosensitive glasses have been developed. These are generally based on silver compounds introduced into the glass. UV exposure results in free silver atoms being released in the exposed areas. The glass is then heat treated so that it crystallizes around the free atoms. The result is that the exposed glass etches at up to 20 times faster than unexposed glass in 10:1 HF. The process is outlined in Figure 3.17. Note that the depth of the structure produced in the glass will be dependent on the etching time and etching rate of the exposed glass, and the sidewall quality will depend both on the quality of the illumination (divergence, for instance) and the selectivity ratio of exposed to unexposed areas.

There are two problems that presently limit the wider uptake of this process. The first is that the etch rate and final results depend strongly on process parameters, and it can take some time to set up a reliable system. The second is that heat treatment of the glass may take several hours. Although it only needs to be held at an elevated temperature of between 500°C and 600°C for 1 to 2 h to effect crystallization, considerable time may be required to ramp up and down to these temperatures to ensure goodresults.

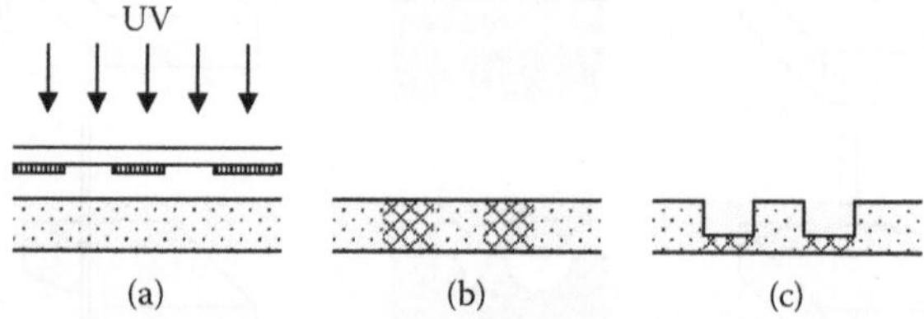

FIGURE 3.17 Process for patterning photosensitive glass (FOTURAN): (a) the glass is exposed to UV light through a mask, (b) following heat treatment, the exposed glass crystallizes, (c) the crystallized glass is then etched in 10:1 HF.

FIGURE 3.18 Cross section of channel etched in photostructurable glass (FOTURAN), 1-mm-thick substrate, 1 degree slope on wall, 100-μm-wide channel (approximate). (Image courtesy of mgt mikroglas technik AG, Mainz, Germany. www.mikroglas.com.)

Nonetheless, the process can achieve quite remarkable results given the relatively simple equipment requirements. Figure 3.18 shows the cross section of a high-aspect-ratio channel etched in photostructurable glass.

3.9 PRECISION ENGINEERING

Various techniques developed under the banner of precision engineering have either been adapted directly as microengineering processes (CMP and EDM, for example, or wafer-dicing techniques from Chapter 2) or fall into the categories of microengineering and nanotechnology by dint of the results that they are capable of achieving. Many of the tools involved in precision engineering have been discussed elsewhere; they include:

- Solid cutting or abrasive tools (e.g., diamond saw blades used in wafer dicing Chapter 2, section 2.8.1)
- Free abrasives (in fixed abrasive processes; e.g., CMP Chapter 3, section 3.2)
- Scanning tip tools (e.g., STM and AFM Chapter 10, section 10.3)

- Energy beam tools (e.g., lasers and ion beam milling Chapter 3, section 3.5 and Chapter 10, section 10.4.3, respectively)
- Measuring probes

One of the more widely used techniques that is capable of ultraprecision machining of a variety of materials is single-point diamond turning. These systems are commercially available and are often used in the optics industry. They are capable of machining glasses and ceramic in the ductile regime, so surfaces will be free from damage due to brittle fracture of the material. Consequentially, the surface finish is of very high quality (smooth), and the finished structure may be usable without any subsequent grinding or polishing.

CNC (computer numerically controlled) machines for single-point diamond turning possess a number of specific features to enable them to achieve the required precision. They will almost certainly be operated in a very-tight-temperature-controlled environment (to better than ±0.25°C), and the position of the tool will be controlled close to nanometer accuracy through use of feedback control systems; often laser interferometry or holography will be employed. Some of the capabilities of ultraprecision turning centers, available both industrially and in research environments, are given in Table 3.2.

Exactly what can be achieved depends on the tool, and Table 3.2 suggests what could be expected. A typical commercial machine, for instance, may have a programming resolution of 10 nm over 300-mm travel and achieve a straightness of 300 nm over the full travel. The surface finish is very dependent on the conditions and the material being machined; the 7 Å rms figure would be under ideal conditions, whereas tens or several tens of angstroms would be more usual. Materials can include nonferrous metals, ceramics, polymers, and crystals.

3.9.1 Roughness Measurements

These have been given here as angstroms rms. One Angstrom (1 Å) is 1×10^{10} m, or 0.1 nm. Here, rms stands for "root, mean, square," and this describes how the number was obtained. A rough surface will feature a number of peaks and troughs (Figure 3.19) of random heights and depths. The distance between the highest peak and deepest trough is not necessarily a good measure of the roughness, as they may appear only once on the entire surface. A better measure would be to find some sort of representative average. This can be performed by measuring an area of the

TABLE 3.2
Parameters of Ultrahigh-Precision Turning Centers

Tool positioning resolution	1.25–10 nm
Tool travel	0.1–2 m
Accuracy over full range of travel	100–600 nm
Surface finish (rms)	7Å–300 Å rms

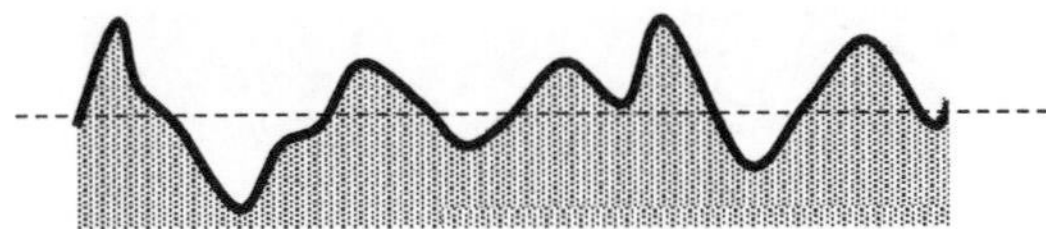

FIGURE 3.19 A rough surface in cross section. The dotted line indicates the average level of the surface; the rough surface deviates from this perfectly flat surface in both directions. Surface roughness can be calculated as the rms deviation from this.

surface (Figure 3.19 shows a single line; normally the surface would be scanned in the x and y directions) and finding the line about which the average displacement (i.e., the average of all peaks and troughs) is 0; this is where the surface would be if it was perfectly flat. The roughness is then calculated as the deviation from this imaginary perfect surface. At all measured points, the deviation is first squared (i.e., multiplied by itself; hence, negative deviations become positive), the average (mean) of these squares is found, and then the square root of this number is taken. This gives a reasonable estimate of the surface roughness, but note that the actual deviation from the imaginary ideal surface will normally have a greater amplitude.

Surface roughness can be readily measured using scanning probe microscopes (Chapter 10), particularly the AFM, and many AFMs are supplied with software that allows the surface roughness to be readily measured.

It is also necessary to be aware that surface roughness will be different when measured on different scales. A 100-μm-deep hemispherical pit may be atomically smooth when measured with an AFM, but a surface covered in such pits would appear to be quite rough to touch. Another analogy would be a road that appears very smooth to drive over, but very rough when touched by the hand.

3.10 OTHER PROCESSES

The purpose of this chapter has been to provide a brief survey of alternative (nonsilicon) micromachining processes, and not all of these have been presented. It should be borne in mind that alternatives are available and that a silicon-based process is not necessarily the best solution. The glass filters shown in Figure 3.20 were produced by drilling glass blanks and then drawing them out. A number of other processes are commonly referred to in discussion of microengineering problems. These include:

- Ultrasound — used for drilling (even welding) as well as to assist in cleaning and etching.
- Sand blasting.
- Electroless plating — as with electroplating, a thin film of metal is deposited on the surface of an object immersed in solution containing metal ions. However, the process by which the metal ions are reduced (brought of solution to be deposited) is chemical and not electrical, so

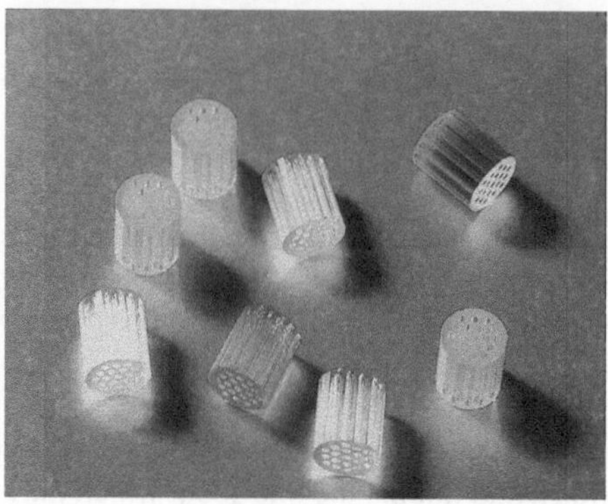

FIGURE 3.20 Glass filters. (Image courtesy of mgt mikroglas technik AG, Mainz, Germany. www.mikroglas.com.)

metal films can be electroless-plated onto nonconducting materials (plastics, glasses, etc.).

- EDM-related drilling processes — shaped tube electrolytic milling (STEM), capillary drilling (CD), and electrostream drilling (ESD).

REFERENCES

1. Effenhauser, C.S., Bruin, G.J.M., Paulus, A., and Ehrat, M., Integrated capillary electrophoresis on flexible silicone microdevices: analysis of DNA restriction fragments and detection of single DNA molecules on microchips, *Anal. Chem.*, 69, 3451–3457, 1997.
2. Jackman, R.J., Brittain, S.T., Adams, A., Prentiss, M.G., Whitesides, G.M., Design and fabrication of topologically complex, three-dimensional microstructures, *Science*, 280, 2089–2091, 1998.
3. Qin, D., Xia, Y., Rogers, J.A., Jackman, R.J., Zhao, X.-M., Whitesides, G.M., Microfabrication, Microstructures and Microsystems, in Manz, A. and Becker, H., Eds., *Microsystem Technology in Chemistry and Life Sciences*, Springer-Verlag, Berlin, Heidelberg & New York, 1998, chap. 1.
4. Ikuta, K., Maruo, S., Kojima, S., New Micro Stereo Lithography for Freely Movable 3D Micro Structure, *Proceedings of the IEEE 11th Annual International Workshop on Micro Electro Mechanical Systems*, MEMS 98, Heidelberg, Germany, 290–295, 1998.

4 Mask Design

4.1 INTRODUCTION

The process of photolithography has already been introduced in Chapter 1 of this volume, and many of the micromachining techniques introduced in Chapter 2 and Chapter 3 make use of this process. The present chapter presents the process by which such masks are designed. In this chapter, the following assumptions have been made:

- Designs are fabricated by a laboratory or foundry that has developed its own processes and can supply relevant process details (required alignment marks and design rules).
- The micromachined devices have relatively large (micrometer) minimum feature sizes.

Generally speaking, the design of masks for MEMS is a fairly straightforward process. All that is required is some suitable CAD (computer aided design) software and a platform on which to run it.

4.2 MINIMUM FEATURE SIZE

The concept of a *minimum feature size* is important in mask design (readers will note that it has already been described in Chapter 1). The minimum feature size is the width of the smallest line or gap that appears in the design. When the mask is produced, information on the minimum feature size will be required to set up the equipment. A small minimum feature size will mean that the mask creation process takes a long time and becomes increasingly expensive; with very small features (submicron) it may be necessary to perform additional processing of the mask for optical correction purposes. This should not happen for most micromachining applications: one would expect to have minimum feature sizes in the region of a few micrometers.

4.3 LAYOUT SOFTWARE

The basic CAD software required is a *layout editor*, which resembles a drawing package with additional features to facilitate the creation and editing of IC designs. Figure 4.1 shows screen shots of two different layout editors: Tanner's L-Edit Pro and Static Free Software's Electric. Electric will be used for the examples in this chapter.

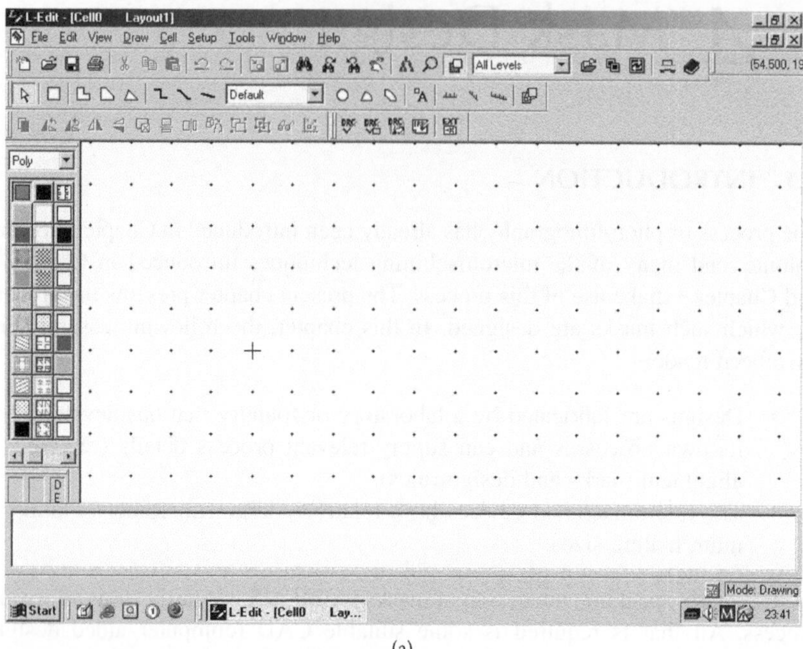

(a)

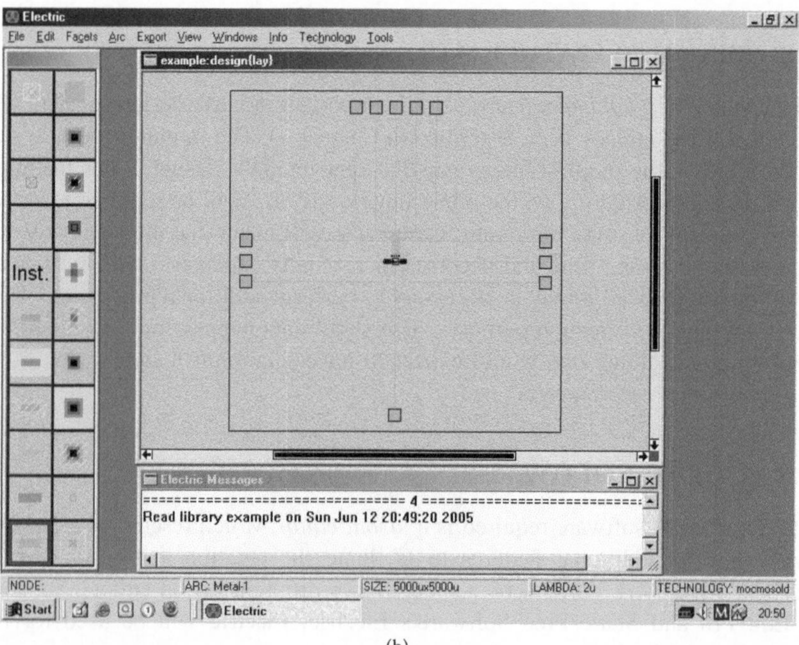

(b)

FIGURE 4.1 Mask design software: (a) Tanner's L-Edit (www.tanner.com), (b) Electric from Static Free Software (www.staticfreesoft.com).

The layout editor enables the placement of different polygons onto different layers of the design. Each layer corresponds to a different mask that will be used in the fabrication of the devices. The polygons will be translated into opaque (or clear) areas on the mask plate.

As will become apparent, the design can be created hierarchically. Polygons can be grouped together into *structures* or *cells* (Electric terms them *facets*), which can then be placed adjacent to each other to create a larger or more complex design. The other essential feature of a layout editor is the facility to import and export designs in file formats that can be used in the mask-making process.

Many software packages will offer additional features. It is also possible to use some drawing packages (AutoCAD, for example) provided that the files created can be translated into an appropriate format. Anyone familiar with a CAD or a drawing package should have few problems using a layout editor.

4.3.1 File Formats

There are two file formats that are commonly accepted by most mask makers: CIF (CalTech Intermediate Form) and Calma GDSII. Any layout editor worth its salt should be able to import and export designs in both of these formats (editors usually use their own format for storing designs that are being worked on). There are relatively few differences between the two formats that are apparent to the user. Documentation supplied with the software should describe how conversions between the two formats are handled and point out any potential problems. One thing to note is that GDSII does not draw true circles; it uses multisided polygons. The number of sides used is normally set by default in the software, but this setting should be accessible for the user to change.

CIF is an ASCII (text) format, whereas GDSII is a binary format. This means that it is possible to view CIF files with a text editor if desired. This can be useful when dealing with foreign files that have no associated technology files (see following text), as layer names can be extracted from the design file itself (otherwise it is necessary to monitor the errors occurring when the file is imported and act accordingly).

Differences between CIF and GDSII

CIF	GDSII
• ASCII format	• Binary format
• Layers referred to by names of up to three letters long	• Layers numbered (GDSII stream numbers)
	• Does not support true circles — uses multisided polygons instead

4.3.1.1 Technology Files

In addition to the design file, which stores details of the mask designs, layout editors also require technology files. These essentially store the "preferences" for the design, such as:

- Layer names used in the design
- Layer numbers used in the design (layers may well be named and numbered)
- GDSII stream numbers (to or from which layer names and numbers are translated when writing or reading a GDSII file)
- CIF layer names (as in the preceding point, but for CIF)
- What layers are made available for drawing
- How to color the different layers on screen
- The units to be used
- How to draw wires (essentially long rectangular polygons, but made easy to draw as interconnections between other elements of the design)

Other things that may be included in the technology file are how to scale certain drawing objects, design rules, and physical parameters to enable the extraction of electrical properties from the final design. For some fabrication processes, vias (holes in an insulating layer that enable one conducting layer to be connected to another below) may only be allowed to be of a fixed size; to connect larger areas or allow the passage of higher currents, several vias rather than one large one may be used. In this case, the ability to specify how certain objects scale can be useful and make the design process a little easier. Design rules are discussed in greater detail in the following text. But allowed overlaps and spacings between polygons must be set.

Parameter extraction is very useful in IC design as it allows the designer to extract the electrical parameters (e.g., resistance and capacitance of interconnects) of the design as laid out on the chip. The design can then be simulated, with these parameters entered, to verify its operation. For MEMS, some specialized software exists for modeling; however, there are generally so many unknowns that it is not possible to extract physical parameters from a given layout to enable verification. The closest approximation is where structures with known properties are included in the design — the remaining electrical parameters can then be extracted and simulated along with appropriate models of these structures.

Technology Files and Imported Files

Note that when importing a CIF or GDSII file, it will be necessary to tell the layout editor which technology file it should use. It will be necessary to create one if a suitable one does not exist.

4.3.1.1.1 Units

The thing that may come as a novelty to users of conventional drawing packages is the concept of *user units* and *internal units* (or design units). The user units of interest for micromachining are micrometers (microns, μm). Other options may be mils (1/1000 in., or 25.4 μm) and lambda (λ), which is used for scalable IC designs.

The user units are the units that the design will be created in, i.e., entering 25 into the software will cause a 25-μm-wide line to appear. The internal units are the units used by the software for its calculations. It is normal to set 1000 internal units to every user unit. Internal units are generally invisible to the user, but because the software normally has a fixed memory (i.e., internal unit allocation), the overall size of the design will be affected by the number of internal units or user units that one specifies. In other words, if the software is set for 2000 internal units or user units, the total area that can be used to create the design will be half of that using 1000 internal units or user units.

Lambda units are used for scalable IC design, and design rules may well be set in terms of lambda units if the fabrication process is based or partially related to an IC process. Lambda is half the minimum feature size, i.e., for a 4-μm minimum feature size design, λ will be set to 2 μm. Figure 4.2 shows a dialog

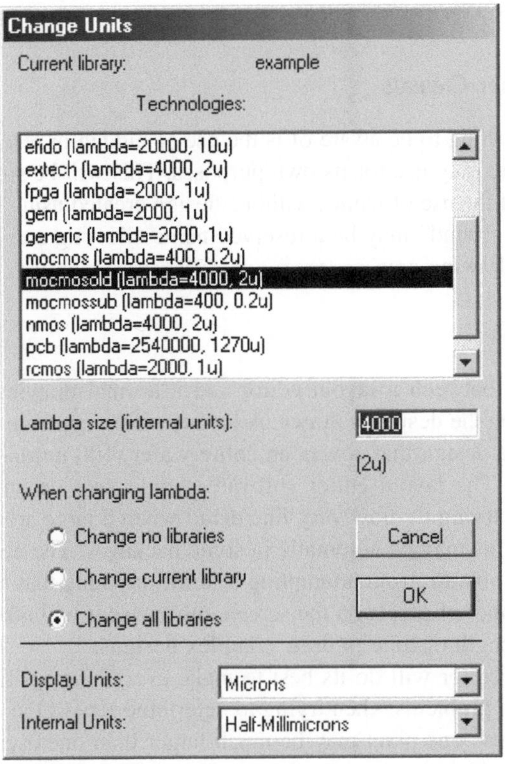

Change Units
Current library: example
Technologies:
efido (lambda=20000, 10u)
extech (lambda=4000, 2u)
fpga (lambda=2000, 1u)
gem (lambda=2000, 1u)
generic (lambda=2000, 1u)
mocmos (lambda=400, 0.2u)
mocmosold (lambda=4000, 2u)
mocmossub (lambda=400, 0.2u)
nmos (lambda=4000, 2u)
pcb (lambda=2540000, 1270u)
rcmos (lambda=2000, 1u)
Lambda size (internal units): 4000
(2u)
When changing lambda:
Change no libraries
Change current library
Change all libraries
Cancel
OK
Display Units: Microns
Internal Units: Half-Millimicrons

FIGURE 4.2 Electric dialog box to set up units.

box for setting up display (user) units and their relationship to lambda and the internal units (half millimicrons, i.e., one internal unit will be 0.5 nm).

It is usually possible to create a design using either micrometers or lambda units. The latter are generally used for circuitry and not for micromachined devices or structures. When moving to a new IC fabrication process with a smaller feature size, the value of lambda is scaled appropriately so that the design need not be recreated from scratch. The software may offer an option to indicate which parts of a design can be scaled and which cannot (e.g., bonding pads). In such a case, it may be possible to create scalable designs that combine microstructures with circuitry, but the process is generally difficult to implement.

Units in Summary

Internal units	Used by the software
User units	Used to create the design, usually in micrometers, and usually set to about 1000 internal units
Lambda units	Half the minimum feature size

4.3.1.2 Further Caveats

One additional thing to be aware of is the use of hidden or reserved layer names that the software may use for its own purposes. This will be software dependent and will restrict the use of names within a design created using that package. For instance, "background" may be a reserved name allowing one to vary the background of the drawing area.

4.3.2 Graphics

One difference between a layout editor and a normal drawing software is the scale over which the design is drawn. A layout will range from features as small as 1 μm up to a design that covers an entire wafer (100 mm) — a difference of 100,000 times. The layout editor software should provide an option that will prevent it from trying to draw very fine detail when a large area of the design is being viewed; this may be automatic in some packages. The normal approach is to prevent the software from attempting to draw anything that takes up less than a specified number of pixels on the screen, otherwise it will take the machine an unreasonable length of time to draw complex designs.

The layout editor will do its best to make everything visible on the screen. This can lead to problems when trying to determine size or the distance between two items because one pixel may be much larger than one user unit, depending on the scale at which the design is being viewed. There will normally be a ruler or cursor system in the software package to allow distances to be measured.

In order to get an accurate measurement, the design should be viewed at an appropriate scale — normally, the highest possible magnification.

4.3.3 Grid

A grid is a useful feature of any drawing package and more or less indispensable for mask design. Layout editors generally feature grids that can be turned on or off (i.e., displayed or hidden) with variable mesh dimensions. Points and drawing elements will be locked to the intersections of the gridlines (i.e., snap to grid), and it is often useful to set the minimum spacing on the grid to the minimum feature size being employed. The usefulness of the grid cannot be overemphasized.

Polygons can also be edited by specifying the coordinates of their vertices (corners), as illustrated in Figure 4.3. Very complex polygons can be created in this manner, but a lot of preplanning is required. Because MEMS normally involve fairly complex polygons, it is worth planning the coordinates that these will require beforehand.

4.3.4 Text

Most packages allow text to be typed onto one layer of the design. This will not normally appear on the mask itself when fabricated. Most layout editors do not allow text to be entered in such a manner that it will appear on the mask (sometimes it is supplied as an advanced option). When placing text on a mask

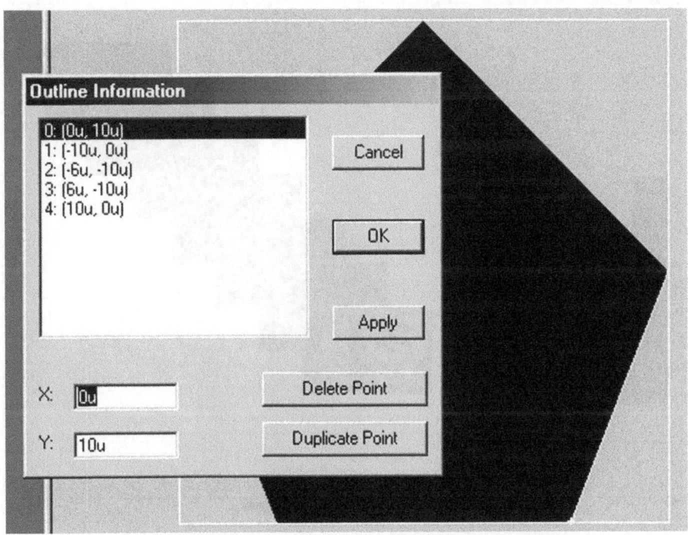

FIGURE 4.3 Setting vertices of a polygon by editing points directly.

design (layer names, mask serial number, etc.), it is normally necessary to create it as complex polygons.

4.3.5 Other Features

Other things that may help when creating a design include: turning different layers on or off to make them accessible or inaccessible for modification and options for selecting an item to be modified. These will obviously be software specific.

4.3.6 Manhattan Geometry

Manhattan geometry refers to a restricted form of layout in which designs may only be composed of horizontal and vertical lines. Complex polygons may also be restricted. This is usually a feature of older CAD packages and is somewhat restrictive in MEMS designs. The way around this is to put simple rectangles up against one another to create complex shapes: any two polygons on the same layer that touch or overlap will appear as a single contiguous area on the mask. Diagonal bars can be approximated by an array of overlapping squares or rectangles. Tools for placing arrays of design features (either simple polygons or more complex structures) are a common component of layout editors. Figure 4.4 illustrates this.

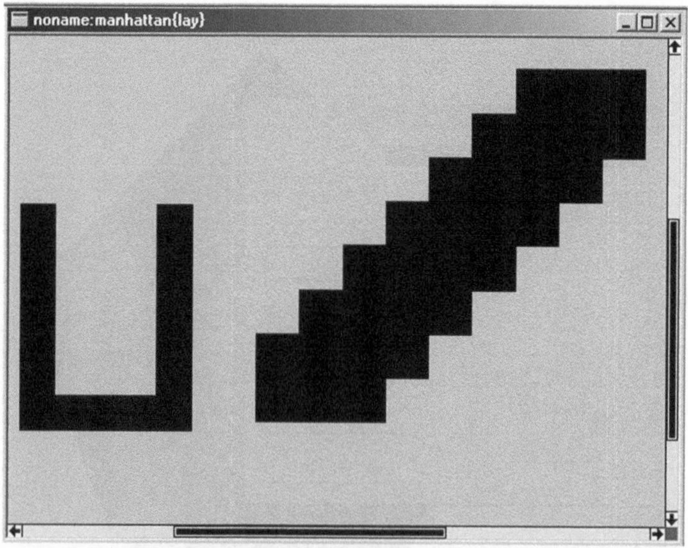

FIGURE 4.4 Creating complex polygons from simple squares and rectangles.

Graphics and Drawing Hints

- Use the grid.
- Plan the design in advance, including the coordinates for complex shapes.
- Measure distances and sizes, and make sure that the display is zoomed in as much as possible.
- Adjacent or overlapping polygons on the same layer will become a single contiguous shape on the mask.
- Most packages allow text to be placed on the design but it will not appear on the mask (check the manual or a CIF file to see if this is true for a particular package).

4.4 DESIGN

If the concept of creating a design by depositing and machining successive layers is comprehended, the process of designing the required masks should be straightforward. It only requires the necessary mental visualization to see how the two-dimensional mask designs relate to the final structure.

In order to explain the basic concepts of mask design, the device outlined in Figure 4.5 will be used as an example. It is not intended to be a real device but is a simple design that could represent a microbiosensor. There are a set of exposed

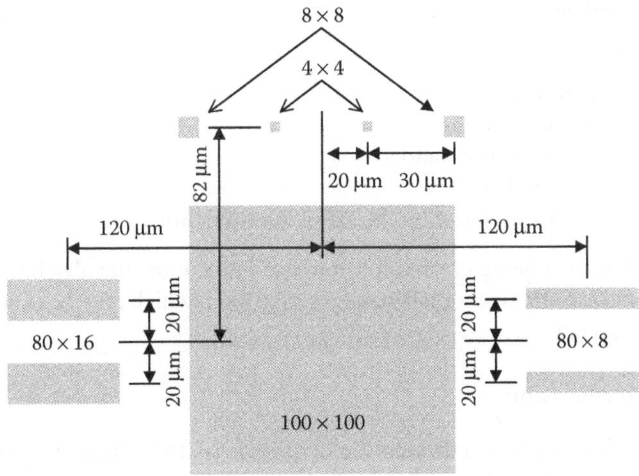

FIGURE 4.5 Electrode layout for the device used in the examples in this chapter. These are metal electrodes; the passivation layer is etched to create the electrode pattern illustrated. This is to be sited at the center of a 5 × 5 mm die. Each electrode is to be connected to a 180-μm square bond pad at the edge of the die.

metal electrodes at the center of a 5 × 5 mm silicon die (chip). The large central electrode represents a reference (ground or 0 V) electrode. The two long thin electrodes to either side could be some form of active sensor, possibly functionalized by electroplating with a conductive polymer combined with a bioactive compound. The four electrodes at the top could be used for electrical impedance measurements (two to apply a signal and two to sense it). A small drop of the sample could be placed in the center of the chip and bonding pads at the edge of the chip could connect it with appropriate circuitry.

The design consists of a silicon substrate and three layers: an insulating layer (CVD or thermal oxide) deposited on the wafer but not patterned, a metal layer (nominally M1) to form the electrodes' bonding pads and conducting tracks between them, and a top insulating layer of oxide (e.g., PECVD) that would be patterned to open vias (holes) above the electrodes and bonding pads so that electrical connections could be made to these; this layer would nominally be V1.

This is a three-mask process — one (A1) to cut alignment marks into the silicon substrate, M1 to pattern the metal, and V1 to pattern the vias. The mask set will be called EX1 and will employ a 4-μm minimum feature size (although such a simple design could be created with a larger minimum feature size).*

A number of different structures or cells will be created and stitched together to create a wafer-sized design.

4.4.1 The Frame and Alignment Marks

Each individual chip design will be drawn within a frame. This is usually supplied by the foundry or laboratory that will manufacture the devices and will normally include the following:

- Scribe lane
- Alignment marks
- Test structures
- Layer identification marks
- Mask set number
- Other marks required by the fabrication facility

If the design is being created for in-house fabrication, the absolute minimum requirements are alignment marks and a scribe lane, which may be useful, if only to assist in layout.

4.4.1.1 Scribe Lane

The scribe lane is an area around the chip that has to be kept free of circuitry or components. This defines the grid along which the wafer will be cut into

* The example uses Electric and the mocmosold technology file (library) that it is supplied with; an old double-metal CMOS process that was available through MOSIS [1]. The design will not be suitable for fabrication through MOSIS, and the library itself is no longer supported by MOSIS. It was chosen only because it has a relatively large value of lambda. For this example, the technology is not important.

FIGURE 4.6 "`scribelane`."

individual dies. Options for dicing up the wafer include the use of a dicing saw, a diamond-tipped scribe (which scores a groove across the surface of the wafer where it will be broken), and laser cutting (which burns a series of pits along the scribe lane). All these methods of dicing have disadvantages when dealing with wafers that incorporate micromechanical structures. The saw and scribe induce mechanical vibration that can damage components, and the laser induces thermal damage and can eject material over the surface of the dies. It may be worth considering incorporating a chemical (e.g., KOH) dicing step for some devices.

In the example (Figure 4.6), a 200-μm-wide scribe lane is created on all three layers in use. This is done by creating four 100-μm-wide rectangles on each layer and placing them as illustrated. The outside dimensions of this are 5000×5000 μm, so the area available for the design has been reduced to 4800×4800 μm. This structure has been named, curiously enough, `scribelane`.

4.4.1.2 Alignment Marks

In order to line up one layer with a previously fabricated layer when performing photolithography, it is necessary to incorporate appropriate marks into the mask design. The foundry producing the devices will probably have standard marks that should be used. Furthermore, these may need to be placed in a specific position on the design. Use the standard alignment marks if available. It is not unknown for process technicians and engineers to attempt to align parts of the design itself with the alignment marks appearing in the previous layer when dealing with unfamiliar designs.

A simple alignment-mark scheme consisting of squares and a cross is shown in Figure 4.7. When creating alignment marks it is important to be aware of

FIGURE 4.7 A simple alignment mark scheme. The cross on one layer is aligned inside four squares on another layer.

whether the mask will be dark field or light field. If a dark field-mask has to be aligned with a preceding light-field mask, then the alignment mark has to be larger, so that the previous mark can be viewed through the mask.

This is illustrated with the example in Figure 4.8. The structure `align` consists of a set of square alignment marks surrounded by a frame so that when

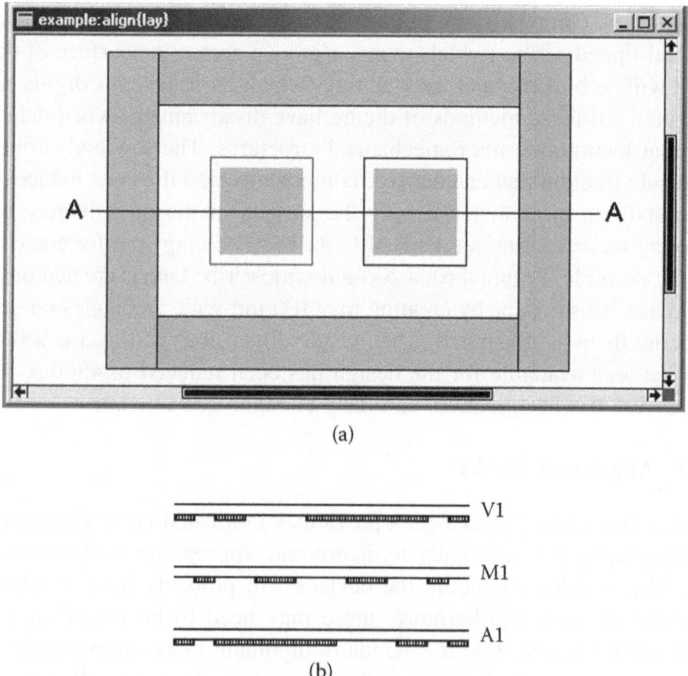

FIGURE 4.8 (a) Example "`align`," (b) outline of the masks in cross section A–A. Note that M1 is to be produced as a light field mask, and V1 and A1 are dark fields.

performing the alignment it is clear which mark is being aligned with which (the frame appears on all layers).

Masks A1 and M1 (A1 defines the alignment marks, etc., that will be etched into the silicon) are light field (i.e., will be opaque where a closed or opaque polygon appears in the design). Mask V1 is dark field (i.e., will be transparent where a closed or opaque polygon appears in the design). M1 will be aligned to A1, and V1 will then be aligned to M1. The alignment marks consist of a small M1 square on the left that will be aligned to the larger A1 square and a larger V1 square that will later be aligned to the smaller M1 square on the right. The squares are 20 μm and 16 μm on a side (large and small, respectively). This may imply a ±2-μm alignment error, but it will be influenced by various factors, such as over- or under-etching and the ability of the person performing the alignment to accurately place one square inside another. Although this range of error has been taken into account in the rest of the design (note the V1–M1 overlap elsewhere), it should be possible to achieve a much smaller error than this.

4.4.1.3 Test Structures

The foundry may wish to incorporate a variety of test structures within the design. For CMOS processes, these may include electronic components (transistors, ring counters, etc.) to verify conformation to process specifications. The example only includes a set of bars as the `test` structure (Figure 4.9).

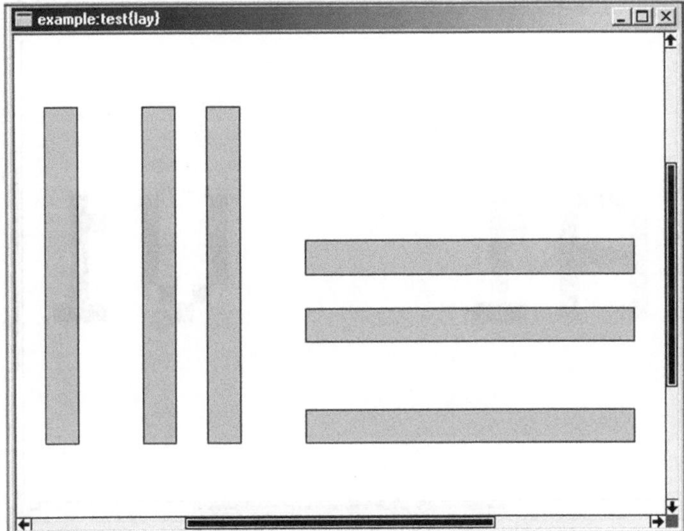

FIGURE 4.9 Example test structures.

A 4-μm minimum feature size has been chosen. The test structure consists of a set of 4-μm bars, spaced 4 μm and 8 μm apart, on each layer. These will enable monitoring of the process to check for such things as:

- Correct exposure
- Under- or over-etching
- Asymmetry between x and y axes (and, thus, the vertical and horizontal bars)

4.4.1.4 Layer and Mask Set Identification Marks

A number of different projects are normally in progress at any one time in a fabrication facility (fabs). For this reason, it is useful to be able to pick up a wafer and determine how far it has progressed and to which particular project it belongs. Therefore, each layer should have the layer name written on it, as illustrated in Figure 4.10a, and a name or number to identify the mask set, as shown in Figure 4.10b. Note that these have been drawn onto each layer as polygons.

4.4.1.5 Putting It All Together

These structures now need to be combined into a single structure that can be used as the basis of any chip design. To do this, a new structure, or a `frame`, is created (Figure 4.11a). The structure's `scribelane`, `align`, `test`, `layernames`, and `maskname` are then placed appropriately (Figure 4.11b). The alignment marks

(a)

FIGURE 4.10 (a) "`layernames`," (b) "`maskname`."

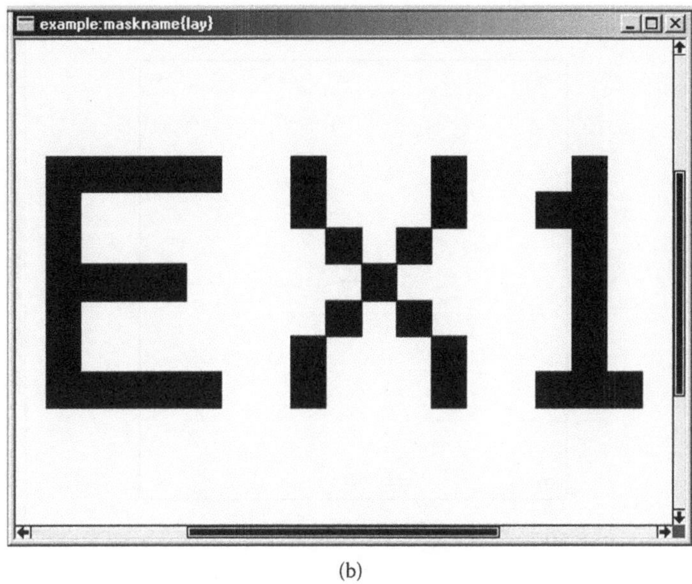

(b)

FIGURE 4.10 (Continued)

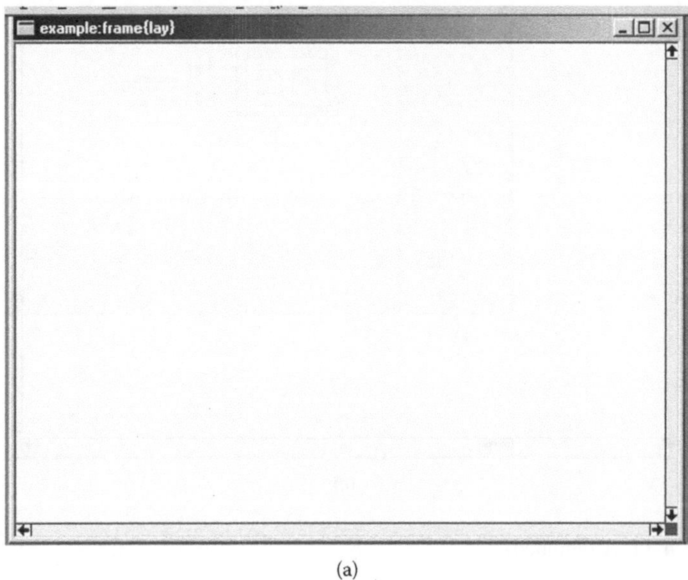

(a)

FIGURE 4.11 (a) Blank structure called "`frame`," (b) "`frame`" completed, (c) "`frame`" expanded and zoomed to lower left corner.

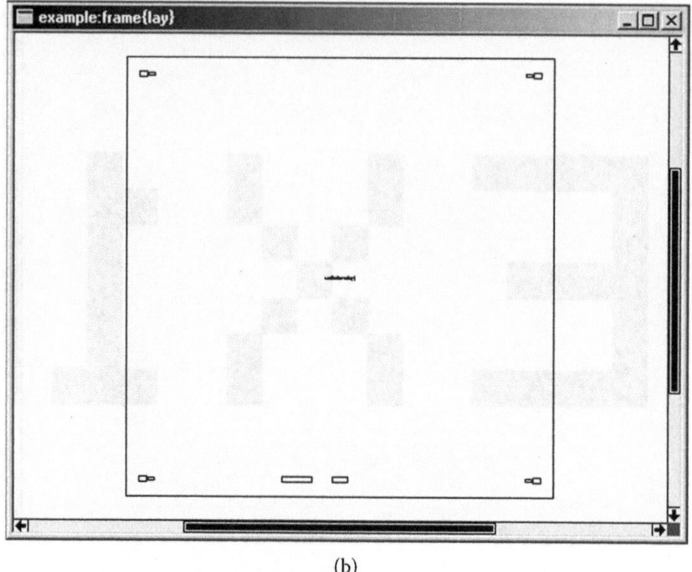

(b)

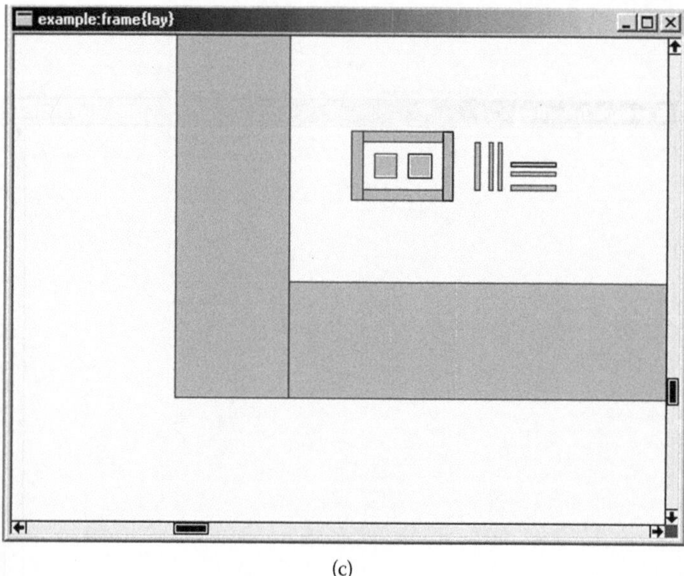

(c)

FIGURE 4.11 (Continued)

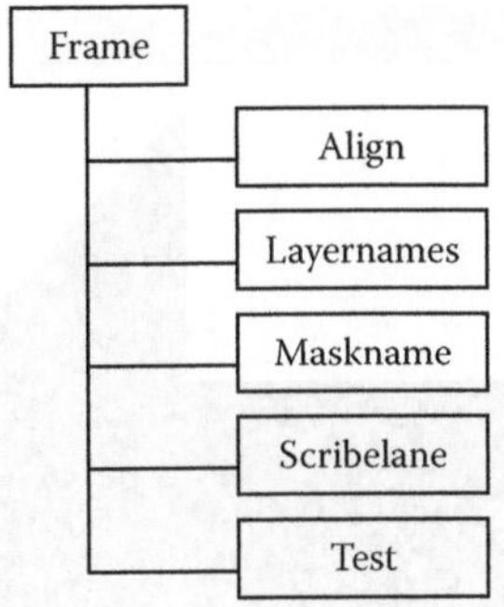

FIGURE 4.12 Design hierarchy for the example "`frame`" structure.

and test structures have been placed in each corner of the scribe lane structure, and the layer and mask names have been placed at the center in the bottom. The software displays these structures as named boxes.

It is possible to view the full design by pushing down (or expanding) on the hierarchy (or instances; each box containing a structure being termed an instance, although the terminology is somewhat software dependent). This has been done in Figure 4.11c and the window has been zoomed in to show the lower-left corner of the design.

It is clear that a design hierarchy is being built up, as illustrated in Figure 4.12. Note that if the mask name is changed in `maskname`, then the change will be reflected in all instances of the `frame`.

4.4.1.6 Another Way to Place Alignment Marks

Notice that the more structures that need to be placed in the frame, the less room there is available for the design. In the example being pursued in this section, it is known that the alignment will be performed by a mask aligner that will view alignment structures that exist along the centerline of the mask. So it is only necessary to place alignment marks in this area.

To do this, a special frame will be created. This will have an alignment mark at the center and a number of arrows to guide the eye to its position. It is only necessary to create one `arrow` structure (Figure 4.13a). This is placed with the correct orientation into `alignmentframe` and duplicated using the array placement command. The resulting structure is shown in Figure 4.13b and Figure 4.13c.

4.4.2 The Device

The sensor design illustrated in Figure 4.5 is shown as a layout in Figure 4.14 `(design)`. The `frame` structure has not been used, as the `alignmentframe` structure, discussed in the previous section, will be used instead. The area of the device has been defined by the use of `scribelane`, and a set of bonding pads have been placed near the edges of the design.

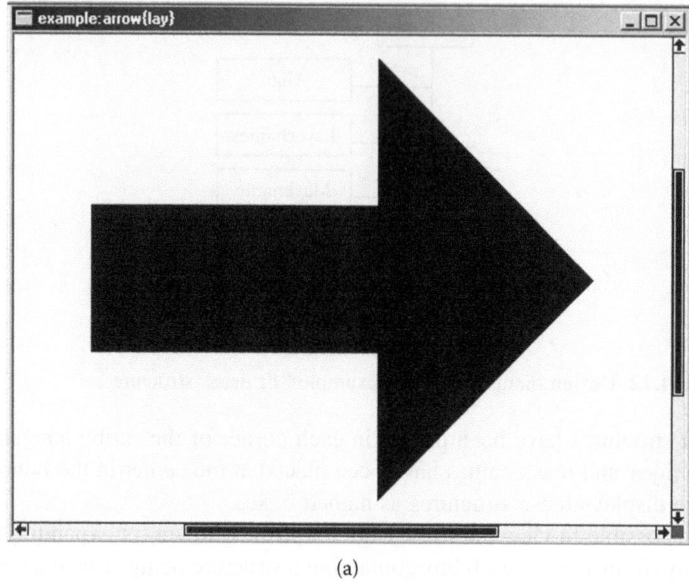

(a)

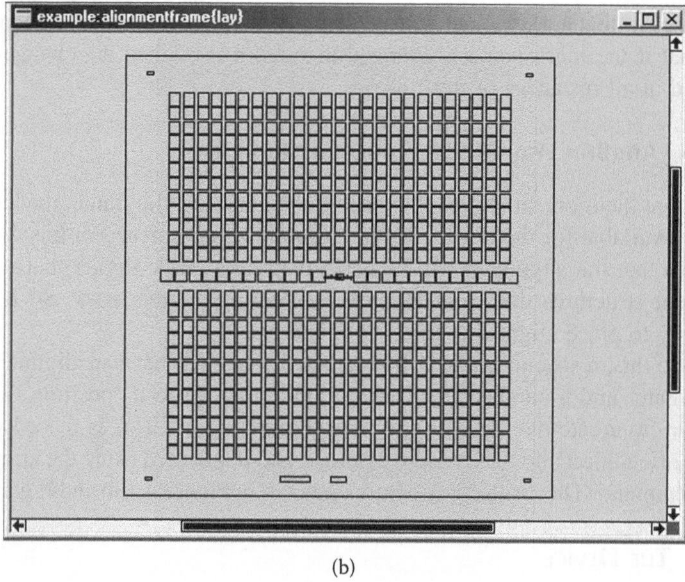

(b)

FIGURE 4.13 (a) "`arrow`," (b) "`alignmentframe`," (c) "`alignmentframe`" expanded and zoomed to center.

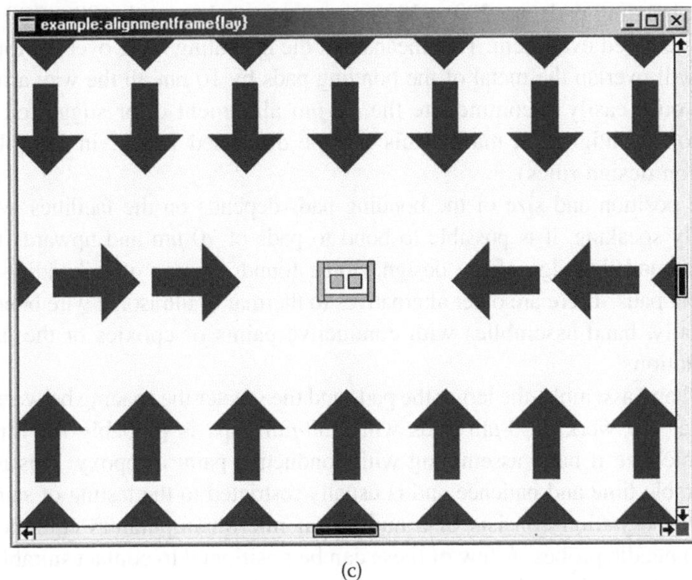

(c)

FIGURE 4.13 (Continued)

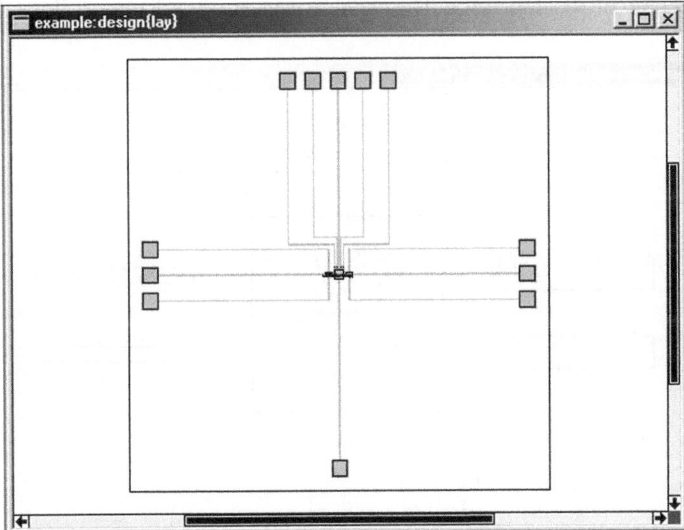

FIGURE 4.14 "`design`" layout

The bonding pads are 200 × 200 μm squares on M1, with 180 × 180 μm V1 squares centered over them. This means that the insulating layer over the top of the device will overlap the metal of the bonding pads by 10 μm all the way around — which could easily accommodate the ±2-μm alignment error suggested by the design of the alignment marks (this will be discussed further in the following section on design rules).

The position and size of the bonding pads depends on the facilities available. Generally speaking, it is possible to bond to pads of 70 μm and upwards that are placed around the edge of the design. Some foundries may offer bonding to less accessible pads. There are other alternatives to thermal or ultrasonic wire bonding — specifically, hand assemblies with conductive paints or epoxies or the use of a probe station.

For hand assembly, the larger the pads and the greater the spacing between them, the easier the work; 200-μm pads with 200-μm gaps is probably the minimum reasonable size if hand assembling with conductive paint or epoxy. This requires considerable time and patience and is usually restricted to the testing of samples.

A probe station consists of a number of micromanipulators equipped with tungsten needle probes. A few of these can be positioned to contact suitably sized pads anywhere within the design, and it is a fairly common piece of test equipment. Pads for probes can be placed anywhere within the area of the layout and, as such, can be useful for breaking out signals that may be important when testing or debugging, but they are not required once the packaged device is in use.

The sensor area is shown in Figure 4.15 with a 4-μm grid superimposed over it. The smallest openings in the V1 layer are 4 × 4 μm (the pair of small electrodes at the center of the top row). The next largest are 8 × 8 μm, followed by the

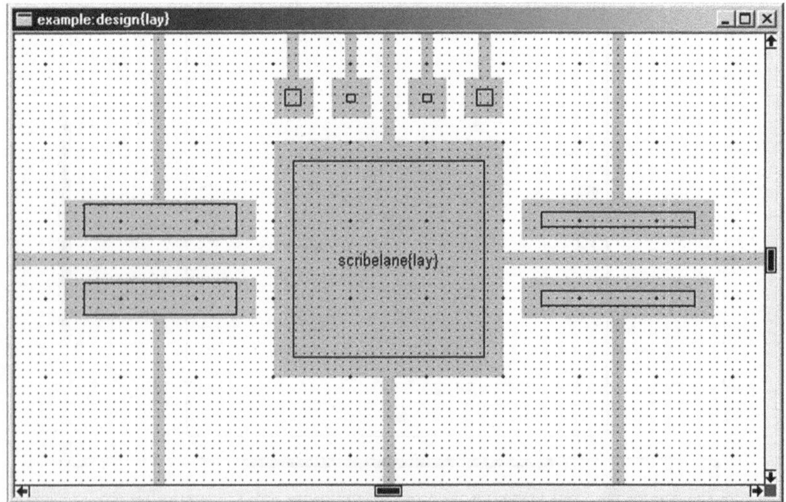

FIGURE 4.15 Sensor area of "`design`" with 4-μm grid superimposed (bold dots are 40 μm apart).

narrow ones on the right-hand side, which are 8 × 80 μm, then the 16 × 80 μm ones and, lastly, the 100-μm square reference electrode at the center. Notice that only three different sized M1 structures have been used: 20 × 20 μm, 20 × 100 μm, and 120 × 120 μm. This gives only a 2-μm V1–M1 overlap for the larger long electrodes. The tracks were drawn using the wire tool and are 6-μm wide.

It is normal for a designer to incorporate structures of several different dimensions into a single design during the development process. It is unlikely that the first process run will produce the desired results, so the more alternative designs that can be fitted into the available area, the better. It would have been wise for the designer in this example to have included another four-electrode impedance sensor at the bottom of the design, as it is quite possible that the 4-μm vias will not be etched properly.

Any departures from process guidelines or design rules should be tested on some area of the wafer, not just structures that are at the minimum feature size. A process optimized for etching 4-μm vias, for instance, may not perform as well when etching larger areas: the excess debris may be distributed over adjacent areas of the design, the etch may proceed more rapidly over larger areas, etc. Similarly, in research establishments in particular, processes may perform optimally for devices near the center of the wafer, so it is worthwhile laying out the wafer with an even distribution of the different designs.

Design and Layout Hints

- Fit as many designs as possible into the area available.
- Include several different approaches — especially if design rules are being broken.
- Do not break design rules unless absolutely essential, and do not expect a good result.
- Distribute different designs evenly across the area of the wafer.
- Include test structures where possible to check the patterning of various films at the dimensions of interest.
- Leave ample overlap between features on different layers.
- Do not make bonding or probe pads too small (70 μm on up; 100 μm preferred).
- Include marks to identify different parts of the design and orientation.
- Keep designs away from the edge of the wafer if a deep etch is to be employed.

Finally, the individual designs need to be laid out for fabrication. In this case, a 4-in. (100-mm)-diameter wafer will be used. The structure `wafer` is created and, to simplify the layout, an annulus (ring) of 100-mm outside diameter and 90-mm inside diameter is created on an unused layer of the design. This will be deleted later or simply not translated to CIF/GDSII.

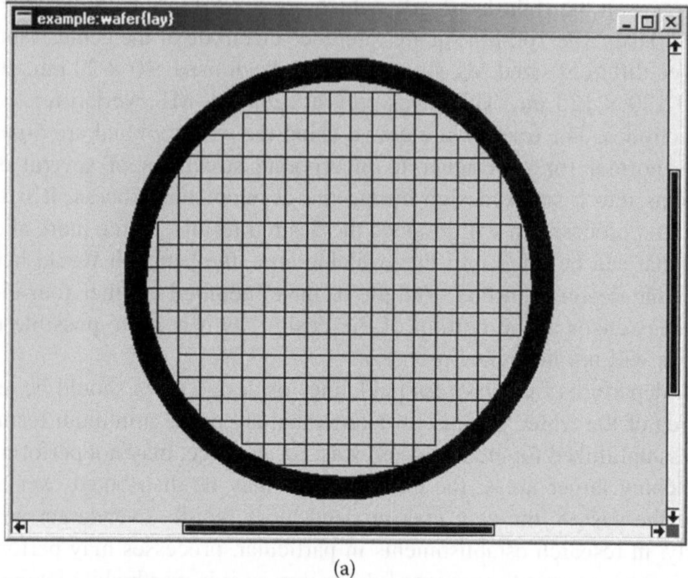

(a)

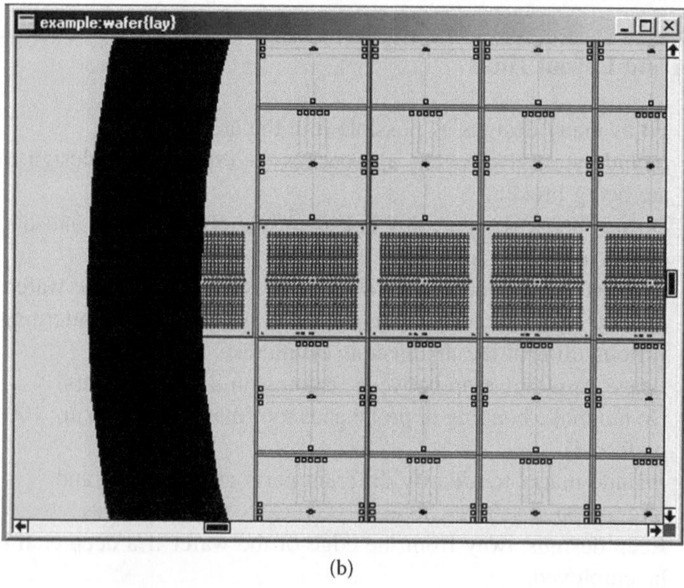

(b)

FIGURE 4.16 (a) "`wafer`," (b) "`wafer`" zoomed in to the left-hand edge and expanded. The black circular border is on a separate non-fabrication layer, and is used to guide the layout of the dies.

An `alignmentframe` structure is placed in the center of the design and repeated across the axis using the array placement option. The process is repeated to fill in the rest of the working area with the `design` structure. Note that, depending on the process, designs within about 5 mm of the edge of the wafer will probably be damaged or only partially fabricated. If a deep etch (KOH or deep RIE, for instance) is to be employed, then this could damage the edge of the wafer, making it difficult to handle and more likely to break. The final layout is shown in Figure 4.16.

4.5 DESIGN RULES

Design rules are a set of rules that must be followed at the masks' design stage to ensure that the outcome of each stage of the fabrication sequence will be as expected. They have to be developed for each process and depend primarily on the accuracy of alignment and upon the processes followed. Design rules normally specify one of several things:

- Minimum feature size.
- Minimum overlap (or underlap) between two layers (i.e., if a via hole is to be aligned over a metal pad, then the pad must be larger than the via hole by a specified amount. If they are both the same size, then a slight misalignment will result in only part of the pad being exposed through the etched via hole.)
- Minimum separation between features on the same layer to ensure that they will appear as separate features on the device. (This is usually the minimum feature size; but for processes with considerable lateral progression, it will be larger — a deep boron diffusion will progress laterally under the mask approximately the same distance that it progresses into the silicon wafer, for example.)
- Minimum separation between features on different layers to ensure that they do not overlap. (Notes on alignment sequence in the text that follows explain why this may be different for different layers.)
- Allowed dimensions of specific features. (When etching via holes in an insulating layer, for example, all of them must be opened cleanly in a single etch step. Because small holes tend to etch a little more slowly than larger holes, all via holes may have to be the same size in order to ensure even etching across the device; to make a via capable of carrying more current, several holes are used instead of one large hole. Other process properties may also come into play to restrict the maximum allowable dimensions.)

These are the most common design rules, but the list is by no means exhaustive. There can be any number of design rules relating to process-specific aspects.

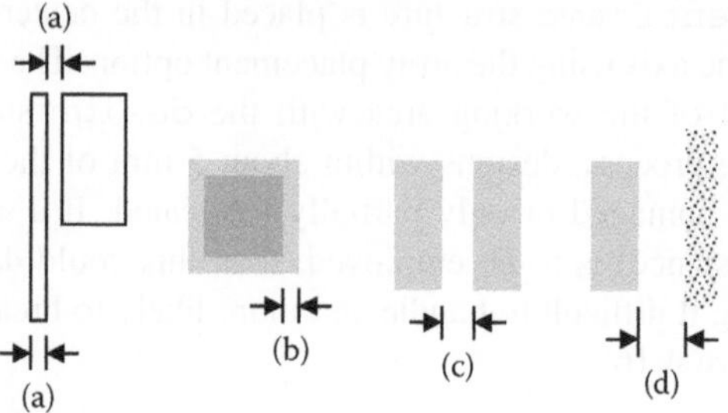

FIGURE 4.17 Design rule illustrations: (a) minimum feature size: line width or gap width, (b) minimum overlap between two layers to ensure that a via (dark gray) opens up over a metal pad (light gray), (c) minimum distance between two features on the same layer to ensure that they are separate when fabricated; for two parts of the same feature (e.g., a U shape) this may differ; it may also be different for two electrically connected features, (d) distance between features on two different layers to ensure that they will not overlap when fabricated.

When using a surface micromachining process, for example, it is necessary to ensure that the sacrificial material is completely etched away during the release stage. It would be reasonable, therefore, to suggest a rule that states the largest area of structural material that can be placed. To create a larger structure, it would be necessary to introduce holes within the structural material layer to enable the sacrificial material to be removed; these holes would need to have a minimum dimension and a maximum spacing.

Figure 4.17 illustrates the design rules outlined above. Figure 4.18 shows some design rules for the MOSIS (www.mosis.org) scalable CMOS process (SCMOS). CMOS design rules are commonly given in lambda units, whereas design rules for MEMS may well be in microns.

Design Rule Checking

Most CAD packages have a built-in design rule checker (DRC). This may be something that is run upon completion of a design to check that it complies with specified design rules, or it may work interactively and warn the user as soon as a structure that violates the specified design rules is drawn. Microengineering often requires unusual rules that cannot be checked using conventional DRC software or it may involve breaking specific rules (e.g., when creating structures using an electrical layer). Despite its limitations, DRC software is invaluable and should be learned and understood along with the layout editor. Rules that cannot be handled by the software should be checked by at least two people.

Mosis Scmos layout rules–7–metal 1		
Rule	Description	Lambda
7.1	Minimum width	3
7.2	Minimum spacing	2
7.3	Minimum overlap of any contact	1
7.4	Minimum spacing when either metal line is wider than 10 lambda	4

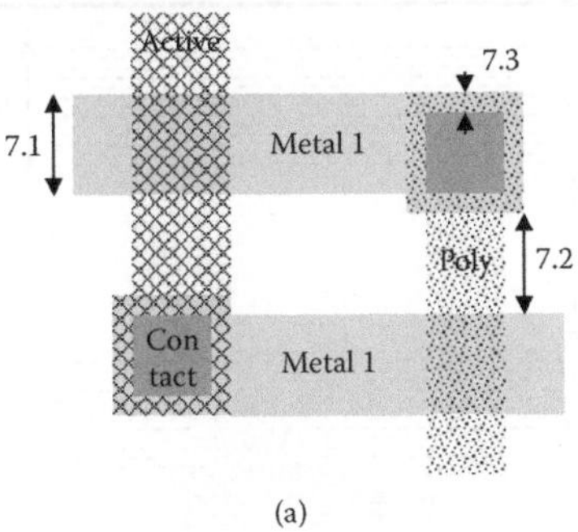

(a)

Mosis Scmos layout rules–8–via		
Rule	Description	Lambda
8.1	Exact size	2 × 2
8.2	Minimum via 1 spacing	3
8.3	Minimum overlap by metal 1	1
8.4	Minimum spacing to contact for technology codes that do not allow stacked vias (SCNA, SCNE)	2
8.5	Minimum spacing to poly or active edge	2

Vias must be drawn orthogonal to the grid of the layout.
Non-Manhattan vias are not allowed.

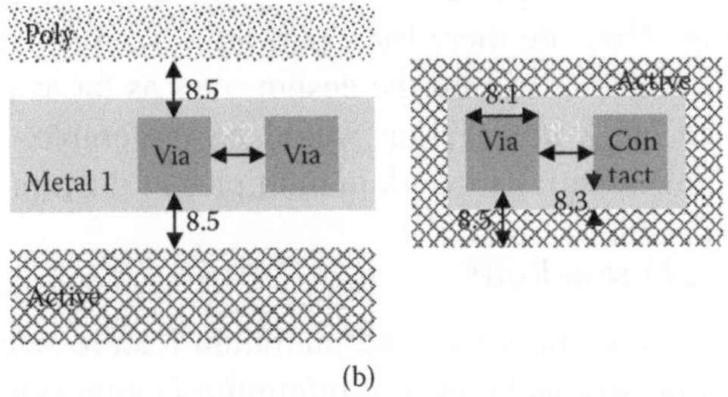

(b)

FIGURE 4.18 (a) Metal 1 design rules for the MOSIS SCMOS process. (b) Via design rules for the MOSIS SCMOS 2 metal process. Under certain circumstances additional design rules may apply; with rule 5 (contact to poly) it is stated that: "On 0.50-μm process (and all finer feature size processes), it is required that all features on the insulator layers (contact, via, via2) must be of the single standard size; there are no exceptions for pads (or logos, or anything else); large openings must be replaced by an array of standard-sized openings. Contacts must be drawn orthogonal to the grid of the layout. Non-Manhattan contacts are not allowed." Notice, also, that the contact is given as an "exact size," not a minimum size. (c) Metal 2 design rules for the MOSIS SCMOS 2 metal process. Courtesy of MOSIS, Marina del Rey, CA (www.mosis.org).

Mosis Scmos layout rules–9–metal 2		
Rule	Description	Lambda
9.1	Minimum width	3
9.2	Minimum spacing	3
9.3	Minimum overlap of via 1	1
9.4	Minimum spacing when either metal line is wider than 10 lambda	6

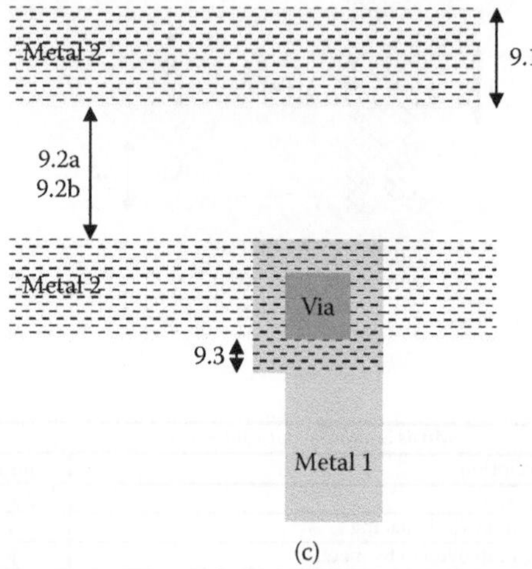

FIGURE 4.18 (Continued)

Note: Do not break design rules unless you are absolutely sure you know what you are doing. They are there for a purpose.

It is usually desirable to exceed the design rules as far as possible. If a 4-μm overlap is specified, an 8-μm overlap would be preferable. Furthermore, it is helpful, although not essential, to work in multiples of the minimum feature size.

4.5.1 Developing Design Rules

The first principle is to comply with the minimum feature size. No structures or gaps should have dimensions below the minimum feature size. This will set the absolute minimum width and separation rules (Figure 4.19a).

Process characteristics contribute to separation rules. Diffusion processes proceed laterally under the mask as well as into the silicon. The minimum separation between diffused impurities (P-well or N-well) is likely to be larger (Figure 4.19b).

Taking the process characteristics into account, overlap requirements are set principally by the accuracy of alignment during photolithography. If it is possible to align one layer with another to ± 1 μm, then an overlap of at least 2 μm would

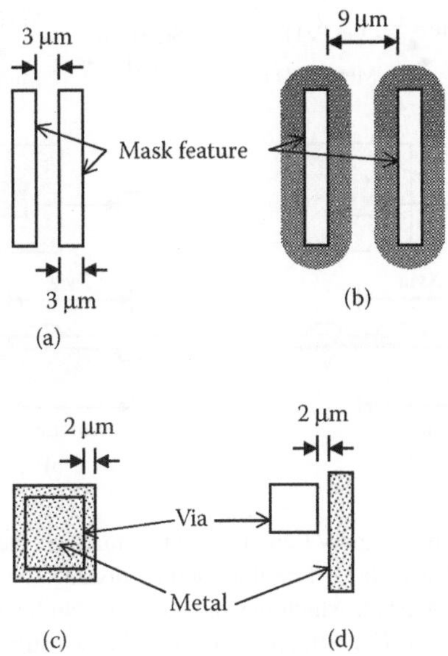

FIGURE 4.19 Building design rules: (a) minimum width and separation set by minimum feature size, (b) separation between 3-μm-deep diffusions needs to be larger, because the diffusion (shaded area) extends beyond the mask features, (c) if a via is to open over a metal pad, there must be a minimum overlap dependent on the alignment error (± 1 μm in this example), (d) if the via is not to open over the metal, it must be spaced from it by at least the same amount.

be required (Figure 4.19c). A similar argument holds for spacing between features on different layers (Figure 4.19d).

The next thing to be considered is the *alignment sequence*. Not all layers will be directly aligned to the one below; indeed, one layer may have to be registered with respect to features on two layers. An example of this is in cases in which a single etching step is used to open contact holes to more than one layer — the same etch mask may be used to cut vias from metal 1 to both poly1 and the n diffusion region (source or drain) of a transistor.

Figure 4.20 illustrates two different alignment sequences and the resulting errors. An initial set of alignment marks is etched into the substrate. In Figure 4.20a, M1 and V1 are both aligned to within ± 1 μm with the mark etched into the substrate. As a result, V1 could be misaligned with M1 by as much as ± 2 μm. In Figure 4.20b, M1 is aligned to the mark etched in the substrate and V1 is aligned to M1 — in both cases to within ± 1 μm. Now, V1 is aligned to M1 with an accuracy of ± 1 μm but could be misaligned with the substrate by as much as ± 2 μm.

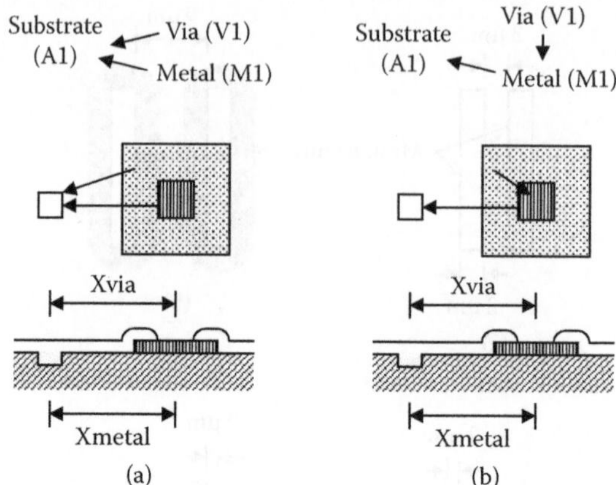

FIGURE 4.20 In (a) both M1 and V1 are aligned to a mark on the substrate, with an error of ±1 μm. This means that Xvia will be the nominal distance on the mask design, X, ±1 μm. The same is true of Xmetal, which means that Xvia can be as much as X+1 μm and Xmetal as little as X – 1 μm. The difference between the two structures on the final device will then be as great as 2 μm. This is addressed in (b), where M1 is aligned to A1 and V1 to M1. While Xmetal is still X±1 μm, and the alignment error between V1 and M1 is now ±1 μm, Xvia is now X ± 2 μm.

4.6 GETTING THE MASKS PRODUCED

Assuming that the masks are to be produced by an independent supplier or brokerage, it will be necessary to provide the design files, obviously, and also additional information regarding the physical format of the mask.

4.6.1 Mask Plate Details

The specification of the blanks to be used to create the set of masks will either be relatively simple for most silicon-related micromachining processes, where the demands are not great, or quite complex for odd requirements such as x-ray lithography masks. Normally, it would be necessary to obtain the latter from a specialist provider; for UV lithography, the masks will normally be chrome on low-expansion glass or quartz.

Mask plates should, clearly, be larger than the wafer being processed: a 2-in. (50-mm)-diameter wafer would require 2.5-in. (63.5-mm) square blank, and a 4-in. (100-mm)-diameter wafer would require a 5-in. (125-mm) square blank. Table 4.1 lists some common plate dimensions.

The chrome may either be "bright chrome" or have a low-reflection coating (LRC) applied. This can help to some extent with optical aberrations during the lithography process. Quartz (fused silica) glass is the preferred material for stability, but low-expansion glass is a cheaper substrate that can be used; usually, MEMS dimensions are not so critical as to require quartz. In the last instance,

TABLE 4.1
Plate Sizes

Wafer[a]	Size (mm)	Thickness (mm)
2"	6.3.5	1.5
4"	125	2.5
6"	178	3

[a] 2": 63.5 × 1.5 mm plate; 4": 125 × 2.5 mm plate, 6": 175 × 4 mm plate.

the foundry where the micromachining is being performed will usually have a preferred format and supplier.

For x-ray lithography (for LIGA), more exotic mask materials are required. These are often specific to a particular organization but are generally formed using a gold film (to absorb the x-rays) that has been patterned on a thin membrane (e.g., an etched silicon/nitride membrane).

4.6.2 Design File Details

Normally, the mask manufacturer will be supplied with either a GDSII or CIF design file. The name and language of the design file should be clearly identified. Furthermore, for GDSII files in particular, it is necessary to identify the structure at the top of the hierarchy. In the case of the example given in earlier sections of the chapter (see Figure 4.16a), this would be `wafer`.

It is common practice to submit design files via the Internet, but tapes and floppy disks have also been used in the past.

4.6.3 Mask Set Details

The result of mask manufacture will be a set of mask plates, each corresponding to a layer of the design file. Firstly, it is necessary to identify the layer name (CIF) or number (GDSII) that will be produced on each plate. It is advisable to ensure that the file submitted for design does not contain data for layers that are not to be produced as masks.

It is at this point that each mask layer should be identified as either light field or dark field; this could alternatively be identified as positive or negative resist. The minimum feature size, i.e., the smallest line or gap width, needs to be specified, and if it is necessary to specify line width control beyond that normally achieved by the mask manufacturer selected for the requested feature size, then this must also be noted for each layer that requires it. Line width control refers to the tolerance to be applied to the physical dimensions across the entire design. If the design is particularly sensitive to variation in physical dimensions, it may be worth revisiting this before trying to take it into production.

4.6.4 Step and Repeat

If the design is of a single chip, to be repeated across the entire mask (see Figure 1.2 in Chapter 1), then details need to be supplied. Care should be taken to ensure that the information given matches the desired pitch (spacing) of the chip designs and that the entire design can fit onto the blank selected.

4.6.5 Placement Requirements

It is normal practice to place all designs symmetrically about the center of the plate. However, it is possible to specify that the design be mirrored about the vertical axis. There are several reasons why this may be required:

- If the design is double sided, then masks for the back side of the wafer will need to be mirrored (with respect to the other layers).
- As seen in Chapter 1, the design is written onto the chromium side of the plate, but when performing lithography this is the side that is in contact with the substrate. This may result in the design appearing as a mirror image following photolithography. Some mask manufactures will normally mirror layers so that the final design comes out as it appears in the design file, whereas others do not do so. It is advisable to check beforehand and to place features (e.g., text) in an obvious place on the design so that the orientation can be verified. This will not normally have functional implications but can mean that bonding pads, etc., appear on the opposite side of the design to that expected.

Note that by specifying which layers are to be mirrored at this stage, rather than using the layout software, it is possible to verify the alignment of features in double-sided designs before going on to mask manufacture.

4.7 GENERATING GERBER FILES

Chapter 1 introduced the reader to some cheaper alternatives to chrome masks. These options included the use of printed circuit board (PCB) artwork and the use of laser-cut stencils. The common design file current in the PCB industry is the Gerber file format. Modern PCB design software can readily cope with multiple layers and can perform design rule checks similar to those performed by layout software. There is no confusion between internal units, lambda units, and user units. The design is created in either millimeters or mils (thousandths of an inch). The only thing to be aware of is that when generating the final Gerber files, it is normally possible to set the precision of the dimensions — the number of places before and after the decimal point; the software will automatically round any dimensions to the specified precision. In the author's experience, PCB layout tools are slightly more difficult to come to grips with than simple polygon pushers.

At the time of writing, there are a number of PCB design packages that are available in one form or another for downloading from the Internet; see Table 4.2.

TABLE 4.2
PCB Design Packages for Download

Name	Web address	Company	Comments
Target 3001	http://www.ibfriedrich.com/	Ing.-Büro FRIEDRICH Am Schwarzen Rain 1 D-36124 Eichenzell Germany	Limited demo version available for download
GC-Prevue	http://www.graphicode.com/	GraphiCode A Division of ManiaBarco 6608 216th Street SW, Suite 100 Mountlake Terrace, WA 98043 U.S.	Free Gerber data viewer. Useful for checking Gerbers generated by other packages; some companies prefer GC-Prevue files to Gerbers
Eagle	http://www.cadsoftusa.com/	CadSoft Computer, Inc. 801 South Federal Hwy., Suite 201 Delray Beach, FL 33483-5185 U.S.	Limited demo version available for download
Protel	http://www.protel.com/	Altium Inc. 17140 Bernardo Center Drive, Suite 100 San Diego, CA 92128 U.S.	Fully functional version available for 30-day trail
Easy PC	http://www.numberone.com/	Number One Systems Oak Lane, Bredon Tewkesbury, Glos. GL20 7LR, U.K.	Limited trial versions available for download

Film Mask Parameters

- Resolution: 8000, 16000, or 32000 dpi (dots per inch)
- Minimum feature size: 6 μm

(These were available at the time of writing from JD Photo-Tools Ltd., Unit 4a, Meridian Centre, King St., Oldham, OL8 1EZ, U.K., www.jdphoto.co.uk.)

4.8 MASK DESIGN — KEY POINTS

1. The design will be created using a layout editor in two dimensions:
   a. The output will be CIF or GDSII.
   b. Each layer will be numbered (GDSII) or named (CIF).
   c. Each individual layer of the design will be transferred to a single mask plate.
2. The layout editor will make use of a technology file that outlines the design (layer names, numbers, colors for rendering, etc.):
   a. Units will be design units (used by the software), microns, or lambda units (half the minimum feature size).
   b. Plan the design beforehand.
   c. Use the grid for help when drawing the design.
   d. Text may not appear on the design layer unless drawn as a polygon or track.
3. The design will be hierarchical.
4. The design will normally include a number of standard features specified by the foundry:
   a. Frame
   b. Alignment marks
   c. Scribe lane
   d. Bonding pads
   e. Test structures
5. Each layer of the design should include a mask set identifier and layer identifier.
6. Always work to the design rules given by the foundry, and use the layout editor's DRC where possible to check:
   a. Minimum feature size
   b. Minimum overlap
   c. Minimum separation
   d. Exact dimensions where required
7. Masks are normally chrome on quartz or chrome on low-expansion glass:
   a. The mask plate should be about 1 in. (2.5 mm) larger than the design.
8. When the mask set is produced, it is possible to individually specify each layer:
   a. As being light or dark field
   b. As being mirrored

Part II

Microsystems

II.1 INTRODUCTION

A microsystem can be considered to be a device or unit made up of a number of microengineered components. A convenient model of a microsystem is that of a control system (Figure II.1); many proposed microsystems take this form. Microsensors detect changes in the parameter to be controlled; electronic control logic then operates microactuators based on information from the sensors to bring the parameter to be controlled within the desired limits.

An example of such a system would be the refreshing of a medium in a small cell culture dish. Sensors could detect changes in pH, pO_2, or pCO_2, and a micropump could deliver new culture medium from a reservoir as required. Not all devices need follow this control-system scheme. For instance, an accelerometer designed to inflate an air bag in the event of a car crash may not only incorporate a micromachined acceleration sensor but also electronics to condition the signal and detect a rapid deceleration, and microactuators that put a force on the sensor, allowing the device to be tested before the driver moves off.

Microsystems may be constructed from component parts produced using different technologies on different substrates, which are then assembled together, i.e., a hybrid system. For example, a silicon chip would be used to implement control circuitry, whereas the actuators it controlled could be micromolded in plastic or electroplated metal (using the LIGA technique, perhaps). Alternatively, all components of a system could be constructed on a single substrate using one technology (a monolithic system). Hybrid systems have the advantage that the most appropriate technology for each component can be selected to optimize system performance. This will often lead to a shorter development time because microfabrication techniques for each component may already exist, and compromises will not have to be made to ensure that each component can be fabricated without damaging components already existing on the substrate. Monolithic devices will typically be more compact and more reliable than hybrid devices (fewer interconnections that can go wrong, for example). Further, once the fabrication process has been developed, they can be manufactured more cheaply because less assembly is required. Various microsensors and microactuators are discussed in Chapter 5 and Chapter 6.

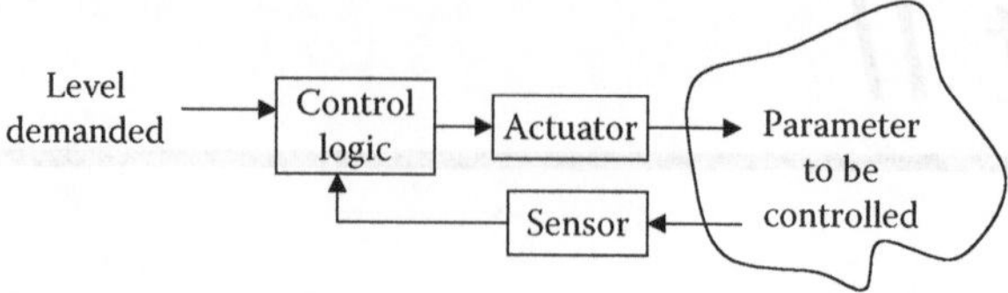

FIGURE II.1 Model of a microsystem

A significant problem facing microengineers is that of assembling many microscopic components. Potential solutions include self-assembling systems and desktop factories staffed by microrobots. At some point many microsystems will have to interact with macroscopic systems. Often it may be that only a critical component of a system has to be microengineered and supported by a complex system produced using more conventional engineering techniques. It is easy to underestimate the problems involved in mounting and packaging microdevices and integrating them with macroscopic supporting systems. Chapter 9 deals with some issues regarding assembly and packaging of microsystems.

Although microsensors and microactuators are given a fairly general treatment within this section, there are some microsystems and related technologies that have become so complex as to require deeper treatment. Microelectronics is one example, and this has been left to the vast body of existing literature on the subject. Similarly, radio frequency (RF) MEMS have not been addressed here, nor have millimeter wave technologies (MMIC). However, some effort has been made in Chapter 7 and Chapter 8 to introduce micro total analysis systems and integrated optics.

It should be recalled that many of the techniques required for other technologies have been addressed in Part I of this volume. In terms of RF MEMS, for example, one technique is to fabricate coils suspended within the wafer, on an oxide or nitride membrane.

Physical dimensions of devices are constantly reducing because of ongoing development. Chapter 10 introduces some aspects of nanotechnology, which, on some scales, overlaps with microtechnology, and in other areas represents the next step.

II.1.1 Microsystem Components

As can be seen from Figure II.1, a microsystem typically comprises components from one or more of three classes: microsensors to detect changes in the system's environment, an intelligent component that makes decisions based on changes detected by the sensors, and microactuators by which the system changes its environment. Obviously, a system need not be a closed loop as illustrated in Figure II.1. Even if it is, some elements may be implemented in macroscopic technology; for instance, microsensors may be used as cheap disposable components to control large actuators.

The intelligent component of the system would be implemented using microelectronic components or a computer. Although microelectronics itself is not addressed, Part III gives a good introduction to techniques that may be employed to interface sensors and actuators to a computer or microcontroller.

Some of the microsensors described in Chapter 5 are micromechanical devices that have to be driven into resonance. This is an additional role for the microactuators described in Chapter 6.

5 Microsensors

5.1 INTRODUCTION

A *transducer* is a device that converts one physical quantity into another. The change in refractive index of some crystals under an applied magnetic field is one example of how this occurs (magnetooptic effect). Deformation of a piezoelectric crystal under an applied electric field is another. Sensors and actuators are special types of transducers. In the present context, a *sensor* is a device that converts one physical or chemical quantity to an electrical one for processing by the microsystem. Similarly, an *actuator* is a device that converts an electrical quantity into a physical or chemical one.

Many of the sensors described in this chapter have been developed within the microelectronics industry and do not involve any special micromachining techniques. However, some of these sensors can be enhanced by the use of micromachining techniques (e.g., for thermal isolation).

5.2 THERMAL SENSORS

There are a number of different types of thermal (temperature) sensors. Two of the most common types are thermocouples and thermoresistors (thermistors).

5.2.1 Thermocouples

When two dissimilar metals (e.g., copper and iron) are brought together in a circuit and the junctions are held at different temperatures, a small voltage is generated and an electrical current flows between them.

A working thermocouple is shown in Figure 5.1. It consists of a sensing junction at temperature Ta and a reference junction at temperature Tb. The voltage developed by the thermocouple is measured with a high-resistance voltmeter. The open circuit voltage (i.e., as measured by an ideal voltmeter with infinite input impedance) is related to the temperature difference ($Ta - Tb$), and the difference in the Seebeck coefficients of the two materials ($Pa - Pb$) is

$$V = (Pa - Pb)(Ta - Tb) \tag{5.1}$$

V will typically be on the order of millivolts or tens of millivolts for metal thermocouples with temperature differences on the order of 200°C.

Semiconductor materials often exhibit a better thermoelectric effect than metals. It is also possible to integrate many semiconductor thermocouples in

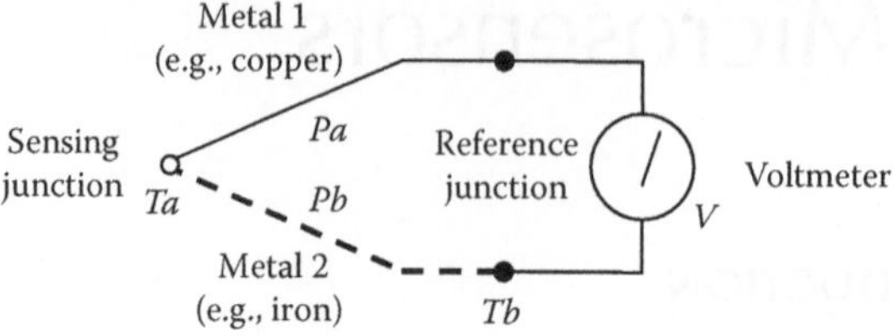

FIGURE 5.1 Thermocouple.

series to make a thermopile, which has a larger output voltage than a single thermocouple on its own. However, the high thermal conductivity of silicon makes it difficult to maintain a large temperature gradient ($Ta - Tb$).

Therefore, one application of silicon micromachining is to thermally isolate the sensing element from the bulk of the silicon wafer. This may be done by fabricating the device on bridges or beams machined from silicon.

5.2.2 Thermoresistors

The electrical resistivity of metals varies with temperature. Above –200°C, the resistivity varies almost linearly with temperature. In this approximately linear region, the variation of resistivity (r) with temperature (T) can be adequately described by a quadratic equation:

$$r = R(1 + aT + bT^2) \tag{5.2}$$

where R is the resistivity of the material at a reference temperature (0°C), and a and b are constants specific to the metal being used. Platinum is often used as its resistance variation is particularly linear with temperature (i.e., b is particularly small).

As metal thermoresistors generally have relatively small resistances and their rate of change of resistance with temperature (temperature coefficient of resistance, or TCR) is not particularly large, they require the use of a resistance bridge network to detect the signal (Chapter 11, Section 11.4).

Semiconducting thermoresistors (or *thermistors*) can be formed from metal oxides or silicon. These are generally not as accurate or stable as platinum thermoresistors but are cheaper to manufacture and are potentially easier to integrate with microelectronic circuitry on the same substrate. The TCR of a thermistor is highly nonlinear and negative and quite dependent on the power being dissipated by the device. The nominal resistance is typically expressed relative to the resistivity at 25°C with no power being dissipated by the device, and can range from 500 Ω to 10 MΩ.

Owing to the negative TCR, it is possible for the resistor to go into a self-heating loop: current flowing through the resistor heats it up, the resistivity drops, more current flows, it gets hotter, etc. However, a large TCR does make it possible to couple thermistors directly to amplifier circuits without the requirement for a bridge configuration. The nonlinearity would typically be dealt with by calibrating the device.

Microengineering techniques can be used in a variety of ways to enhance thermal sensors. As mentioned earlier, they can be used to thermally isolate the sensing element from the rest of the device. Also, arrays of sensors can be produced to give signals that are larger than what one sensor on its own would produce. If the device is small and thermally isolated, then its response time (the time the sensor takes to heat up or cool down in response to changes in the temperature of the environment) can be quite fast. With silicon-based devices, there is, of course, all the potential benefits that could come if electronics were integrated onto the chip (e.g., calibration done on-chip, self-testing).

5.2.3 Thermal Flow-Rate Sensors

There are a number of ways by which the flow rate of gases (and liquids, although clogging of the sensor may be more of a problem) can be monitored by the use of thermal sensors. One can measure the temperature of a fluid as it enters and then leaves the sensor having been passed over a heating resistor; the temperature difference will be proportional to the mass flow rate. Another possibility is to maintain the sensor at a constant temperature (using heating resistors with thermal sensors for feedback control) and measure the amount of power required to maintain the temperature. Again, this will be proportional to the mass flow rate of material over the sensor. This type of sensor is typified by the use of a platinum resistor both as the sensing and heating element.

Equation 5.2 gave the relationship between temperature and resistivity for a metal-sensing element (for platinum, a is approximately 3.9×10^{-3} and b is approximately 0.58×10^{-6}, so the response is roughly linear). If a heating current, I, is flowing through the element, then this can be related to the mass flow rate by

$$I^2 = p + q\sqrt{M} \tag{5.3}$$

where p is a constant related to the heat loss under zero-flow conditions, and q is a constant dependent on the geometry of the system and the fluid. Note that these constants will also be dependent on the temperature difference between the fluid and the heating element. In the case of a MEMS sensor, the element will be situated on a thermally isolating structure such as a bridge, otherwise heat will be dissipated through the silicon (or another) substrate. In order to make a measurement, the element can be incorporated into a Wheatstone bridge (see Chapter 11, Section 11.4) as shown in Figure 5.2.

In Figure 5.2, a bridge with two sensing arms is shown. When the bridge is in balance ($V_{dif} = 0$), one arm has a resistance of approximately 11000 Ω, whereas the other has a resistance of approximately 278 Ω. This implies that the 100-Ω platinum element (100 Ω referring to its nominal resistance at 0°C) has been

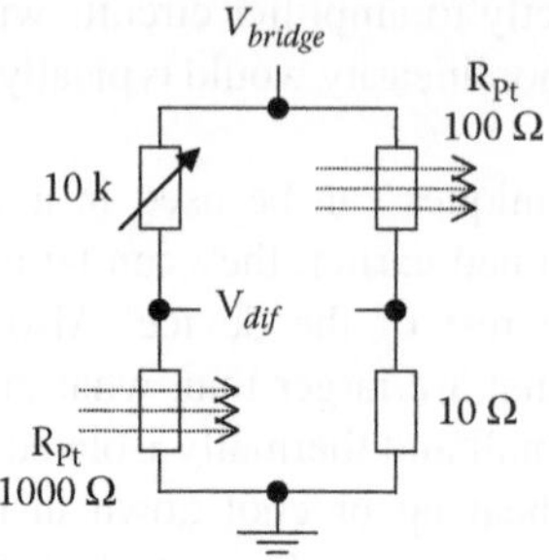

FIGURE 5.2 Thermal mass-flow-rate sensor and bridge circuit. Dashed arrows indicate flow over the sensing elements.

heated to approximately 100°C by the current flowing through the bridge. The heating is due to current flowing through the platinum element. Because this side of the bridge has a much lower resistance than the other, much more current will flow, resulting in more power being dissipated by the element (power dissipated being proportional to the square of the current).

The use of two sensing elements means that ambient temperature is compensated for to some extent, because the 1000-Ω element will not heat up. To obtain a signal that is related to the mass flow rate past the sensor, the current flowing through the 100-Ω element is measured and controlled to keep the bridge in balance. The system can then be calibrated to give a mass flow rate based on Equation 5.3. (Alternatively, if the fluid and thermal aspects of the system can be reasonably well simulated, then the constants p and q can be estimated or even calculated.)

5.3 RADIATION SENSORS

There are a variety of radiation sensors for different types of radiation, including nuclear radiation as well as visible light, infrared, and ultraviolet. Only a few of the most common ones will be considered here: the photodiode and phototransistor, charge-coupled devices (CCDs), and pyroelectric sensors.

5.3.1 PHOTODIODES

The simplest photodiode is a reverse-biased pn (diode) junction. When no light falls on the device, only a small amount of current flows (the *dark current*). When light falls on the device, additional carriers are generated, and more current flows.

Photodiodes typically work in the visible light — near the infrared region of the spectrum. They are high-impedance devices, and operate at relatively low currents (typically 10 μA dark current, rising to 100 μA when illuminated).

They have fairly linear responses to increasing illumination, and they generally have very fast response times.

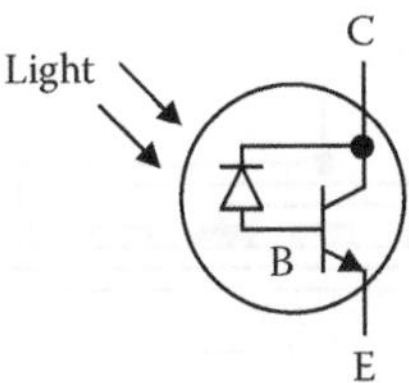

FIGURE 5.3 Phototransistor.

It is worth remembering that, although photodiodes are carefully designed to have specific response characteristics, light falling on any pn junction can potentially induce charge carriers. This will appear as unwanted noise in a signal, and it is, therefore, desirable to ensure that sensitive electronics are packaged so as to prevent light from falling on them.

5.3.2 Phototransistors

The phototransistor has a much higher current output than a photodiode for comparable illumination levels. However, it does not operate as fast as photodiodes (about 100 kHz being the top limit) and also has higher dark current.

The phototransistor is essentially a transistor with the base current supplied by the illumination of the base–collector junction; it can be considered to be similar to a photodiode supplying the base current to a transistor (Figure 5.3). Normal transistor action amplifies the small base current.

5.3.3 Charge-Coupled Devices

CCDs can be built as large linear and two-dimensional arrays; the latter are often used in small video cameras. They consist of a large number of electrodes (gates) on a semiconductor substrate. A thin insulating layer is situated between the metal gates and the semiconducting substrate.

The operation of a CCD is shown schematically in Figure 5.4. The substrate has been doped so that the main current carriers are positive (i.e., "holes" — the term originates from semiconductor physics). When a positive voltage is applied to every third gate (V1), the majority carriers are repelled from the region beneath (Figure 5.4a), leaving "wells." When light falls on the device, additional carriers are generated (as with photodiodes). The positive carriers are repelled from the gate, but the negative charge carriers (electrons) are attracted to the gate and fill the wells (Figure 5.4b). After the carriers accumulate, the entire array may be read out by shifting the carriers from one well to the next, the number of carriers being proportional to the amount of light that fell on each well. The electrical potential on the gates to one side of those already biased (V2) is increased; thus the charge is now shared between wells under two gates (Figure 5.4c). Then the first potential (V1) is then switched off. The charge is transferred to the adjacent well (Figure 5.4d), and so on.

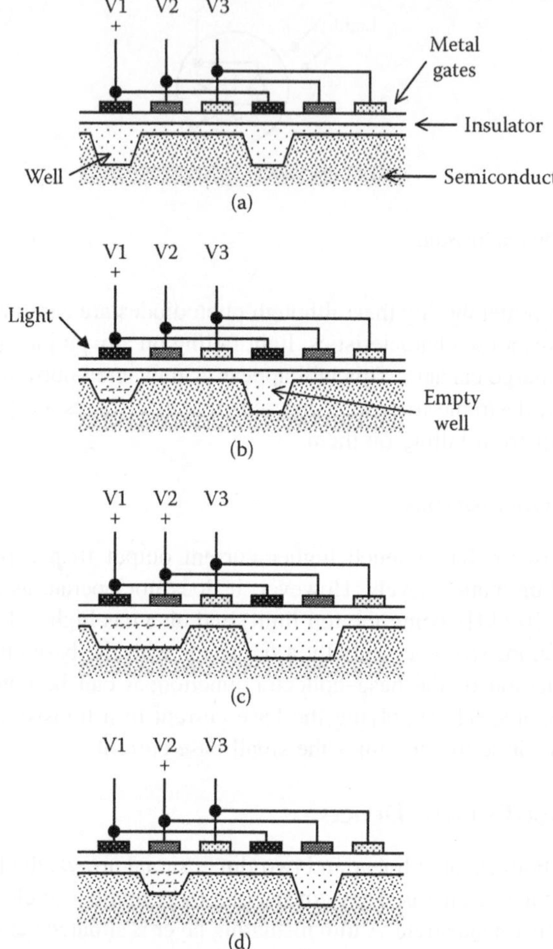

FIGURE 5.4 Operation of a charge-coupled device (CCD).

5.3.4 Pyroelectric Sensors

These devices operate on the pyroelectric effect in polarized crystals (e.g., zinc oxide). These crystals have a built in electrical polarization level that changes with the amount of incident thermal energy.

These are generally high-impedance devices and so are often buffered using field effect transistors. They can be made to automatically bring themselves to the ambient temperature, so they only respond to rapid fluctuations. One potential problem is that crystals that exhibit the pyroelectric effect may also exhibit piezoelectric effects (see the discussion on mechanical sensors in the following text), thus pyroelectric sensors need to be designed to avoid strain on the crystal.

Zinc oxide for piezoelectric or pyroelectric applications is usually applied by spinning on a solgel and then baking. This is discussed further in the section on piezoelectric actuators in Chapter 6.

One common application of these devices is in motion detectors for intruder alarms. A lens cuts the sensor's field of view into discrete sections. As someone moves across the field of view, thermal radiation from their body falls on the sensor, resulting in discrete pulses as the person moves from one part of the field of view to the next. It is thus possible to build relatively cheap motion detectors that can be tuned to respond to a particular rate of motion.

5.4 MAGNETIC SENSORS

There are many ways of sensing magnetic fields. Optical sensors can be based on crystals that exhibit a magnetooptic effect or on specially doped optical fibers. Coils can be used, although microfabricated coils are generally two dimensional, which are often not useful for many applications. The continuing development of high-temperature superconductors is also broadening the possibilities for sensors based on superconducting quantum interference devices (SQUIDs), which are capable of detecting the magnetic fields in the heart or brain. There are also a variety of other devices. Many measurements can be made, however, using Hall effect sensors. These are very common and are outlined in the following text.

A Hall effect sensor is shown diagrammatically in Figure 5.5. The sensor consists of a conducting material, usually a semiconductor, and a current is passed between two contacts on opposite sides of the device. Two sensing contacts are placed on two other sides of the device opposite each other and perpendicular to the current flow. A magnetic field perpendicular to the plane of the contacts causes a deviation in the current flow across the device. This in turn is detected as a potential difference between the two sensing contacts.

Hall effect sensors operate typically in the range 0.1 mT to 1 T (the Earth's magnetic field is about 0.05 mT). Hall effect IC packages that typically give an output of about 10 mV per mT are available.

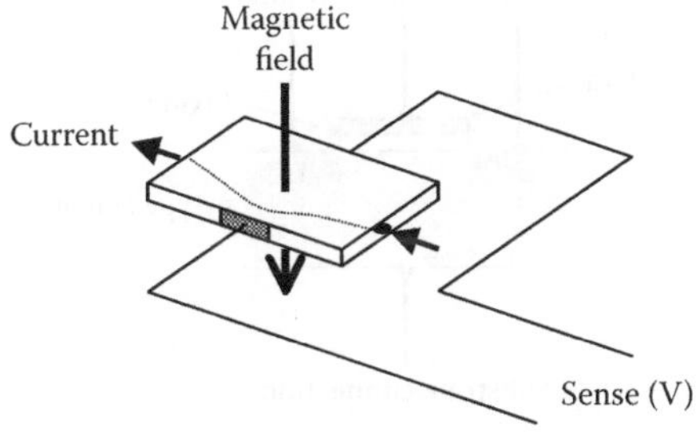

FIGURE 5.5 Hall effect sensor.

5.5 CHEMICAL SENSORS AND BIOSENSORS

There is a wide variety of different chemical sensors, especially if one includes biosensors as a subclass of chemical sensors, as is done here. A large proportion of chemical sensors are based on metal-oxide-semiconductor field effect transistor (MOSFET) devices. For this reason, the ion-sensitive field effect transistor (ISFET) will be discussed in this section. ISFET devices have been around for some time now (since about 1970), so there is quite a bit of literature on them.

The term *biosensor* refers to any sensor that uses an active biological (or sometimes biologically derived) component in the transduction process. This may be a sensory cell taken from a living organism and mounted on an electrode. Alternatively, antibodies may be used, which will lock onto the material of interest and hold it in an appropriate position for sensing. Another option is to use an enzyme that catalyzes a reaction that can be detected by suitable means. As there is considerable interest in monitoring blood glucose levels (to provide closed-loop control of blood glucose by means of an artificial pancreas for diabetics), blood glucose sensors have received much attention. One of these, based on the glucose oxidase enzyme, will be outlined.

One thing to note is that a lot of research has gone into these sensors — biosensors and blood glucose sensors in particular. Although progress has been made, there are still a lot of problems to be solved. One big problem in this area is that the sensor performance drifts or degrades over time, often in unpredictable ways. So the device has to be calibrated regularly or just before use. Clearly, a blood glucose sensor that only gives reliable readings over a period of 100 d cannot be used in an implanted artificial pancreas. Thus, although there are many potential uses for chemical sensors, their use is often complicated by calibration requirements.

5.5.1 ISFET Sensors

ISFETs sense the concentration (activity level) of a particular ion in a solution. These devices are generally based on the enhancement-mode MOSFET structure, shown in Figure 5.6.

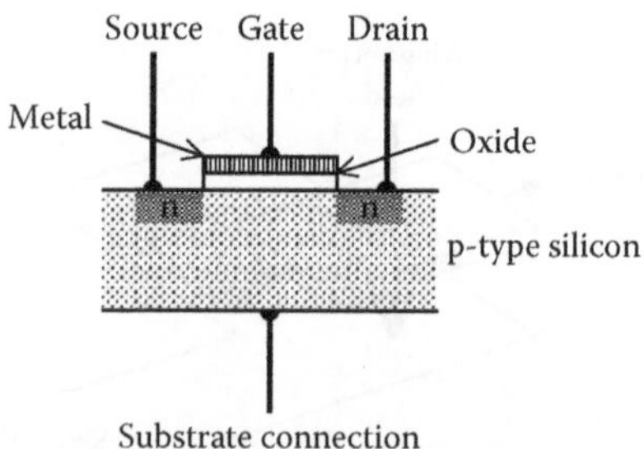

FIGURE 5.6 n-channel enhancement-type MOSFET structure (not to scale).

A MOSFET has a metal *gate* electrode, insulated from the semiconductor (silicon) wafer by a thin layer of silicon dioxide (oxide). The bulk of the semiconductor (i.e., the substrate) is doped with impurities to make it p-type silicon; in this material current is carried by positive charge carriers called holes. On either side of the gate are small areas of silicon doped with impurities so that negatively charged electrons are the main carriers in these n-type silicon regions: the *source* and the *drain*. Both n-type and p-type silicon are used to form diodes; current will flow from p-type to n-type, but not vice versa. Thus, to keep the bulk of the silicon substrate from interfering with the transistor (gate, drain, and source), this is connected to the most negative part of the circuit (often connected inside the transistor package to the source; although the device may appear symmetrical, it should be connected according to the pinout on the data sheet).

When in use, a positive voltage is applied to the gate. This repels holes from the region near the gate and attracts electrons, forming, between the drain and source, a small channel in which the majority charge carriers are electrons. Current can flow through this channel, and the amount of current that can flow depends on how large the channel is and, thus, on the voltage applied to the gate.

In the ISFET, the gate metal is replaced with an ion-selective membrane (Figure 5.7), and the device is immersed in a solution. Ions in the solution interact with the ion-selective membrane. When there is a high concentration of positive ions in the solution, a lot of them will accumulate on the gate, widening the channel between the source and drain. With a low concentration of positively charged ions, the channel will be narrow.

In order to ensure that the FET channel is biased to an optimum size about which sensing can take place, the solution is maintained at a reference potential by an electrode placed in it. Generally, the reference potential is

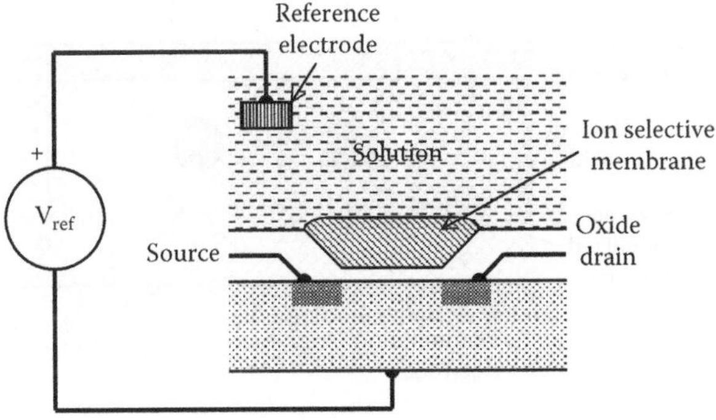

FIGURE 5.7 Ion-sensitive field effect transistor (ISFET) structure (not to scale).

adjusted to maintain a constant current flowing from the drain to the source, so the ionic concentration will be directly related to the solution reference potential with respect to the substrate potential (in the circuit shown in Figure 5.7).

One significant problem in the design and fabrication of ISFETs is ensuring that the ion-selective membrane adheres to the device. If the integrity of the membrane is compromised, then the device is useless; this problem has considerable effect on the yield (percentage of functioning devices on a wafer) of the fabrication process. The simplest ISFET device is a pH sensor, which employs a glass (oxide) as the "membrane."

5.5.2 Enzyme-Based Biosensors

Enzymes are highly specific in the reactions they catalyze. If an enzyme can be immobilized on a sensing substrate and the reaction products detected, then one has the basis of a highly selective biosensor. The enzyme-based biosensor described in the following text is for monitoring glucose levels; this application has been investigated considerably because glucose is important in diabetes and in many industrial fermentation processes.

The operation of a glucose-oxidase-based sensor is shown schematically in Figure 5.8. The enzyme is immobilized on a platinum electrode and covered with a thin polyurethane membrane to protect the enzyme layer and reduce the dependence of the sensor on blood oxygen levels. Glucose oxidase, in its oxidized form, oxidizes the glucose entering the sensor to gluconic acid, resulting in the conversion of the enzyme to its reduced form. The enzyme does not remain in

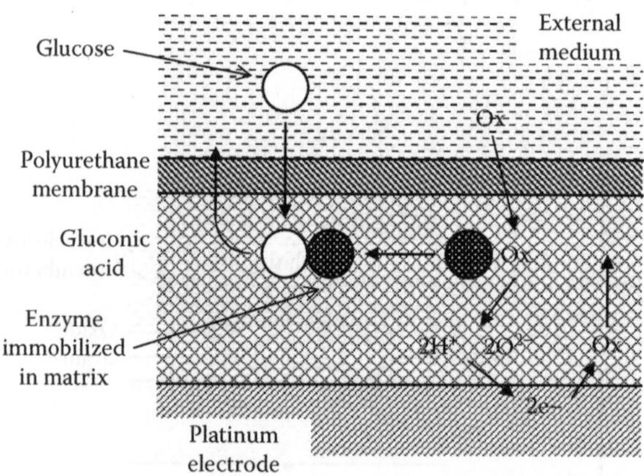

FIGURE 5.8 Glucose oxidase sensor.

this form for long. It interacts with the oxygen entering through the membrane. The products of this interaction are the oxidized form of the enzyme, two hydrogen ions, and two oxygen ions. When the platinum electrode is biased to the correct potential, it will reduce one of the oxygen ions such that the end products are oxygen and water. The resulting electrode current can be measured and will be proportional to the concentration of glucose in the external medium. (Note that this is a simplified explanation; there are also many other ways to monitor the reaction).

One thing to note is that because the various molecules have to physically move through the materials of the sensor, such biosensors can be quite slow to respond to changes in the external medium.

5.6 MICROELECTRODES FOR NEUROPHYSIOLOGY

Microelectrodes of fine-wire or electrolyte-filled micropipettes have been used for some time now to study the nervous system on a cellular (individual neuron) basis. These, in particular the metal wire microelectrodes, are prime targets for the application of microengineering techniques. The small signal amplitudes involved (in the region of 100 μV) and high interface impedances (1 to 10 MΩ at 1 kHz) between the metal and the tissue mean that it is advantageous to place the amplifier as close as possible to the recording site. In addition, the characteristics of microfabricated devices can be more reproducible than those of handmade metal wire microelectrodes, and their small size enables the accurate insertion of many recording sites into small volumes of tissue to study networks of neurons or for neural prosthesis applications.

The microelectrodes operate by detecting the electrical potential generated in the tissue near an active nerve fiber because of action potential currents flowing through the fiber membrane. There are three common types of micromachined microelectrode (Figure 5.9). Array-type microelectrodes (Figure 5.9a) are used to form the floor of cell culture dishes: signals are recorded from neurons that are placed or grown over these. Probe-type microelectrodes (Figure 5.9b) have recording sites on a long thin shank that is inserted into the tissue under investigation. Regeneration electrodes (Figure 5.9c) are placed between the ends of a severed peripheral nerve trunk; nerve fibers then regrow (regenerate) through the device.

These microelectrodes can be quite difficult to use. For array microelectrodes, appropriate cell culture methods have to be developed and practiced. Probe types have to be mounted on amplifier boards, and different situations require probes of many different shapes and size. Regeneration electrodes have to be fixed to the stumps of the nerve trunk and are required to be connected to the outside world. All devices can potentially generate huge amounts of data that have to be collected and analyzed.

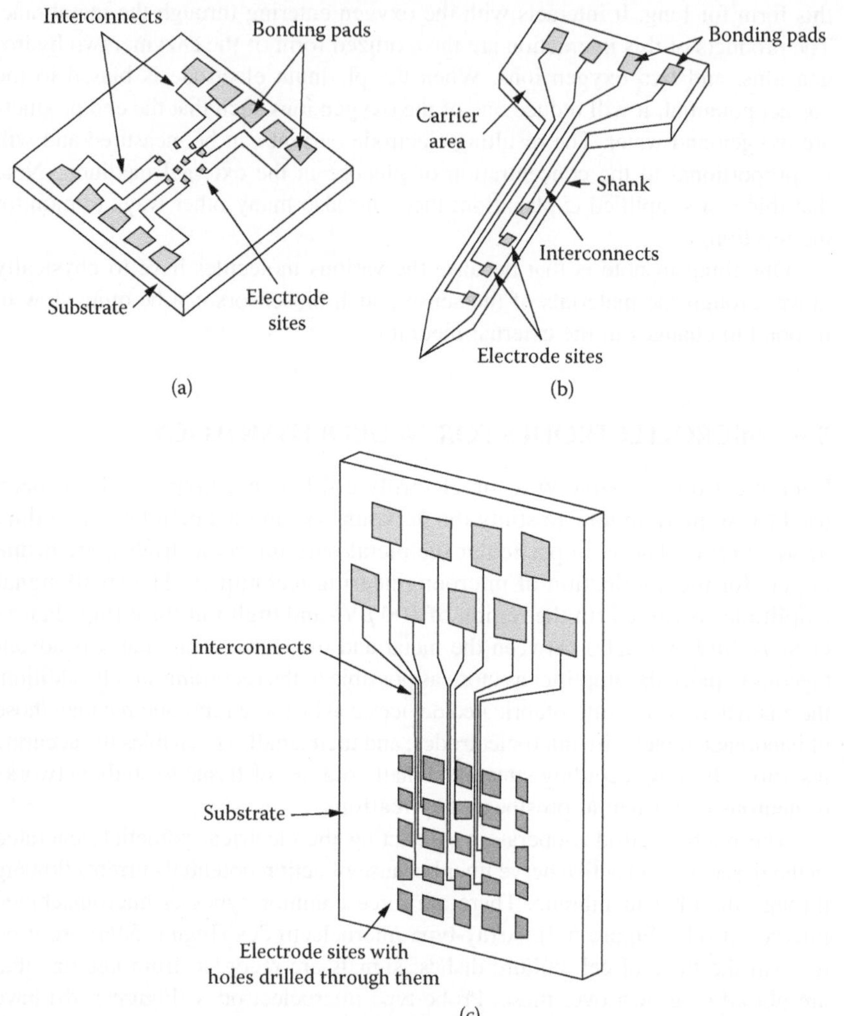

FIGURE 5.9 Microelectrodes for neurophysiology: (a) designed to form the bottom of a cell culture dish, recording and stimulating activity in cells growing over the electrodes in the center, (b) a probe-type microelectrode designed to have the shank, with the electrode sites, inserted into nervous tissue; the larger "carrier area" projects outside the tissue and provides a support for handling the device and for wire bonding, (c) a regeneration electrode sewn between the severed ends of a peripheral nerve; the nerve fibers then grow through it making contact with the electrodes (not to scale).

5.7 MECHANICAL SENSORS

Two different types of mechanical sensors will be discussed here. The first uses physical mechanisms to directly sense the parameter of interest (e.g., distance and strain). The second uses microstructures to enable the mechanical sensors to detect parameters of interest (e.g., acceleration) that cannot be measured directly with the first type of sensor.

5.7.1 PIEZORESISTORS

The change in resistance of a material with applied strain is termed the *piezoresistive effect*. Piezoresistors are relatively easy to fabricate in silicon, being just a small volume of silicon doped with impurities to make it n type or p type.

Piezoresistors also act as thermoresistors, so to compensate for changes in ambient temperature they are usually connected in a bridge configuration of some sort with a dummy set; this is illustrated in Figure 5.10. Figure 5.10 illustrates part of an accelerometer mass suspended by a beam from the bulk of the silicon

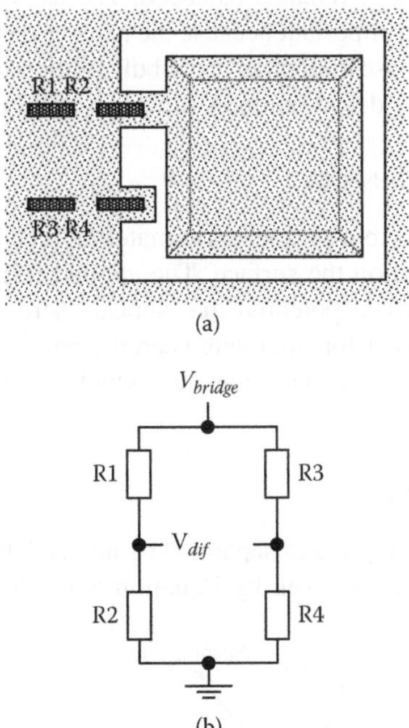

FIGURE 5.10 Piezoresistor configuration: (a) piezoresistors (dark rectangles) are implanted into the beam suspending an accelerometer mass, seen from above, and a reference pair are implanted on a dummy beam; (b) the four resistors are connected in a bridge circuit. If all resistances match, then V_{diff} will be 0 V. If the beam bends as a result of acceleration, then R2 will change its resistance and the bridge circuit will no longer be balanced.

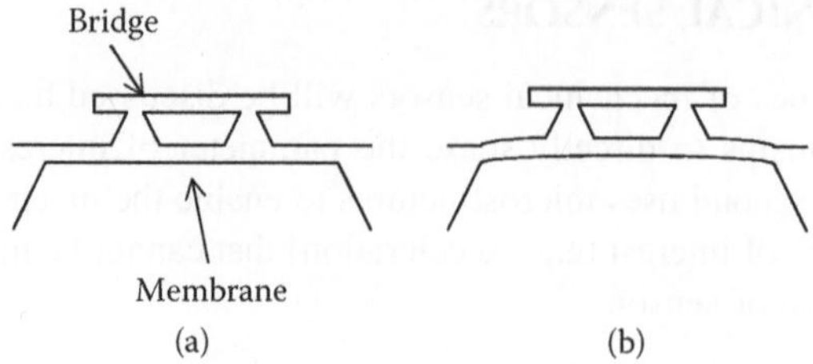

FIGURE 5.11 (a) A resonant bridge can be used to sense deformation, (b) when the membrane is deformed, the resonant frequency of the bridge will change.

wafer (compare this with Figure 5.12, which shows a similar design in cross section). Next to the suspension beam is a second beam, which is not attached to the mass. Resistors have been implanted into both beams and the bulk of the silicon substrate. These have been connected to form a bridge circuit (Chapter 11, Section 11.4 explains how this operates). Should the temperature change, the close proximity of the four resistors ensures that they all experience the same temperature and the bridge remains balanced. The use of the dummy beam not connected to the mass is important because the resistor on the beam is in a slightly different thermal than those implanted in the bulk of the wafer. Thus, only changes in resistance induced by deformation of the beam will be registered.

5.7.2 Piezoelectric Sensors

When a force is applied to a piezoelectric material, a charge proportional to the applied force is induced on the surface. The applied force can thus be deduced by measuring the electrical potential that appears across the crystal. Common piezoelectric crystals used for microengineered devices include zinc oxide and PZT (PbZrTiO3 — lead zirconate titanate), which can be deposited on microstructures and patterned.

5.7.3 Capacitive Sensors

For two parallel conducting plates separated by an insulating material, the capacitance between the plates is given by Equation 5.4, where A is the area of the

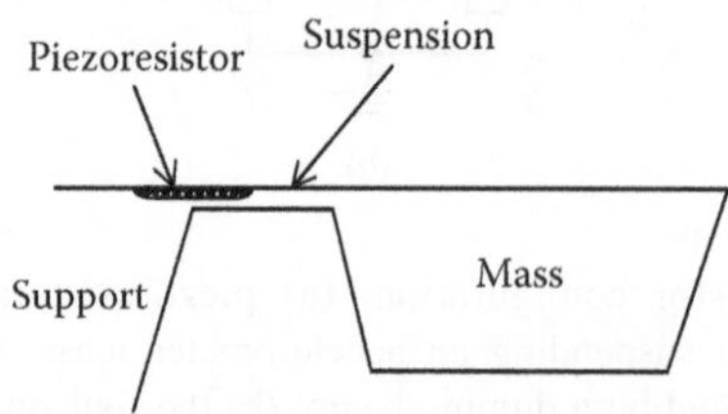

FIGURE 5.12 Bulk micromachined accelerometer in cross section — compare with Figure 5.10 (not to scale).

plates, d the distance between them, and ε a constant depending on the material between the plates (this assumes that the circumference of the plates is much larger than the distance between them, so what happens at the edges of the plates can be neglected).

$$C = \varepsilon \frac{A}{d} \tag{5.4}$$

(For air, ε is approximately $8.9 \times 10 - 12$ F/m.) From this it can be seen that the measured capacitance is inversely proportional to the distance between the two plates. It is possible to use this technique to measure small displacements (microns to tens of microns) with high accuracy (subnanometer); however, the instrumentation required to measure capacitance changes can be a little complex. The capacitor bridge, for example, is dealt with in Chapter 11, Subsection 11.4.1.

5.7.4 Optical Sensors

Silicon is a reflective material, as are other materials used in semiconductor device fabrication (e.g., aluminum). Thus, optical means may be used to sense displacement or deformation of microengineered beams, membranes, etc. A laser is directed at the surface to be monitored in such a way that interference fringes are set up. By analyzing these fringes, displacement or deformation may be detected and quantified. One area in which optical sensing is often employed is atomic force microscopy — to monitor the deflection of the beam upon which the sensing tip is mounted (see discussion on scanning probe microscopy in Chapter 10).

5.7.5 Resonant Sensors

These are based on micromachined beams or bridges that are driven to oscillate at their resonant frequency. Changes in the resonant frequency of the device would typically be monitored using implanted piezoresistors, capacitive, or optical techniques.

Figure 5.11a shows a bridge driven to resonance on a thin membrane. The resonant frequency of the bridge is related to the force applied to it (between anchor points), its length, thickness, width, mass, and the modulus of elasticity of the material from which it has been fabricated. If the membrane that the bridge is mounted on is deformed (Figure 5.11b), for instance, there is greater pressure on one side than the other. Then the force applied to the bridge changes, and hence the resonant frequency changes.

Alternatively, a resonant device may be used as a biosensor by coating it with a material that binds to the substance of interest. As more of the substance binds to the device, its mass will be increased, again altering the resonant frequency.

5.7.6 Accelerometers

Microengineered acceleration sensors, accelerometers, consist of a mass suspended from thin beams (Figure 5.12). As the device is accelerated, a force (force = mass × acceleration) is developed, which bends the suspending beams. Piezoresistors situated where the beams meet the support (where strain is greatest) can be used to detect acceleration. Another alternative is to capacitively sense the displacement of the mass.

5.7.7 Pressure Sensors

Microengineered pressure sensors are usually based on thin membranes. On one side is an evacuated cavity (for absolute pressure measurement), and the other side is exposed to the pressure to be measured. The deformation of the membrane is usually monitored using piezoresistors or capacitive techniques.

6 Microactuators

6.1 INTRODUCTION

Microactuators are required to drive resonant sensors (see Chapter 5, Subsection 5.7.5) to oscillate at their resonant frequency. They are also required to produce the mechanical output required of particular microsystems: this may be to move micromirrors to scan laser beams or switch them from one fiber to another, to drive cutting tools for microsurgical applications, to drive micropumps and valves for microanalysis or microfluidic systems, or these may even be microelectrode devices to stimulate nervous tissue in neural prosthesis applications.

In the following section a variety of methods for achieving microactuation are briefly outlined: electrostatic, magnetic, piezoelectric, hydraulic, and thermal. Of these, piezoelectric and hydraulic methods seem to be most promising, although the others have their place. Electrostatic actuation runs a close third and is possibly the most common and well-developed method, but it does suffer a little from wear and sticking problems. Magnetic actuators usually require relatively high currents (and high power) and, on the microscopic scale, electrostatic actuation methods usually offer better output per unit volume (the limit is somewhere in the region of 1-cm^3 to few-cubic-millimeter devices depending on the application). Thermal actuators also require relatively large amounts of electrical energy, and the heat generated has to be dissipated.

When dealing with very smooth surfaces, typical of micromachined devices, sticking or cold welding of one part to another can be a problem. These effects can increase friction to such a degree that all the output power of the device is required just to overcome it, and they can even prevent some devices from operating. Careful design and selection of materials can be used to overcome these problems, but they still cause trouble with many micromotor designs. Another point to be aware of is that when removing micromachined devices from wet-etch baths, the surface tension in the liquid can be strong enough to stick parts together.

6.2 ELECTROSTATIC ACTUATORS

For a parallel plate capacitor, the energy stored, U, is given in Equation 6.1 (where C is the capacitance and V is the voltage across the capacitor).

$$U = C\frac{V^2}{2} \tag{6.1}$$

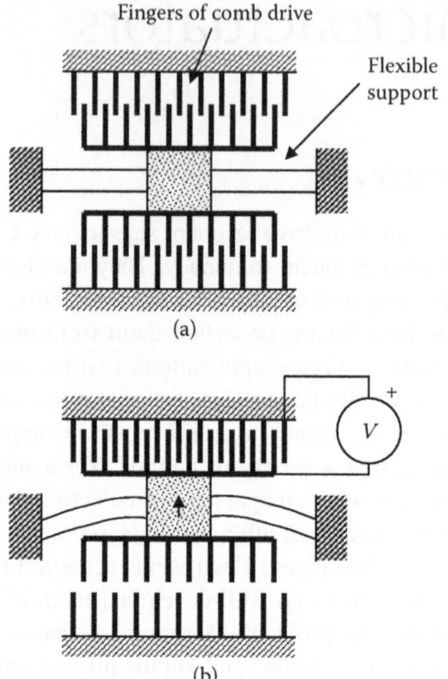

FIGURE 6.1 Schematic of comb drive operation.

When the plates of the capacitor move toward each other, the work done by the attractive force between them can be computed as the change in U with distance x. The force can be computed using Equation 6.2:

$$F_x = \frac{V^2}{2}\frac{\partial C}{\partial x} \tag{6.2}$$

Note that only attractive forces can be generated in this instance. Also, to generate large forces (which will do the useful work of the device), a large change of capacitance with distance is required. This has led to the development of electrostatic comb drives (Figure 6.1).

6.2.1 Comb Drives

These are particularly popular with surface-micromachined devices. They consist of many interdigitated fingers (Figure 6.1a). When a voltage is applied, an attractive force is developed between the fingers, which move together (Figure 6.1b). The increase in capacitance is proportional to the number of fingers; so to generate large forces, a large number of fingers are required. One problem with this device is that if the lateral gaps between the fingers are not the same on both sides (or if the device is jogged), then it is possible for the fingers to move at right angles to the intended direction of motion and stick together until the voltage is switched off (and in the worst-case scenario, they will remain stuck even after that).

For a comb drive with N electrodes (or rather, $2N$ gaps between the fingers), the capacitance is approximately:

$$C \approx 2Nhx\frac{\varepsilon}{g} \tag{6.3}$$

where h is the depth of the structure, g the gap between two electrode fingers, x the overlap of the two combs, and ε the permitivity. Thus:

$$\frac{\partial C}{\partial x} = 2Nh\frac{\varepsilon}{g} \tag{6.4}$$

6.2.2 Wobble Motors

Wobble motors are so called because of the rolling action by which they operate. Figure 6.2a and Figure 6.2b show a surface-micromachined wobble motor design. The rotor is a circular disk. In operation the electrodes beneath it are switched on and off one after another. The disk is attracted to each electrode in turn, the edge

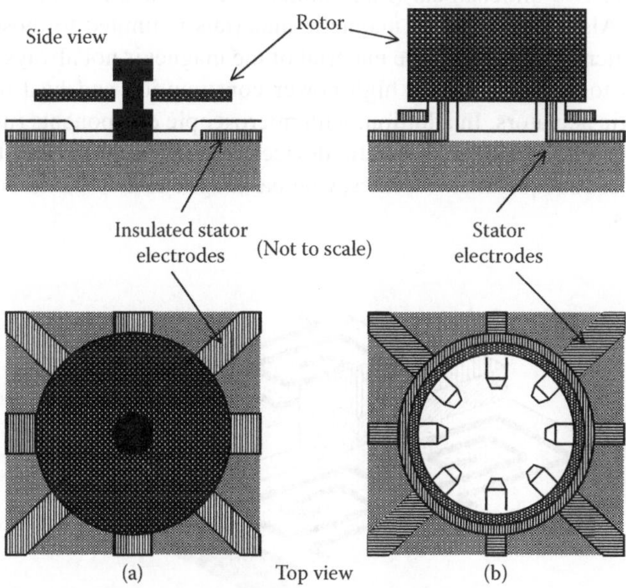

FIGURE 6.2 Wobble motors: (a) a surface-micromachined type, (b) use of LIGA to achieve a larger overlap between rotor and stator electrodes.

of the disk contacting the insulator over the electrode. In this manner it rolls slowly around in a circle, making one revolution to many revolutions of the stator voltage. Problems can arise if the insulating materials on the stator electrodes wear rapidly or stick to the rotor. Also, if the rotor and bearing are not circular (this is possible because many CAD packages draw circles as many-sided polygons), then the rotor can get stuck on its first revolution.

A problem with surface-micromachined motors is that they have very small vertical dimensions, so it is difficult to achieve large changes of capacitance with the motion of the rotor. LIGA techniques can be used to overcome this problem — for instance, the wobble motor shown in Figure 6.2c and Figure 6.2d, where the cylindrical rotor rolls around the stator.

6.3 MAGNETIC ACTUATORS

Microstructures are often fabricated by electroplating techniques, using nickel. This is particularly common with LIGA. Nickel is a (weakly) ferromagnetic material and so lends itself to use in magnetic microactuators. An example of a magnetic microactuator is the linear motor shown in Figure 6.3. The magnet resting in the channel is levitated and driven back and forth by switching current into the various coils on either side of the channel at the appropriate time.

From Figure 6.3, one common problem with magnetic actuators is clear: the coils are two dimensional (three-dimensional coils are difficult to microfabricate). Also, the choice of magnetic materials is limited to those that can be easily micromachined, so the material of the magnet is not always optimum. This tends to lead to a rather high power consumption and heat dissipation for magnetic actuators. In addition, with microscopic components (up to about millimeter dimensions), electrostatic devices are typically stronger than magnetic devices for equivalent volumes, whereas magnetic devices excel for larger dimensions.

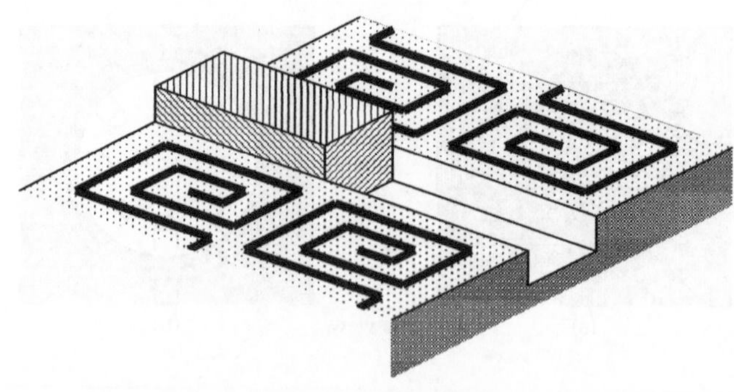

FIGURE 6.3 Magnetic actuator.

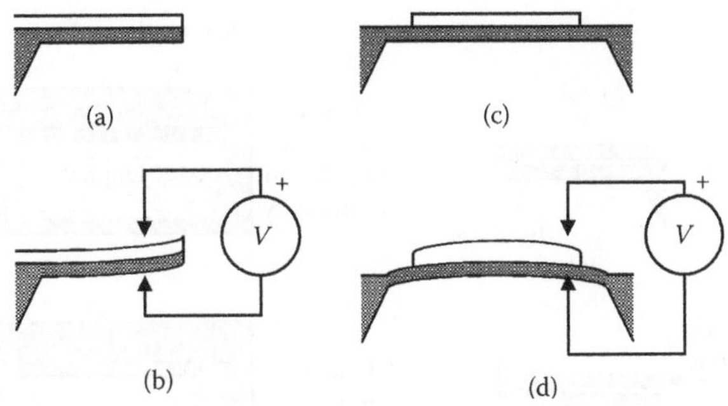

FIGURE 6.4 Piezo actuators: (a) and (b) a cantilever beam; (c) and (d) an actuated membrane.

6.4 PIEZOELECTRIC ACTUATORS

The piezoelectric effect mentioned previously for use in force sensors also works in reverse. If a voltage is applied across a film of piezoelectric material, a force is generated. Examples of how this may be used are given in Figure 6.4. In Figure 6.4a, a layer of piezoelectric material is deposited on a beam. When a voltage is applied, the stress generated causes the beam to bend (Figure 6.4b).

The same principle can be applied to thin silicon membranes (Figure 6.4c). When a voltage is applied, the membrane deforms (Figure 6.4d). This, when combined with microvalves, can be used to pump fluids through a microfluidic system.

When fabricating piezoelectric devices, it is necessary to ensure that the films are suitably thick so that high enough voltages can be applied without dielectric breakdown (sparks or short circuits across the film).

6.5 THERMAL ACTUATORS

Thermal microactuators are commonly either of the bimetallic type or use the expansion of a liquid or gas.

In Figure 6.5a, a beam is machined from one material (e.g., silicon) and a layer of material with a different coefficient of thermal expansivity (e.g., aluminum). When the two are heated, one material expands faster than the other, and the beam bends (Figure 6.5b). Heating may be accomplished by passing current through the device, thus heating it electrically.

Figure 6.5c shows a cavity containing a volume of fluid with a thin membrane as one wall. The current passed through a heating resistor causes the liquid in the cavity to expand, deforming the membrane (Figure 6.5d). The most effective method of actuation is critical point heating. A liquid with a suitably low boiling point is chosen and actuation is effected not merely through thermal expansion of the liquid but by heating it to its boiling point. The large volume change that

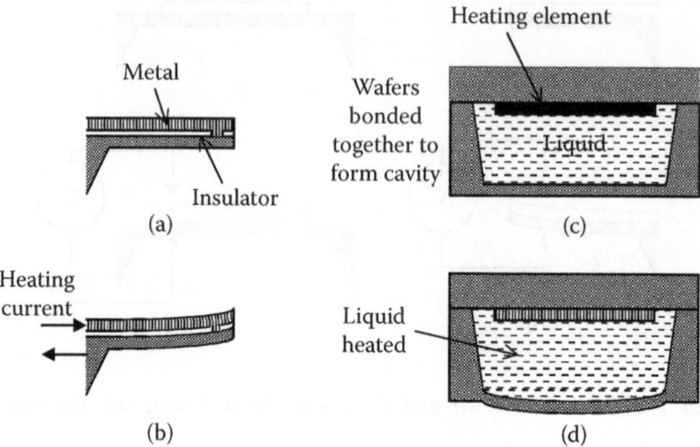

FIGURE 6.5 Thermal actuators: (a) a bimetallic bar form, (b) when a heating current is passed through the structure, the metal and silicon expand with different coefficients of thermal expansion, and the beam bends, (c) a cavity filled with a liquid, (d) when the liquid is heated, it expands deforming the membrane.

accompanies the state change from liquid to gas can be used to create a very powerful actuator.

Thermally actuated devices can develop relatively large forces, and the heating elements consume fairly large amounts of power. Also, the heated material has to cool down to return the actuator to its original position; so the heat has to be dissipated into the surrounding structure. This will take a finite amount of time and will affect the speed at which such actuators can be operated. There is also a trade-off: the faster it cools, the more energy used to heat the device to actuation point.

6.6 HYDRAULIC ACTUATORS

Despite problems associated with leaky valves and seals (a problem in many microfluidic systems), hydraulic actuators have considerable potential as quite a lot of power can be delivered from an external source along very-narrow-diameter tubes. This has potential in areas such as catheter-tip-mounted microsurgical tools.

LIGA techniques can be used to fabricate turbines (as in Figure 6.6), which can deliver power to cutting tools.

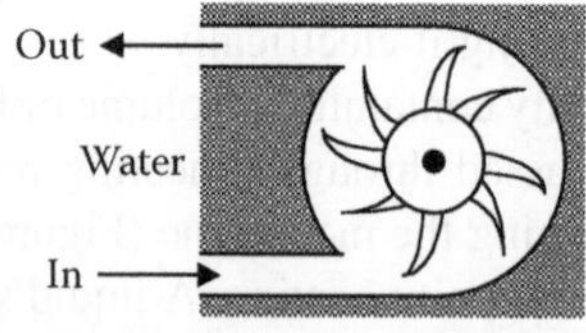

FIGURE 6.6 Hydraulic actuators; turbine-like structures can be created using LIGA.

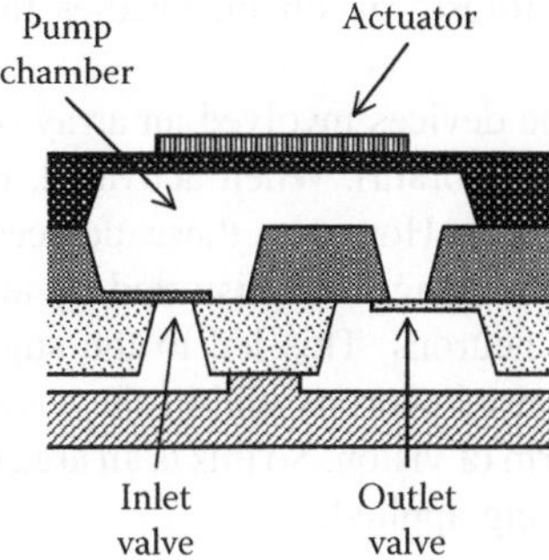

FIGURE 6.7 Microfluidic device constructed by bonding four bulk machined wafers (cross section, not to scale); the inlet and outlet valves are formed by cantilever beams, and a pump is created by deforming a membrane to change the volume of the pump chamber.

6.7 MULTILAYER BONDED DEVICES

Surface micromachining (Chapter 2, Subsection 2.7.2) provides an obvious way to build complex multilayered structures. The disadvantage of this approach is that the vertical dimensions of the structures are usually quite small (on the order of microns rather than hundreds of microns). A similar approach is available for bulk micromachined devices, whereby bulk micromachined wafers are stacked and bonded together. There are obvious problems with registration and alignment, and devices constructed in this way are generally quite large. Figure 6.7 gives a cross section through a hypothetical microfluidic device constructed using this method.

6.8 MICROSTIMULATORS

One further method of actuation is illustrated by the use of microelectrode devices to electrically stimulate activity of nerves and muscles. Common designs for these devices have already been discussed in Chapter 5 (Section 5.6). The use of microelectrode devices facilitates highly specific stimulation of individual nerve fibers compared to other methods of stimulation; this should allow finer control of the stimulation provided enough electrode sites can be inserted into the tissue.

As relatively large stimulating currents have to be passed through the electrode sites, microelectrodes for stimulation generally have geometrically larger electrode sites than those for recording (500 μm^2 and above compared to 16 μm^2 and above). This is necessary, otherwise the currents involved will damage the electrode sites.

One area in which silicon microengineering is being applied in the hope that it will result in a considerable improvement over more conventional electrodes is the area of visual prosthesis — providing rudimentary vision for the blind.

One project involves a "forest" of silicon needles that will be inserted in the visual cortex.

Early visual prosthetic devices involved an array of electrodes placed on the surface of the visual cortex (brain). When activated, blind volunteers could see points of light (phosphenes). However, these devices required relatively high currents to operate, and the image was distorted by afterdischarges and interactions between groups of neurons. This led to the suggestion that a method for more selective stimulation of neurons within the visual cortex was required to provide any functional form of vision. So this is an area is which microengineering technology is recently being applied.

7 Micro Total Analysis Systems

7.1 INTRODUCTION

The term *micro total analysis system* (μTAS) was originally coined by a group of researchers to describe the kind of chemical processing system outlined in Figure 7.1. This goes beyond chemical sensors or biosensors as it is modeled on the analytical procedures often found in industry:

- Go to the site of investigation.
- Take samples.
- Return the samples to the laboratory.
- Prepare the samples: filter, centrifuge, and split up for different processes.
- Perform analysis — this may involve a series of chemical reactions and may even be performed by robots.
- Record examination results.
- Adjust the industrial process if necessary.
- Repeat as frequently as required.

Obviously, this is not always very convenient. For instance, water companies want to know the results as soon as possible when there has been a pollution incident — before the polluted water reaches the consumer and preferably before there has been any major environmental damage. Furthermore, regular sampling and analysis can be a very expensive process. Highly trained staff have to be employed in an expensively equipped laboratory, and people must go regularly to the appropriate sampling points, which may be long distances apart in the case of large companies or for some systems. Many processes involve hazardous chemicals that are very difficult to sample, transport, and dispose of. The same may be said of the chemical reagents used to analyze the sample, and these may also be very expensive. In the worst situation, the entire manufacturing process may need to be shut down to enable it to be investigated properly.

Given all these challenges, it is not surprising that microengineering techniques were applied to the area of chemical analysis quite early on. In fact, one of the first commercial applications (developed in the 1970s) was the Stanford gas chromatograph, developed by Terry, Jerman, and Angell. This provided a system for separating and detecting gases in a sample on a single silicon wafer.

The main focus in this chapter will be on the core elements of μTAS — the microfluidic, electrophoretic separation systems that it employs, and the detection

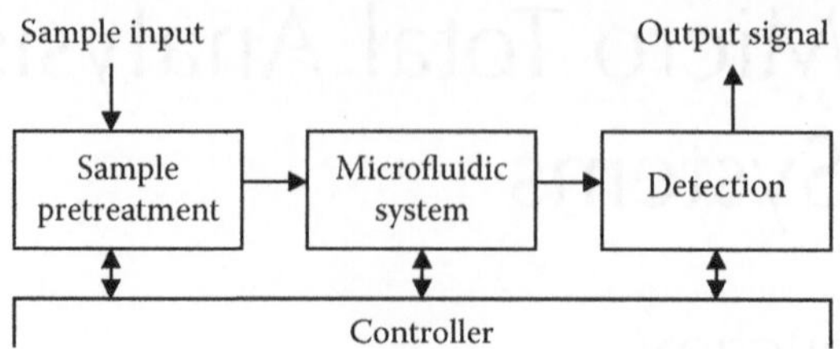

FIGURE 7.1 Micro total analysis system outline. (After van den Berg & Lammerink, in Manz & Becker, "Microsystem Technology in Chemistry and Life Sciences," Springer, 1998).

systems. Other aspects will, however, also be explored. These include a more general treatment of so-called "biochips" and a very brief introduction to chemistry.

7.2 BASIC CHEMISTRY

For the purposes of this book, it is convenient to divide chemistry into three groups: inorganic chemistry, organic chemistry, and biochemistry. Inorganic chemistry deals with the chemical reactions and properties of small and relatively simple molecules. Organic chemistry deals essentially with carbon chemistry. Carbon is a special element, which forms complex long-chain molecules; it is the basis of polymers (plastics, oils, etc.) and life. Biochemistry is a subset of organic chemistry and deals specifically with the molecules found in living organisms: proteins, lipids, carbohydrates, nucleic acids, and their component parts.

Group I	II	III	IV	V	VI	VII	VIII
1 H 1 Hydrogen							2 He 4 Helium
3 Li 6.9 Lithium	4 Be 9 Beryllium	5 B 10.8 Boron	6 C 12 Carbon	7 N 14 Nitrogen	8 O 16 Oxygen	9 F 19 Fluorine	10 Ne 20.2 Neon
11 Na 23 Sodium	12 Mg 24.3 Magnesium	13 Al 27 Aluminum	14 Si 28.1 Silicon	15 P 31 Phosphorous	16 S 32.1 Sulphur	17 Cl 35.5 Chlorine	18 Ar 39.9 Argon
19 K 39.1 Potassium	20 Ca 40.1 Calcium	31 Ga 69.7 Gallium	32 Ge 72.6 Germanium	33 As 74.9 Arsenic	34 Se 79 Selenium	35 Br 79.9 Bromine	36 Kr 83.8 Krypton

FIGURE 7.2 Part of the Periodic Table. The upper number in each cell is the atomic number (the number of protons in each atom). The lower number is the mass number. This represents the number of protons and neutrons in an atom; it is not twice the number of protons in some cases because the naturally occurring substance is made up of different isotopes (one or two extra or missing neutrons). Metals have been shaded.

7.2.1 Inorganic Chemistry

Any exploration of chemistry starts with the Periodic Table of Elements, such as that shown in Figure 7.2. For simplicity, Figure 7.2 only shows a subsection of the Periodic Table.

Matter is composed of atoms of the different elements, which combine (react) according to known rules. Atoms are composed of three types of subatomic particles: protons and neutrons, which are found in the center (nucleus) of the atom, and electrons, which form a cloud around the nucleus. The protons in the nucleus each carry one unit of positive charge, and the electrons each carry one unit of negative charge. The *atomic number* (Figure 7.2) of the element is the number of protons in the nucleus. In a normal atom, the number of electrons in the atom is equal to the number of protons, and the atom carries zero charge. In certain circumstances, it is possible to add or remove electrons from the atom; the resulting charged particle is called an *ion*.

Amount of Substance

Elements also have an associated mass number. This is the number of protons plus neutrons in a sample of the element as it naturally occurs. It is not always exactly twice the atomic number, as may be expected, because many elements exist as different isotopes with slightly different numbers of neutrons. The relative abundance of these isotopes gives rise to noninteger atomic masses. The atomic mass of an element expressed in grams is one molar mass (M) of the element and contains 6.02×10^{23} atoms (this is the Avogadro constant). 1 M dissolved in 1 l of water gives a solution of 1-mol concentration.

The electrons crowd near the nucleus, and as the number of electrons in the atom increases, not all of them get an equal share of the nucleus. It is convenient to consider the electrons as existing in different energy shells — related to the amount of energy required to remove them from the atom. For an atom with a lot of electrons, it is relatively easy to remove the first one (ionize the atom), but as more and more electrons are removed, more and more energy is required.

The basic rule of chemistry is that an atom would like to fill its outer shell with electrons (this being, energetically, the most preferable state to be in). However, if it were to simply add or shed electrons at a whim, matter would be electrically charged. But matter (on the macroscopic scale and, in general, over time) remains electrically neutral. It has already been noted that it takes energy to ionize an atom, and in the absence of that energy, electrons will simply flow from negatively charged areas to positively charged ones so that both become neutral. The atom, therefore, has two options to achieve this neutral state. Either it can borrow electrons from another atom (or atoms), in which case the resulting ions must stay close together to ensure their charges cancel out and give an overall appearance of neutrality, or it can share electrons with another atom (or atoms), in which case the atoms must stay so close together that a single electron cloud can envelop them.

TABLE 7.1
Number of Electrons Required to Complete the Outer Shell

Group	Electrons in the Outer Shell	Gain (+) or Lose (–) to Get a Full Outer Shell	Notes
I	1	–1	Hydrogen sometimes gains one
II	2	–2	
III	3	–3	
IV	4	+4 or –4	
V	5	+3	
VI	6	+2	
VII	7	+1	
VIII	8	0	Helium has, and requires, only two; it does not readily react with anything

This gives rise to two kinds of bonds: the *ionic bond*, formed between ions, and the *covalent bond*, formed when electrons are shared equally between atoms.

The number of electrons that comprise a full outer shell can be determined from the group in which the element appears in the Periodic Table (Figure 7.2); this has been summarized in Table 7.1. The notable exception is hydrogen. It normally loses an electron to form a positive hydrogen ion, but it can form a negatively charged hydronium ion.

Figure 7.2 shows two divisions: metals and nonmetals; semimetals can be found along the line on the nonmetal side. Compounds formed when metals react with nonmetals form ionic bonds. Atoms of nonmetallic elements combine by forming covalent bonds. The elements in group I and group VII are generally more reactive than those in the other groups. Elements in group VIII are the most inert, and these are the only gaseous elements that can be found in their atomic state under normal conditions; all the others forms combine into molecules (oxygen atoms pair up, for instance, to form oxygen molecules). In the solid state, atoms are bonded together into some sort of crystalline or amorphous structure.

Pure elemental metals do not form either ionic or covalent bonds. Rather, they exist in a state in which the electrons in the outer shells escape from the influence of the particular nucleus, so the atoms in a metal can be considered as floating in a sea of electrons; this is why metals conduct electricity so well. In semiconductors, there is sufficient energy at room temperature to enable electrons to occasionally escape from the influence of a particular nucleus, but they are soon captured by another; thus, they conduct electricity but not as well as metals do.

Similarities between Elements

This section has been written in a way that suggests that the chemical properties of an element follow from its position in the Periodic Table. The historical truth is that the table was originally developed by grouping elements that exhibited similar chemical properties. It was only later that these were formally related to atomic structure in the manner described.

7.2.1.1 Bond Formation

Chemical reactions between elements in group I or group II with elements in group VII will form ionic bonds.

Common table salt, sodium chloride, is formed from one atom of sodium combining with one atom of chlorine. As suggested from Table 7.1, the sodium atom gives up an electron to become a sodium ion (denoted as Na^+), and the chlorine atom gains an electron, becoming a chlorine ion (Cl^-); the two ions combine to form the compound, thus:

$$Na + Cl \rightarrow NaCl \tag{7.1}$$

Combination of Sodium and Chlorine

In more detail, showing the formation of sodium and chloride ions and the transfer of electrons, Equation 7.1 develops as follows:

$Na \rightarrow Na^+ + e^-$ sodium ion formation

$Cl + e^- \rightarrow Cl^-$ chloride ion formation, using the electron given up by Na

$Na^+ + Cl^- \rightarrow NaCl$ two charged ions combine into a neutral compound

Magnesium, however, loses two electrons (becoming Mg^{2+}), and these ions are accepted by two chlorine atoms:

$$Mg + 2Cl \rightarrow MgCl_2 \tag{7.2}$$

The movement of electrons is not normally explicitly stated in chemical equations unless they are provided from an unusual source, such as an electric current from an electrode, or if the equation needs to be broken down for some reason.

The metals of group I burn well in air and react violently with water (H_2O, a combination of hydrogen and oxygen); for this reason, they are stored under oil. The relevant equations for potassium are:

$$2K + O \rightarrow K_2O \tag{7.3}$$

and

$$2K + 2H_2O \rightarrow 2KOH + H_2 \tag{7.4}$$

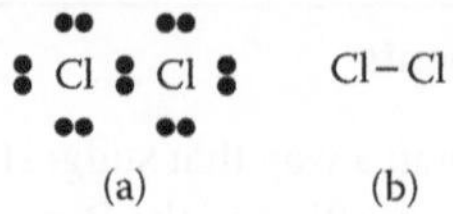

:Cl:Cl: Cl–Cl

(a) (b)

FIGURE 7.3 (a) Two chlorine atoms form a bond by sharing a pair of electrons, (b) this bond can also be drawn as a straight line.

Notice that in Equation 7.4, a hydrogen molecule (H_2) has been given off by the reaction. The other product of the reaction, potassium hydroxide, is formed from the combination of a potassium (K+) ion and a hydroxide (OH^-) ion. Where elements naturally occur in molecular form, then it is normal to show this in the chemical equation. Equation 7.1 to Equation 7.3 thus become:

$$2Na + Cl_2 \rightarrow 2NaCl \tag{7.5}$$

$$Mg ++ Cl_2 \rightarrow MgCl_2 \tag{7.6}$$

$$4K + O_2 \rightarrow 2K_2O \tag{7.7}$$

Covalent-bond formation arises due to the sharing of electrons. In Equation 7.5 and Equation 7.6 it has been suggested that chlorine forms a molecule of two chlorine atoms. Because chlorine is a nonmetal, this involves the formation of a covalent bond. From Table 7.1 and Figure 7.2, it can been seen that chlorine has seven electrons in its outer shell; it would like to have eight. It can achieve this by sharing a pair of electrons with another chlorine molecule, with each atom donating one of the pair, as shown in Figure 7.3. (Obviously, it is not possible to tell which electron came from which atom, as illustrated.)

This sharing of a pair of electrons forms a single covalent bond, which is graphically depicted by a single line (Figure 7.3). Two nitrogen atoms, on the other hand, have to share three pairs, which results in a very strong triple bond; carbon burns in air to form carbon dioxide, with two double bonds (see Figure 7.4).

Because the electrons are shared between atoms in a covalent bond, they cannot be everywhere at once. As a result, at any one time, one part of the resulting molecule will be slightly negative and another part will be slightly positive. This gives rise to a very weak form of intermolecular bonding, termed *van der Waals bonding*. When the atoms in the molecule are of different elements, carbon and chlorine for example, then it may well be that one component of the molecule is more attractive to electrons than the other. In the case of carbon tetrachloride

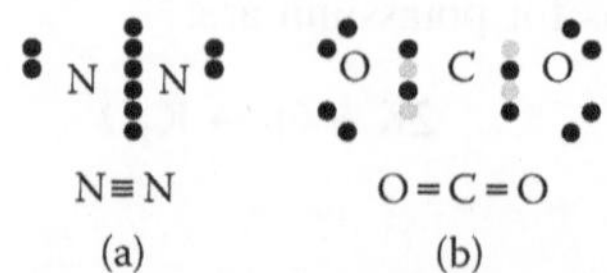

FIGURE 7.4 (a) Nitrogen molecule, (b) carbon dioxide molecule.

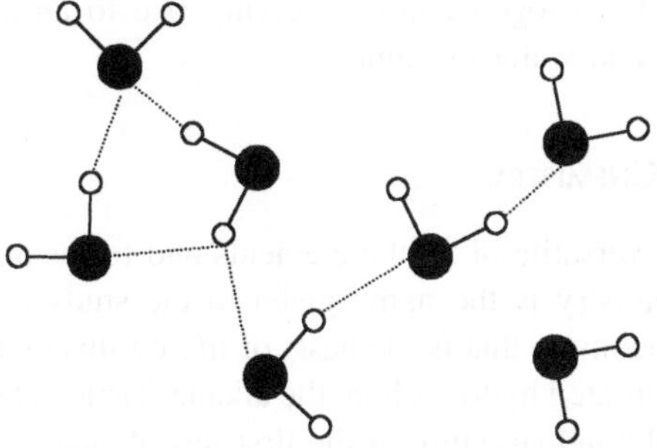

FIGURE 7.5 Hydrogen-bonding in water. The small hydrogen atoms in a molecule form weak bonds with larger oxygen atoms in adjacent molecules.

(CCl_4), there will be a dipole setup with the chlorine atoms being slightly more negative than the carbon atoms, i.e., the bonds will be polarized.

This effect is very important in water, where the bonds between hydrogen and oxygen are polarized. This gives rise to hydrogen bonds (H-bonds) between the water molecules (Figure 7.5), which means that water is a liquid at room temperature and pressure, whereas the much heavier carbon dioxide molecule is a gas.

7.2.1.2 pH

In an aqueous solution (i.e., when substances are dissolved in water), things are never constant. Water molecules themselves are continually breaking up and recombining, although most of them will remain as molecules for the majority of the time. As with all chemical equations, however, an equilibrium is set up:

$$H_2O \leftrightarrow H_3O^+ + OH^- \tag{7.8}$$

The degree to which this reaction is shifted to the right-hand side of the equation is shown by the pH, which is a logarithmic measure of the concentration of hydrogen ions in the solution. Pure water has a pH of 7. A pH of less than 7 means that the solution is hydrogen-ion-rich and therefore acidic. A pH of more than 7 means that it is hydrogen-ion-poor (and thus rich in hydroxide, OH^-, ions) or alkaline. Compounds that dissolve readily in water, giving up hydrogen ions, therefore form acids. Hydrogen chloride (HCl) is one example of a gas that forms a strong acid when dissolved in water.

Potassium hydroxide (KOH, see Equation 7.4), when dissolved in water, releases hydroxide ions and so forms an alkaline solution (alkali-forming compounds are called *bases*). Potassium hydroxide and hydrogen chloride react in solution:

$$\text{KOH (s)} + \text{HCl (aq)} \rightarrow \text{KCl (aq)} + H_2O\text{ (l)} \tag{7.9}$$

(Acid plus base forms a salt plus water.) The states of the different compounds are given in brackets. A solid powder of potassium hydroxide is dissolved in an aqueous solution of hydrogen chloride, giving rise to an aqueous solution of potassium chloride and water (a liquid).

7.2.2 Organic Chemistry

Carbon is the most versatile of all the elements and forms many complex molecules. Organic chemistry is the name given to the study of carbon chemistry, because it is this chemistry that is the basis of life on this planet. The basic form is the long-chain saturated hydrocarbon, the alkane. Table 7.2 shows the chemical formula, name, and bond structure of the first few alkanes.

It is important to note that the structures shown in Table 7.2 are merely two-dimensional representations of three-dimensional structures (Figure 7.6).

TABLE 7.2
Chemical Composition, Name, and Bond Structure — Methane to Hexane

Chemical formula	Name	Bond structure
CH_4	Methane	H-C-H (with H above and below C)
C_2H_6	Ethane	H-C-C-H (with H above and below each C)
C_3H_8	Propane	H-C-C-C-H (with H above and below each C)
C_4H_{10}	Butane	H-C-C-C-C-H (with H above and below each C)
C_5H_{12}	Pentane	H-C-C-C-C-C-H (with H above and below each C)
C_6H_{14}	Hexane	H-C-C-C-C-C-C-H (with H above and below each C)

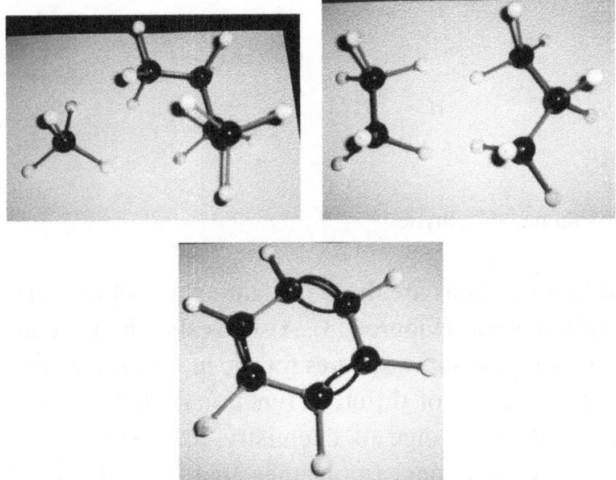

FIGURE 7.6 Stick-and-ball models of methane, butane, ethane, propane, and the six-carbon benzene ring. Note that in the case of benzene, there are three double bonds; in practice, the electrons do not know where to form these bonds, which means that there is a cloud of delocalized electrons around the ring.

Additionally, although stick-and-ball models such as those in Figure 7.6 clearly show the structure, other models are also used. These may show the approximate distributions of electrons about the molecules (electron orbitals) or the charge on the molecule. A variety of different models can be generated by computer for different purposes.

The other important thing to note is that the chemical formula does not clearly define the structure. Figure 7.7 shows two representations of C_4H_{10}. Figure 7.7a is butane but Figure 7.7b is, in fact, 2-methylpropane.

Naming Hydrocarbons

The (IUPAC) standard procedure is:

1. Name the longest carbon chain.
2. Name the substituent groups.
3. Give the positions of the substituent groups.

The groups are numbered, counting from the end that gives the lowest number.

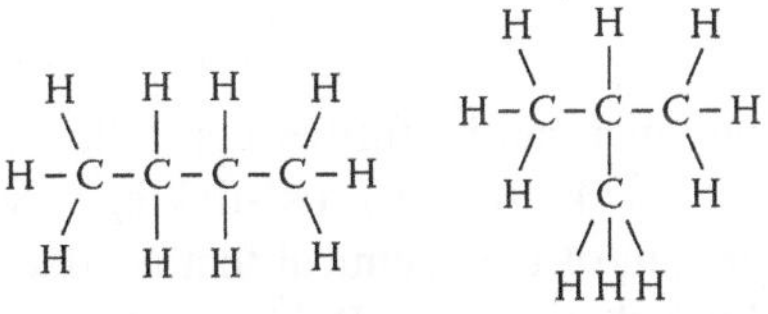

```
                              H   H   H
                              \   |   /
 H   H  H   H               H-C - C - C-H
 \   |  |   /                 /   |   \
H-C -C -C -C-H                H   C   H
 /   |  |   \                    /|\
 H   H  H   H                    HHH

     (a)                          (b)
```

FIGURE 7.7 (a) Butane, (b) 2-methylpropane.

```
 H     H
  \   /
  C = C        H-C≡C-H
  /   \
 H     H
   (a)           (b)
```

FIGURE 7.8 (a) Ethene, (b) ethyne.

Unsaturated hydrocarbons are those containing a double (-ene) or triple (-yne) bond between carbon atoms (Figure 7.8). Also, as shown in Figure 7.7, the carbon atoms can form a variety of ring structures (cyclic molecules) as well as long chains.

Table 7.3 lists a variety of different functional groups that can add on to a hydrocarbon molecule to change its chemistry, and Figure 7.9 illustrates some different kinds of stereoisomers (molecules that have the same chemical and structural formula but different arrangements of bonds in space).

7.2.2.1 Polymers

Carbon compounds can contain very long chains of carbon atoms, plastics being one such example. The molecules of plastics are made of repeats of many smaller identical units. The common plastic polythene (actually, polyethene) is composed of many ethene molecules (Equation 7.10). In the presence of a strong catalyst (which is something that assists in a reaction without being consumed itself), the double bond between carbon atoms is opened out, enabling two molecules to join. This leaves two more bonds at the end of the chain, encouraging more molecules to join, and so on, until something causes the chain to stop growing (running out of ethene molecules, for instance). One consequence of this is that the molecules in a plastic will be of different lengths, depending on when the chains stopped growing:

$$nCH_2 = CH_2(g) \rightarrow -(CH_2-CH_2)-_n \qquad (7.10)$$

Polymers can also form through chemical reactions between two small precursor materials. Nylon 66 is formed by a reaction between a diamine (hexane-1,6-diamine) and a dicarboxylic acid (hexane-1,6-dioic acid):

$$nH_2N(CH_2)_6NH_2 + nHOOC(CH_2)_4COOH \rightarrow$$
$$-(HN(CH_2)_6NHOC(CH_2)_4CO)-_n + 2nH_2O \qquad (7.11)$$

Cross-Linking

The properties of a polymer can be further controlled by adding agents that join chains together; this is known as cross-linking. The physical properties will depend on the amount of the chemical that is added, which controls the amount of cross-linking that occurs. Rubbers are a good example of the application of cross-linking; many are very soft in their pure state, but increasing the crosslinking makes them harder and chemically more resilient.

TABLE 7.3
Different Substituent Groups

-CH_3 -C_2H_5 -C_nH_{2n+1}	methyl ethyl alkyl
$>C=C<$	alkene functional group
-C≡C-	alkyne functional group
-Fl -Cl -Br	Halogens / halide groups $C_2H_4Cl_2$ is dichloroethane.
-OH	hydroxyl group, making alcohols (e.g. C_2H_5OH is ethanol)
O // -C \\ O-H	Carboxyl group, forming carboxylic acids (CH_3COOH is ethanoic acid)
O // -C \\ H	Aldehyde
O // R-C \\ R'	Ketone (R and R' are hydrocarbon chains)
O // R-C \\ O-R'	Ester
R-NH_2 R \\ NH / R' R' \\ R-N / R''	Amines, derived from ammonia (NH_3)
H_2NCHCO_2H | R	Amino acid

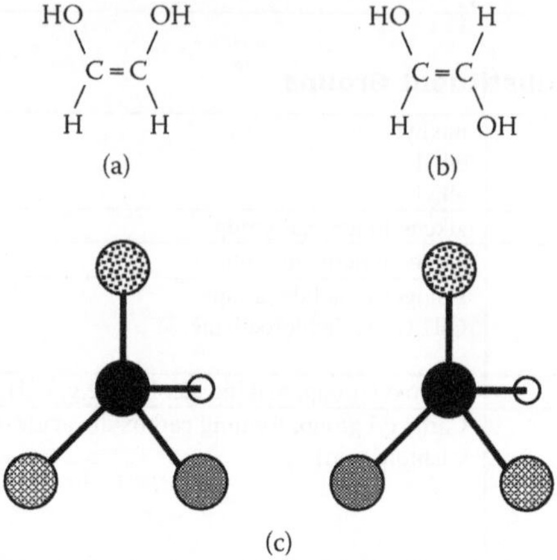

```
HO    OH        HO    H
  \  /            \  /
  C=C             C=C
  /  \            /  \
 H    H          H    OH
  (a)              (b)
```

(c)

FIGURE 7.9 *Cis-* (a) and *trans-* (b) isomers of ethane-1,2-diol; note the importance of the double bond; if this were a single bond, then the carbon atoms could rotate about it, (c) shows stereoisomers of the same four-atom compound; one is the mirror image of the other and cannot be superimposed on it by simple rotation.

7.2.2.2 Silicones

As mentioned earlier, elements in the same group in the Periodic Table have similar properties. Therefore, one may reasonably expect that silicon should be capable of forming long-chain molecules similar to those formed by carbon. Although silicon does not act exactly as carbon does, it can be coerced into forming long-chain molecules with a silicon–oxygen backbone. Carbon side chains on these molecules prevent them from forming three-dimensional structures and control their properties. These polyorganosiloxanes can be liquids (oils), gels, or solids (rubbers). The most common polyorganosiloxane is polydimethylsiloxane (PDMS). The basic repeating unit is:

$$—(CH_3)_2SiO—$$

To control the length of the chain formed, end-blocking molecules are added (see Figure 7.10).

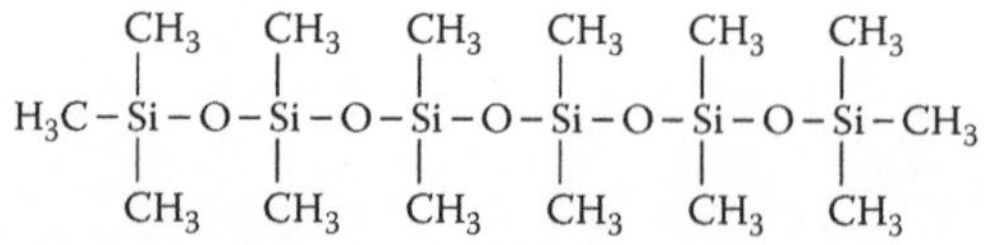

```
      CH3   CH3   CH3   CH3   CH3   CH3
       |     |     |     |     |     |
H3C - Si - O-Si - O-Si - O-Si - O-Si - O-Si - CH3
       |     |     |     |     |     |
      CH3   CH3   CH3   CH3   CH3   CH3
```

FIGURE 7.10 Short PDMS chain.

Silanes

The small molecules of silicon chemistry are silanes (or organosilanes if they involve carbon atoms). The simplest molecule of them all is silane itself:

$$SiH_4$$

Compare this formula with that for methane, CH_4.

Silicones are important engineering materials (not just for MEMS) because they are chemically relatively inert and stable with changes in temperature. Silicones are not naturally occurring polymers, unlike long-chain carbon molecules, and have to be manufactured.

7.2.3 Biochemistry

Because organic chemistry has developed from its original definition as the chemistry of compounds found in living things to encompass the entirety of carbon-based chemistry, biochemistry has developed to encompass this original area: the molecules found in living organisms and how they work together.

Different molecules in the cells of living organisms perform different functions. These include, among others, the following:

- Storage of genetic information (which encodes instructions for making different proteins — it apparently stores all the information required to make an organism)
- Replication of genetic information
- Translation of genetic information into proteins
- Formation of structural elements (membranes and skeletal components, and also molecular machines)
- Energy storage
- Signaling (enabling the cell to react to different environmental conditions)
- Communication (with other cells)
- Catalysts

This section will introduce three of the major classes of large molecules in the cell — proteins, lipids, carbohydrates — and the nucleic acids. The two nucleic acids, deoxyribonucleic acid (DNA) and ribonucleic acid (RNA) are the molecules that store genetic information (DNA) and translate it into proteins (RNA).

Proteins are the workhorses in the cell: they form structures, such as the skeleton, molecular motors, molecular machines, and signaling systems. They also participate in or catalyze the reactions used to build (synthesize) other molecules (anabolism) and break down molecules (catabolism).

One of the main tasks for lipids in the cell is the formation of membranes within the cell as well as those that form the outer skins of cells. The cell is not, as it may first appear, a number of interconnected reaction chambers (analogous to an industrial chemical plant, where chemicals enter at various points and are pumped through pipes into different reaction vessels, at different temperatures and pressures, until the desired chemical product emerges at the other end). It is much more analogous to a manufacturing plant with robots moving various components around and assembling them into larger parts. A lot of this work takes place on membranes (or by proteins embedded in membranes). Additionally, small molecules (or ions) can be pumped across membranes as required, larger molecules can be made to flip from one side of a membrane to the other, and membranes can even be formed into spherical vesicles, which then detach and can carry a particular cargo to another part of the cell. The main components of these membranes are lipids.

Probably the best known carbohydrate is sugar, which is an energy store. The use of carbohydrate as a structural component is most obvious in plants, where the fibrous cell walls are formed from a carbohydrate (cellulose), and the exoskeletons of insects (chitin). Carbohydrates also combine with proteins, forming glycoproteins, which take on various tasks. In humans, they are evident (on a common day-to-day basis) as mucopolysaccharides, which form syrupy or gel-like mixtures in water (mucus).

7.2.3.1 Proteins

Proteins are complex polymers composed of amino acids — small organic molecules that incorporate an amino group ($-NH_2$) and a carboxyl group ($-COOH$) (see Table 7.3). The amino group of one amino acid in the protein reacts with the carboxyl group of the next forming a peptide bond; proteins are, therefore, also called polypeptides. Although there are a vast number of possible amino acids, there are only 20 that are encoded for in DNA and commonly found in proteins (they can, however, be modified after being incorporated into the protein). These are:

- Alanine
- Arginine
- Asparagine
- Aspartic acid
- Cystine
- Glutamic acid
- Glutamine
- Glycine
- Histadine
- Isoleucine
- Leucine
- Lysine
- Methionine
- Phenylalanine
- Proline
- Serine

- Threonine
- Tryptophan
- Tyrosine
- Valine

The basic form of 19 of the common amino acids is shown in Figure 7.11. Proline is the exception to the rule. The fundamental (primary) structure of a protein, then, is in the form of a chain of amino acids, with a backbone of peptide bonds from which project different side chains.

It was noted in Subsection 7.2.1.1 that different molecules are polarized to different extents. This is true of the amino acid side chains in proteins. Related to this, it can be seen that different side chains are more or less hydrophobic (do not mix with water); this is important because water is a very small molecule compared to most biological molecules, and as a result, a single protein may be surrounded by thousands of water molecules. The polarized side chains and hydrophobic side chains cause the protein molecule to fold up on itself, the degree of folding being restricted by the flexibility of the bonds between amino acids in the chain. This gives rise to a further level of structure within the protein molecule itself — the secondary structure.

The features of the secondary structure can generally be classified as either helices, sheets, loops, or turns. A helix occurs when amino acids along the backbone of the protein coil up. A sheet occurs when several lengths of amino acid chains run parallel to one another. Helices and sheets are normally interconnected by short turns or may have longer chains (loops) between them.

It is, however, the way that these features fold in on themselves — the tertiary structure — that gives the biggest clue to what a protein does. This determines the shape of the protein and how charge is distributed over its surface. Proteins may be designed to bond very specifically to particular molecules or to hold two molecules close together in order to catalyze the reaction between them. Alternatively, the addition or removal of a phosphate group may radically alter the shape of the molecule, enabling it to act as a motor, for instance.

There is a further level of structure — the quaternary structure — in which different polypeptides come together to form very complex structures (protein machines) and the individual protein molecules are not covalently bonded together, although they act as one unit. A simple example of this is in the protein coats that some viruses use for protection outside the cell. These are formed of many identical protein subunits that come together to form a regular geometric structure.

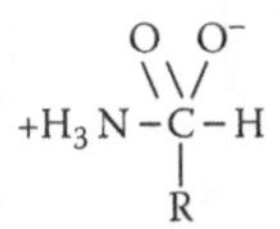

```
       O  O⁻
        \\/
+H₃N - C - H
        |
        R
```

FIGURE 7.11 Form of 19 of 20 common amino acids; R is a unique side chain.

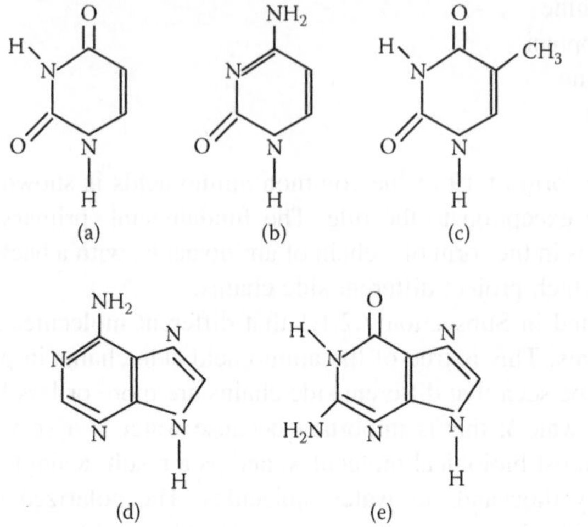

FIGURE 7.12 The five bases: (a) uracil, (b) cytosine, (c) thymine, (d) adenine, (e) guanine.

> **Molecular Mass**
>
> The molecular weights of proteins and other macromolecules, such as nucleic acids, are usually given in *daltons* (Da). A Dalton is $^1/_{12}$ the mass of a ^{12}C atom. Proteins have molecular weights ranging from about 5,000 to over 1,000,000 Da.

7.2.3.2 Nucleic Acids

Nucleic acids are polymers formed of nucleotides. Nucleotides are formed of bases joined to a sugar residue that has a phosphate group attached. The two nucleic acids commonly found in living organisms are DNA and RNA. The component parts of these are:

- Bases: uracil (U), cytosine (C), thymine (T), adenine (A), and guanine (G), Uracil is found in DNA but is replaced by thymine in RNA.
- Sugar: ribose or deoxyribose.

The different bases are termed purines or pyramidines, depending on their structure (see Figure 7.12); ribose and deoxyribose differ by the presence or absence of an oxygen atom (see Figure 7.13). The assembly of the whole into a

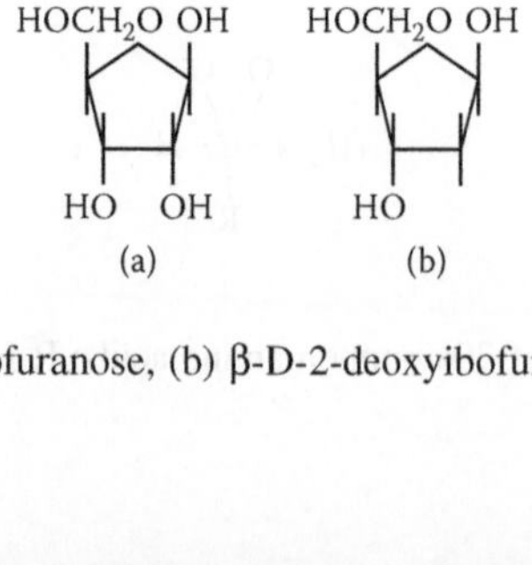

FIGURE 7.13 (a) β-D-Ribofuranose, (b) β-D-2-deoxyibofuranose.

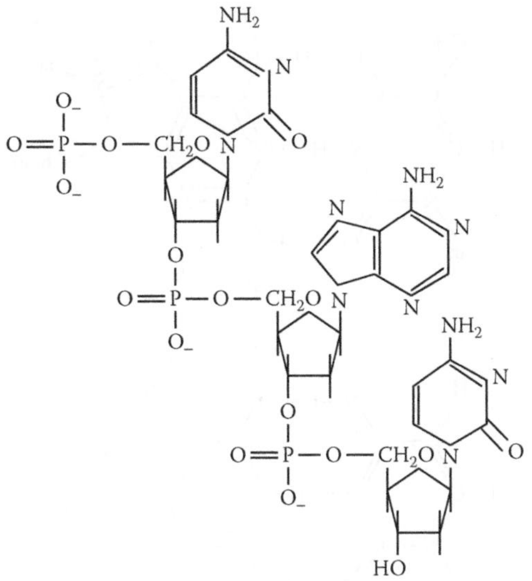

FIGURE 7.14 Nucleic acid chain (DNA).

chain via phosphate residues (PO_4^{3-}) is illustrated in Figure 7.14. Note that in these figures, carbon and hydrogen atoms have been left out for clarity. If bonds have been shown without atoms being labeled, it is assumed to be a carbon atom with the appropriate number of hydrogen atoms. Additionally, the term "base" is also used ambiguously when referring to the complete base–sugar structure in the DNA or RNA molecule itself.

DNA Strand Direction

Single DNA strands have two ends, the 3 end and the 5 end; the terminology arises from the standard nomenclature for the carbon atoms in the sugar ring. When synthesizing a long strand of DNA, the 3 (sugar) carbon atom of the next base will attach to the 5 (sugar) carbon atom of the base at the 5 end of the chain. In double-stranded DNA, the 5 end of one molecule pairs up with the 5 end of the other.

In the cell, information is stored in the DNA, which is chemically more stable than RNA. The key to information storage in DNA is that each purine base forms hydrogen bonds with a particular pyramidine base: cytosine pairs with guanine and uracil (or thymine) pairs with adenine (see Figure 7.15). This means that two complementary strands of DNA can bind together, which they do, forming a double-helical structure. When the cell divides (to reproduce), one strand goes to one daughter cell and the other strand goes to the other daughter cell. The complementary strands are then rebuilt.

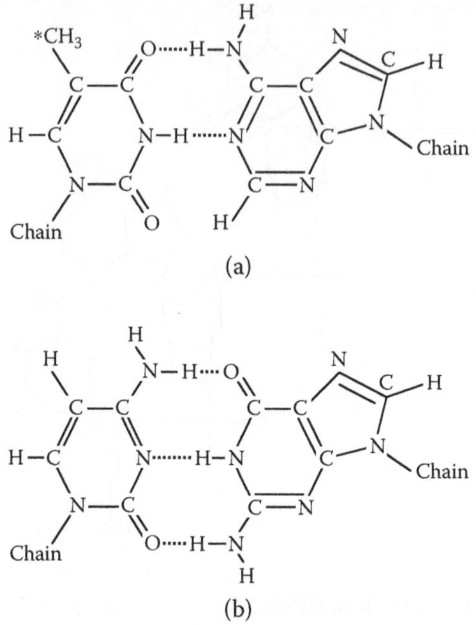

(a)

(b)

FIGURE 7.15 H-bonding between nucleic acid bases (dotted lines): (a) thymine and adenine; if the methyl group marked with an asterisk is replaced by a hydrogen, then this would represent uracil and adenine, (b) cytosine and guanine. Bonds to other atoms in the chain are marked.

Information encoded in the DNA is copied (transcribed) onto shorter strands of RNA. Molecular machines then interpret these and assemble proteins, amino acid by amino acid. Three bases form a *codon*, which is generally translated as shown in Table 7.4. RNAs can form complex structures by hydrogen bonding with themselves and have many other functions in the cell; notably, the machine that assembles proteins is substantially RNA driven.

There are many other sequences in the DNA that code for different things. The most obvious are codons that tell the assembly machinery where to start reading the RNA code and translate it into a protein, and a second sequence is to stop the machinery and make it release the protein. Other sequences in the DNA are designed to allow transcription factors to bind. A stretch of DNA will only be transcribed into RNA (and thence into protein) if another molecular machine can bind to it and start the process of transcription. Because only small parts of the DNA need to be transcribed at any one time, transcription factors bind close to these to enable the transcription machinery to get started.

7.2.3.3 Lipids

Lipids are virtually insoluble in water but can be dissolved in relatively nonpolar solvents such as chloroform. There are many forms of lipids that are of interest in biochemistry; however, this subsection will deal only with fats, phospholipids and, briefly, with cholesterol. These are important in the formation of membranes and,

TABLE 7.4
Codons

Amino acid	Codons
Alanine	GCA, GCC, GCG, GCU
Arginine	AGA, AGG, CGA, CGC, CGG, CGU
Asparagine	AAC, AAU
Aspartic acid	GAC, GAU
Cysteine	UGC, UGU
Glutamic acid	GAA, GAG
Glutamine	CAA, CAG
Histidine	CAC, CAU
Isoleucine	AUA, AUC, AUU
Leucine	UUA, UUG, CUA, CUC, CUG
Lysine	AAA, AAG
Methionine	AUG
Phenylalanine	UUC, UUU
Proline	CCA, CCC, CCG, CCU
Serine	AGC, AGU, UCA, UCC, UCG, UCU
Threonine	ACA, ACC, ACG, ACU
Tryptophan	UGG
Tyrosine	UAC, UAU
Valine	GUA, GUC, GUG, GUU

as such, are doubly relevant to this book because the formation of membranes and artificial vesicles (liposomes) are of interest from the point of view of nanotechnology. Other lipids include waxes, sphingolipids, glycolipids, and lipoproteins.

7.2.3.3.1 Fats

Fats are formed by the reaction of a carboxylic acid with glycerol (Figure 7.16). Any of the three hydroxyl groups may undergo this reaction, forming a monoglyceride (only one), diglyceride (two), or triglyceride (all three). The carboxylic acids that are found to form fats in nature are termed *fatty acids*.

Fats are further classified as saturated or unsaturated, depending on whether they contain only single carbon–carbon bonds or not.

7.2.3.3.2 Phospholipids

Phospholipids are lipids that have one or more phosphate (PO_4^{3-}) residues. The commonest form is the phosphoglyceride (see Figure 7.17). The phospholipids are amphipathic compounds, having a polar head (the phosphate residue) and a nonpolar tail. As a result, in water, the heads tend to dissolve and the tails to aggregate. This gives rise to three basic structures in water (Figure 7.18). The lipid bilayer is the basic form of lipid membrane, with the polar heads facing out toward water and the nonpolar tails in the center. This can fold into a spherical liposome, but if there are very few molecules involved, then a micelle will form instead.

```
  H  H  H          H         H         H
  |  |  |          |         |         |
H-C--C--C-H      H-C---------C---------C-H
  |  |  |          |         |         |
  O  O  O          O         O         O
  H  H  H           \         \         \
                    C=O       C=O       C=O
    (a)
                            (b)
```

FIGURE 7.16 (a) Glycerol, (b) fat, a triglyceride. If there are no double bonds in the long hydrocarbon tails, then the fat is said to be saturated.

The membranes in a cell are based on lipid bilayers. Proteins may be embedded in one side of the bilayer or may form a bridge (or pore) from one side to the other. In cells, there appears to be a tendency for proteins to congregate together in membranes rather than to float around freely bumping into one another.

7.2.3.3.3 Cholesterol

This chemical is particularly well known because of its association with heart disease. Cholesterol is an example of a group of lipids known as steroids, which

```
                       R'
                       |
                       O
                       |
                     O=P-O⁻
                       |
      H        H       O
      |        |       |
    H-C--------C-------C-H
      |        |       |
      O        O       H
       \        \
       C=O      C=O
```

FIGURE 7.17 A phosphoglyceride. R is a polar head group.

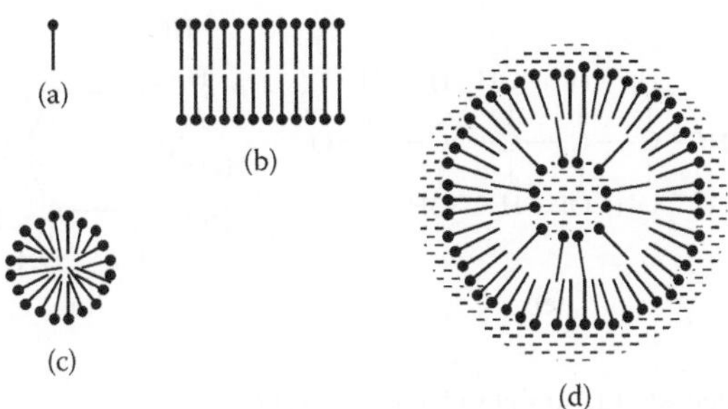

FIGURE 7.18 (a) Stylized phospholipids, (b) lipid bilayer, (c) micelle, (d) liposome.

fulfill many different roles in organisms. They are characterized by the ring system, which can clearly be seen in the cholesterol molecule (Figure 7.19). The cholesterol molecule is much smaller than the tails of the phospholipids that make up biological membranes, and because of the ring system it is quite a rigid molecule. As a result, one of its functions is to stiffen biological membranes.

7.2.3.4 Carbohydrates

Once again, there are many different chemical compounds that are termed carbohydrates (or saccharides). Many, but not all, follow the general formula $C_n(H_2O)_n$. Single monosaccharide units may join together to form disaccharides, oligosaccharides (about 2 to 11 units), or polysaccharides. Glucose is an example of a common monosaccharide (Figure 7.20; note that it can form a ring, or cyclic, structure), and sugar (sucrose) is an example of a disaccharide. Polysaccharides include compounds such as glycogen, which stores energy in animals, and cellulose (Figure 7.21), which is a major structural component of plants. Modified polysaccharides, such as mucopolysaccharides, take on a variety of roles in organisms.

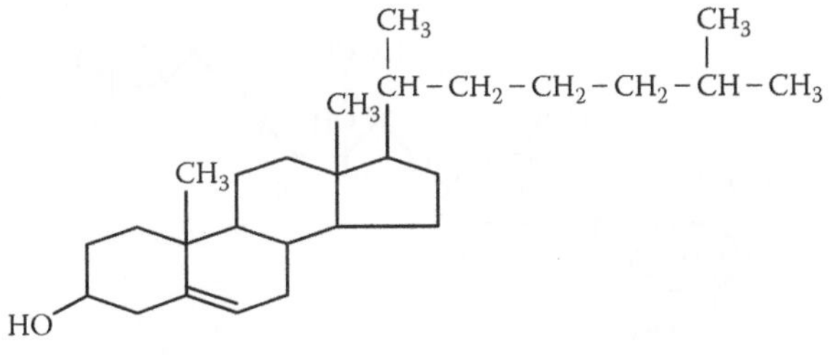

FIGURE 7.19 Cholesterol.

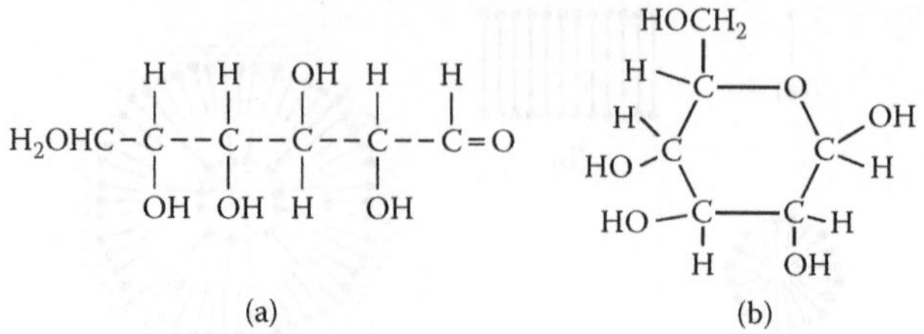

(a) (b)

FIGURE 7.20 Glucose: (a) straight chain, (b) cyclic.

7.3 APPLICATIONS OF MICROENGINEERED DEVICES IN CHEMISTRY AND BIOCHEMISTRY

The characteristics of microengineered devices and systems, relevant to chemical and biochemical applications, are:

- Small size; this means small volumes (so only small quantities of reagent are required), the ratio of surface area to volume is high, laminar flow in small capillaries may cause problems when mixing, small distances for reagents to diffuse, and low thermal mass (rapid heating and cooling).
- Reproducible dimensions and mass production. This raises the possibility of disposable devices.
- Massively parallel systems are possible.
- On-chip processing of data.

Of course, these are not necessarily advantages. A high surface-area-to-volume ratio can be advantageous if part of the process involves molecules becoming attached to the walls of the device but can be a disadvantage if a channel has to be flushed clean during the process.

The following subsections summarize some of the general procedures involved in chemistry and biochemistry that may be helped by miniaturization.

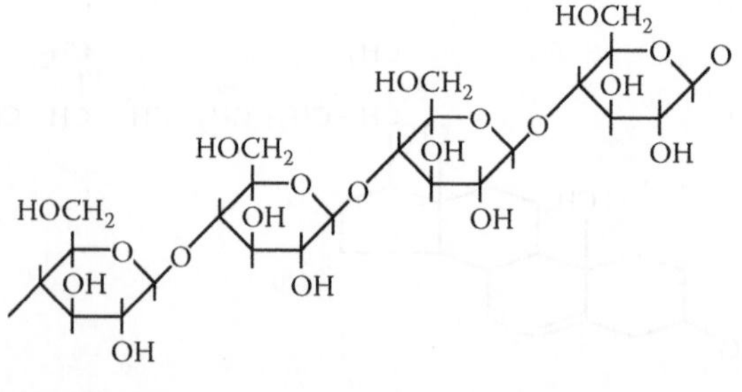

FIGURE 7.21 Part of a cellulose chain.

7.3.1 Chemistry

7.3.1.1 Synthesis

Compounds are produced by combining a number of precursor compounds in a fixed sequence under specific conditions. Microengineered devices can be of use in a number of ways, principally through their size. When expensive reagents are involved, they can cut down on waste. This is also an advantage when hazardous chemicals are required as part of the process. Furthermore, instead of transporting large volumes of hazardous chemicals between industrial sites and the sites where they are required, it can be envisaged that miniature chemical plants could be developed to produce, from the less hazardous precursors, only the quantity required at the point of use. Lower thermal mass and smaller size mean that steep temperature gradients can be developed and maintained, resulting in faster results.

Most industrial chemical processes are large-scale operations, however. One problem in developing new compounds is scaling up from a laboratory procedure to industrial-scale production. Microengineering may be able to help here too: rather than scaling up, many thousands of chips could be run in parallel.

Finally, there is the opportunity to investigate some novel chemistry. Chemical reaction rates can be controlled by moving a plug of one reagent through a stationary volume of a second reagent, for example (see the section on electroosmotic flow). The chemist may also choose to explore the surface-to-volume ratio or other aspects of the systems.

7.3.1.2 Process and Environmental Monitoring

In many production processes, it is necessary to monitor and control many different parameters to ensure the consistency and quality of the final product. There are also requirements for environmental monitoring: identifying toxic chemicals entering the water supply, for example. A further related application is the identification of biological or chemical warfare agents in the environment. For many substances, this process is done off-line: someone visits the site on a regular basis, takes samples, and sends them to a central laboratory where they are analyzed. In many instances it would be preferable to have a continuous real-time reading so that problems could be identified as they occur and as close to the source as possible. Microengineering technologies, by offering the opportunity to perform many processes (i.e., sampling, sample preparation, reaction, separation, detection, and identification), are small compact systems providing a potential solution to this problem.

7.3.2 Biochemistry

Microengineering technologies offer other benefits in addition to those listed earlier. The modern drug development process is a very long and costly process. Once a disease-causing agent or process has been identified, compounds that will interfere specifically with that agent or process are sought. These are normally small molecules because they need to be able to enter cells through their protective outer membranes. The sources of these compounds are diverse; they may come from plants, animals, bacteria, or may be found in existing chemical libraries. Once a promising agent has been identified, it is used as a basis for generating thousands of new chemicals that differ from each other only very slightly. This is *combinatorial chemistry.* Essentially the same sequence of reactions is carried out in an array of small wells on a titer plate, but each well receives a different reagent to all the others at a predefined step in the process.

Each of the new compounds is then tested for binding with the agent of interest (or whatever may be appropriate) and tested on cells in culture. The most promising ones are then tested on animals to see if the organism as a whole reacts as desired (whole kidneys tend to react differently from kidney-derived cells in a cell culture dish, for example). Finally, there are several sets of clinical trials with human volunteers: to see if it is safe to give to humans, what quantities can be given and then to determine if it works.

The early stages of this process, combichem and testing with cell culture, could be speeded up by performing the reactions in parallel on a chip. There is also a further problem that microengineered devices may be able to assist with — the thousands of compounds that are not particularly useful on one occasion may turn out to be of use in other situations, and therefore one would like to store them. Long-term storage is normally performed at –20°C in special freezer units or in liquid nitrogen. Storing titer plates this way is very costly, and microminiaturization would enable more compounds to be stored in the same space.

7.3.3 Biology

The processes that occur in cells are elucidated through a number of different approaches. Essentially, it is necessary to identify what compounds are interacting, where these interactions take place within the cell, and at what point in the growth division cycle the interactions take place. These approaches include: microscopy, radioactive labeling, chromatography, and electrophoresis. Mass spectrometry can be used to determine chemical composition, and because structure plays such a significant part in biological processes, it is desirable to determine what tertiary or quaternary structure a protein (or nucleic acid) takes in a particular situation. For this, x-ray crystallography and nuclear magnetic resonance (NMR) are used.

7.3.3.1 Microscopy

Microengineering techniques have been used for some time to provide tips for scanning probe microscopes (see Chapter 10). These include atomic force microscopes (AFMs), which operate on the principle of attraction between the atoms at the tip of the probe and the atoms in the sample, and scanning tunneling electron microscopes (STEMs), in which the probe tip is brought so close to the sample that an electron is able to vanish from the orbit of an atom in the probe tip and reappear about the nucleus of an atom in the sample. These can, under the correct circumstances, provide atomic-scale resolution. The STEM requires considerable sample preparation, but the AFM can be used in wet biological environments.

One use of nanotechnology in microscopy is in the use of nanospheres (10 nm or so in diameter). These are labeled with antibodies and used in sample preparation for STEMs; the spheres congregate in areas rich in the molecules that the antibodies preferentially bind to, and can be easily seen under the microscope.

A third form of the scanning probe microscope (SPM) is the scanning near-field optical microscope (SNOM). In this the probe tip is an optical fiber coated with an opaque material. The tip is sharpened and a gap that is less than the wavelength of the light illuminating the fiber is formed at the very end. Although light cannot escape from this gap, a very small portion of the electromagnetic field projects beyond the physical end of the waveguide, and this can interact with the material being studied. The position of the tip with respect to the sample is controlled in a manner similar to that for the AFM tip, so it is possible to obtain both topographical and some additional chemical information from this tool.

Although visible light microscopy is not capable of achieving the resolution of electron microscopy or SPM, it should not be neglected. Many techniques have been developed for the identification and manipulation of cells and cell components. Combined with fluorescence labeling, light microscopy and image-processing technologies can be used for a number of processes, such as cell sorting. Fluorescence labeling is becoming increasingly useful. Firstly, small fluorescent molecules (such as rhodamine and fluorescein) can be attached chemically to larger molecules to identify them, hopefully without altering their activity. Secondly, the small jellyfish protein, green fluorescent protein (GFP), and its derivatives (cyan and yellow fluorescent proteins) can now be genetically engineered and combined with proteins of interest in the cell. Thus, it is possible to locate the position of that protein quite precisely (within the limits of the resolution of the microscope) within the cell. For this, the laser confocal microscope is normally used. A laser beam is employed to illuminate the sample (and excite fluorescence), and a pinhole is used to ensure that only light that passes through the focal point of the optical system passes through to the viewer. This enables a three-dimensional view of the sample to be built up in slices.

Related to the confocal microscope is the laser tweezer. These use a laser beam to manipulate very small (micron-scale) objects. These can be chemically primed to investigate surfaces; but of interest to the microengineer is the fact that the forces developed can be quite accurately predicted and, therefore, these tools lend themselves to many other possible uses.

7.3.3.2 Radioactive Labeling

Radioactive labeling can be used in "pulse-chase" experiments to determine how a particular compound is used by an organism. The compound is made up using a radioactive isotope (usually one from ^{32}P, ^{131}I, ^{35}S, ^{14}C ^{45}Ca, ^{3}H, the number being the atomic mass and not the atomic number), and a sample is given to the organism at a particular time (thereafter it is supplied in the nonradioactive form once more). By taking regular samples it is possible to follow the radioactive molecules through the organism.

7.3.3.3 Chromatography

Coupled with the centrifuge, chromatography is one of the major workhorses of modern biology. The centrifuge separates material by mass: it spins liquid samples at very high speeds for long periods of time, following which the most dense material accumulates at the bottom of the sample tube and the least dense accumulates at the top. Chromatography separates liquid samples by passing them through a column packed with something that slows down the passage of the molecules (Figure 7.22). In the most simple form, larger molecules pass through

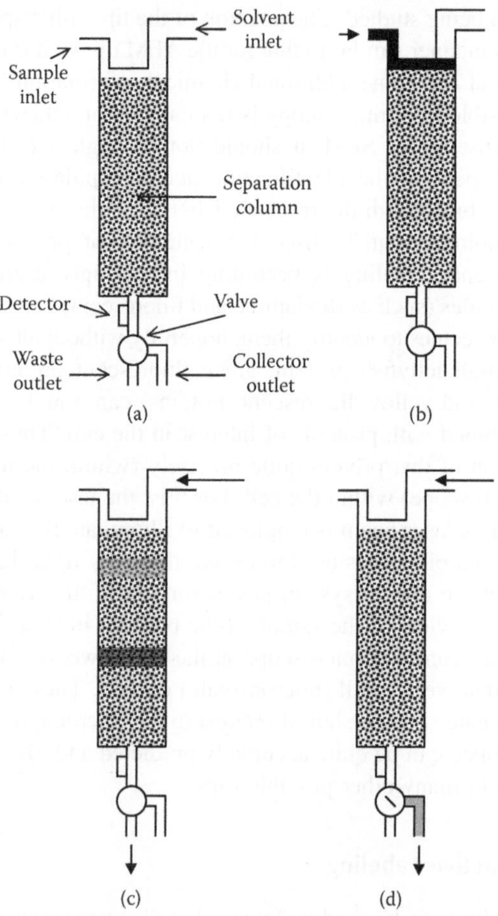

FIGURE 7.22 Principle of liquid chromatography: (a) a separation column is packed with a material designed to retard passage of the compounds of interest; a detector on the outlet enables samples to be diverted for collection and further work as they pass out of the column; the separation is performed under a continuous flow of solvent; (b) sample is injected in a band at the top of the column; (c) different components (compounds) of the sample separate out; (d) collection of band of interest.

TABLE 7.5
HPLC, Capillary Electrophoresis, and Microstructure Separations Compared

	HPLC	Capillary Electrophoresis	Microstructure
Cycle time	90	50	1
Fluid consumption	680,000	90	1

the column more slowly because the packing material slows them down more than it does the smaller molecules. By passing a "calibration" solution through it, one can find out how long it takes for a particular molecule to pass through the column and, with this knowledge, determine if that molecule was present in the sample. When identifying active chemicals in a mixture (e.g., identifying drug candidates from plant extracts), each different band from the column can be analyzed to see if it has the same effect as the mixture and, if so, further work can be carried out to discover exactly what the chemical is.

Additionally, the packing material may be treated to slow down a particular molecule. For instance, if one wishes to find out what proteins bind to a particular DNA sequence, then the inside of the column may be coated with DNA, and proteins that bind to it will be delayed more than others.

The principal advantage of miniaturization is that of speed. High-pressure liquid chromatography (HPLC) requires considerable time to run a single cycle, whereas microstructure separations take seconds or minutes. HPLC also consumes considerable volumes of fluid, which may be especially significant in biological circumstances in which the compound of interest is a relatively rare one within the cell. Table 7.5 compares HPLC, capillary, and microstructure separations (normalized to typical values found for microstructures [3]).

7.3.3.4 Electrophoresis

Electrophoresis extends chromatography; it separates not only by size of the molecule but also by electrical charge (Figure 7.23). In solution, polar elements of macromolecules, such as hydroxide and carboxyl groups, dissociate so that the molecule itself is left with an overall charge. When placed in an electric field, the molecule will migrate toward the electrode that carries the opposite charge. This is usually performed with a gel to slow down and separate the molecules and is, therefore, referred to as *gel electrophoresis*; other methods are mentioned later in this chapter. Although several different types of electrophoresis have been developed, there are three additional points to note: Firstly, because many proteins are folded up into relatively small volumes, they need to be unfolded (denatured) in order for the separation to be effective. This is usually done by using a denaturing agent such as the detergent sodium dodecyl sulfate (SDS). Secondly, some molecules, such as different lengths of DNA, have the same charge-to-mass ratio regardless of

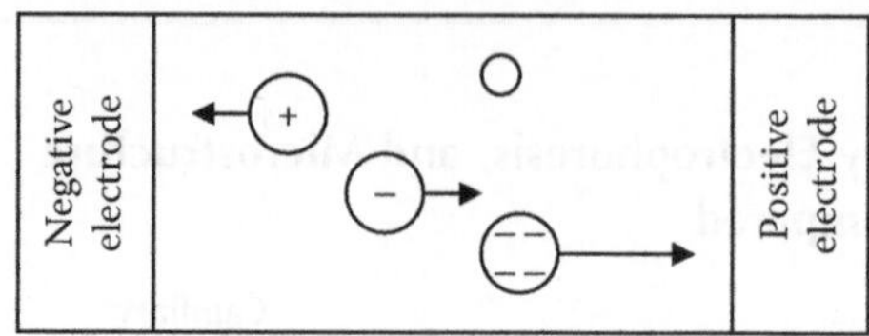

FIGURE 7.23 Principle of electrophoresis. Positively charged molecules move towards the negative electrode (cathode). Negatively charged molecules move towards the positive electrode (anode). Uncharged molecules are unaffected. Highly charged molecules move more rapidly than those with lower charge. Small molecules will move faster than large molecules (depending on the retarding medium and whether or not the molecule is denatured or folded compactly).

their actual length. In such cases, gel electrophoretic separation proceeds purely on the basis of molecular mass. Finally, two-dimensional separation can be performed. The sample is first placed in a pH gradient from very acidic to very alkaline, and an electric field is applied. Molecules will migrate until they reach the pH at which they have no overall charge. Then an electric field is applied in the opposite direction, and the molecules are separated on the basis of mass.

Again, the advantages of microengineering are in improved speed, reduced reagent and sample volumes, and greater accuracy. Reduced dimensions mean that it is possible to apply higher electric fields (heating by electric current being one limiting factor) with smaller absolute voltages (tens, as opposed to thousands, of volts).

7.3.3.5 Mass Spectrometry

Different mass spectrometry (MS) techniques have been developed to identify chemical compounds. The purified sample is introduced into the instrument, broken up into pieces, and ionized. The ions are then accelerated through an electric field and through a magnetic field, which causes their path to curve. The degree of curvature of their path will depend on the momentum (and therefore the mass of the fragment) and the charge. A detector is used to identify the landing sites of different fragments, and a fragmentation pattern can be built up. Although fragmentation is random, careful analysis will reveal the chemical composition and primary structure of the compound. Obviously, portable mass spectrometers would be of immense use in industry and in military and civil defense applications (to identify and give warning of the use of chemical and biological weapons, for example).

7.3.3.6 X-Ray Crystallography and NMR

The arrangement of atoms in a crystal acts as a diffraction grating for x-rays. Shining x-rays through a crystal formed by a particular protein, therefore, creates a particular interference pattern that can be used to determine the tertiary (or even quaternary) structure. The disadvantages of x-ray crystallography are that it is very difficult to produce protein crystals, and it is not possible to observe the proteins changing shape to perform specific tasks. Nuclear magnetic resonance (NMR) offers an alternative approach.

In NMR, the sample is held in a very powerful magnetic field, and radio waves are used to flip the spins of the nuclei in the sample. As they return to their rest state they give off a characteristic signal. This signal is influenced by the bonding situation in which the atom finds itself. The structure of the molecule can then be determined. NMR also has some disadvantages and problems. NMR can be used in a simpler mode as a molecular detector and, as such, could be useful in μTAS.

7.3.3.7 Other Processes and Advantages

Many of the processes involved in understanding cells incorporate the use of biological molecules (as opposed to molecules specifically designed and synthesized by humans). In research terms, it may well be desirable to try and reproduce and control the microenvironments in which studies take place on the same size scale.

Other processes, such as the famous polymerase chain reaction (PCR), which copies lengths of DNA, require thermal cycling. Miniaturization decreases the thermal mass that must be cycled and, hence, the time required for the reaction to be completed.

7.4 MICRO TOTAL ANALYSIS SYSTEMS

At the start of this chapter, Figure 7.1 introduced the concept of an integrated microengineered chemical analysis system. Part I of this book introduced the fabrication techniques by which various elements — channels, valves, micromechanical pumps, and mixers — could be constructed. Chapter 5 and Chapter 6 showed how these elements could be combined and actuated or formed into sensors. This section, therefore, will focus on the principal techniques used for moving chemicals in solution around mirofluidic chips and performing separations — electroosmosis and electrophoresis — and the detection of results.

7.4.1 Microfluidic Chips

Figure 7.24 shows a simplified diagram of a microfluidic chip used for performing separations. The device has at least four ports, one for the carrier fluid (buffer) that will initially be used to fill all the channels on the chip, two for the sample, and one as a waste outlet at the end of the separation channel. If the collection of particular components (fractions) is required, following separation, there may be more than one outlet following the separation channel so that the required fraction can be directed to a specific outlet. The other principal elements are the sample injector, separation channel, and detector. Each port needs to be connected to a reservoir and an electrode that can be switched to control the fluid flow through the channels for separation.

The chips themselves have typically been fabricated by HF etching of quartz glass to form channels. Holes are drilled in a second layer of glass to provide inlet and outlet ports, and this is bonded on top to seal the channels. Glass has

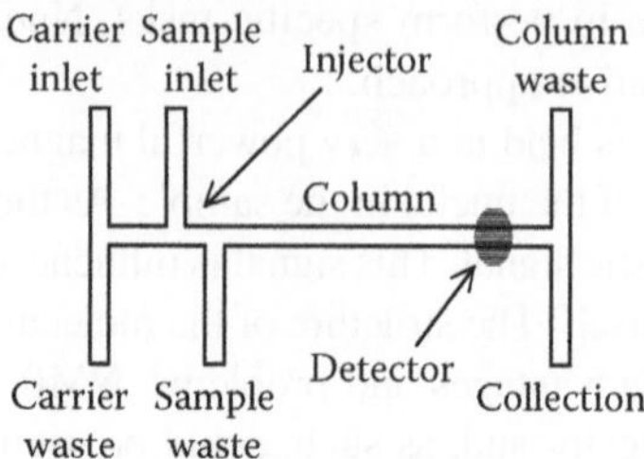

FIGURE 7.24 Typical layout of channels on a microfluidic chip. A carrier solvent is used to prime the column (separation channel, which is typically millimeters in length). The flow in the column can be halted by diverting the flow of carrier to the carrier waste outlet. To inject the sample, carrier flow is directed to carrier waste and sample flow is directed between the inlet and sample waste; the intersection of sample and separation channels form an injector, which enables a precisely defined plug of sample to be injected into the column. For separation, the sample flow is stopped and flow in the column is resumed.

several advantages: the machining processes are well known, it is possible to optically observe progress (fluorescence labeling has been a popular method of detection) and to monitor filling to ensure that there are no air bubbles or obstructions in the channels, and it presents the correct conditions for the formation of electroosmotic flow. It is also chemically relatively inert.

Glass is, however, relatively expensive, and polymers, including PDMS, have been used.

7.4.2 Laminar Flow and Surface Tension

Two characteristics that have to be considered when designing microfluidic chips are that flow is almost exclusively laminar and that surface tension plays a more significant role than in larger systems.

Laminar flow means that turbulent mixing does not occur. Therefore, a variety of mixers have been devised that divide the flow from a single channel into multiple flows and interleave these with a similarly divided flow from a second channel. Various approaches have been used, such as wide channels filled with regularly spaced pillars that divide and redivide the flow (Figure 7.25). Thereafter, diffusion of molecules completes the process.

FIGURE 7.25 Principle of a micromixer, as viewed from above. Pillars divide and redivide the flow.

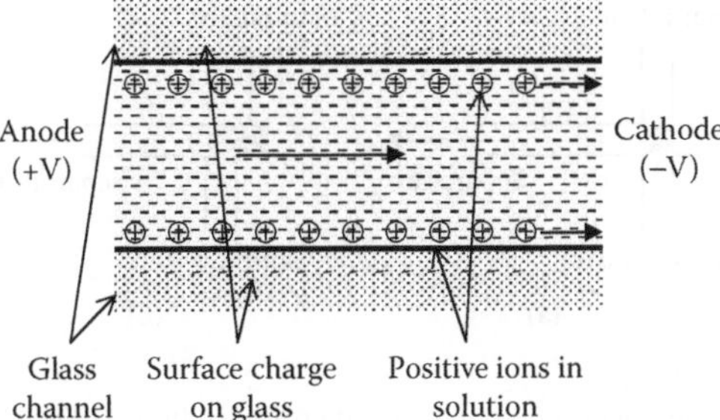

FIGURE 7.26 Principle of electroosmotic flow in a glass microchannel. A negative surface charge on the glass attracts small positive ions in solution to the edges of the channel. Applying a potential difference across the channel causes these to move towards the cathode, dragging the rest of the fluid in the channel with them.

Surface tension has caused problems, for instance, in filling microfluidic channels, but it has also been used to advantage in schemes whereby the direction of fluid flow is controlled by controlling surface tension. Valves have also been formed by introducing restrictions into channels; once the leading surface of the fluid reaches the restriction, additional pressure must be applied to force it past the restriction.

7.4.3 Electroosmotic Flow

Fluids can be relatively easily moved through microchannels by electroosmotic flow. Figure 7.26 illustrates a channel into which an aqueous solution has been introduced. Negative surface charges on the walls of the channel attract small ions from the solution. Applying an electric field across the channel causes the small ions to move towards the negative electrode (cathode). This motion drags the rest of the fluid in the channel toward the cathode. This is known as electroosmotic flow.

The sample to be analyzed is introduced into the channel at the opposite end to the cathode. This will contain neutral particles (molecules) as well as positively and negatively charged particles. Neutral particles will progress through the channel at the same rate as the bulk liquid flow and will only be retarded by mechanical (i.e., fluidic) effects. Positively charged particles will be attracted towards the cathode and so will generally move faster than the bulk buffer solution, whereas negatively charged particles will be retarded because they are attracted toward the positive end of the channel (the anode); however, they will still end up at the cathode as they cannot resist the bulk flow of the buffer solution.

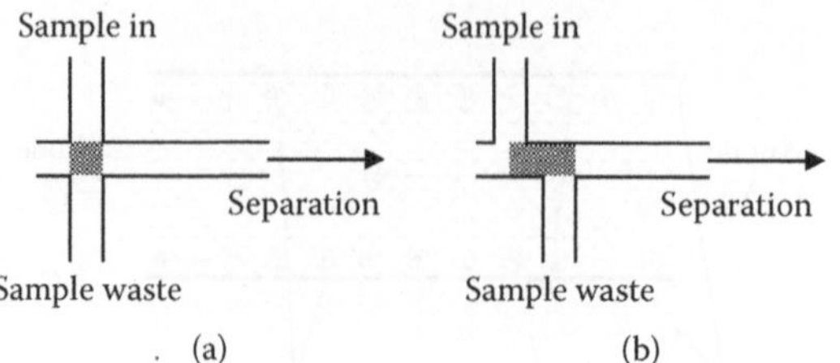

FIGURE 7.27 Sample injector geometries: (a) cross, (b) twin T. The shaded area indicates the shape of the sample plug.

> **Surface Charge Formation**
>
> This is dependant on the ionization of chemical groups on the surface of the channel walls. In the case of quartz glass (SiO_2), it is silanol (SiOH) that is ionized at pH values above about 2. Different materials will have different surface chemistries, and the pH or composition of the buffer solution may have to be adjusted to enable electroosmotic flow to take place.

7.4.4 Sample Injection

Having filled the channels with buffer solution, a plug of sample must be introduced into the separation channel. In glass chips, electroosmotic flow is toward the cathode, so flow through the chip is controlled by switching the electrodes between 0 V and a negative voltage. (If ports are left without being connected to an electrode, there will be no electroosmotic flow toward them.)

Figure 7.27 illustrates two common injection geometries: the cross and the twin T; the latter design enables the plug size to be controlled independent of channel width. The procedure is the same for both geometries. The buffer inlet port and separation waste port are left unconnected, and a negative voltage is applied between the sample inlet port (0 V) and sample waste port (–V). The channel between the two sample ports, and the injector, fill with the sample solution. The voltage is then switched so that the buffer port is most positive (0 V) and the other ports are negative. The sample plug will thus propagate along the separation channel, and the buffer will flow toward the sample ports to prevent additional sample from filling the separation channel.

Controlling the sample plug size and ensuring consistent injection is crucial to obtaining consistent and repeatable results.

7.4.5 Microchannel Electrophoresis

The force acting on a charged particle (charge q C) in an electric field (E V/m) is given by Equation 7.12:

$$F = qE \tag{7.12}$$

When the frictional forces retarding the particle (i.e., coefficient of friction f multiplied by the velocity of the particle) equal the force due to the electric field, then the particle will move with constant velocity. This can be rearranged to give the velocity:

$$v = \frac{q}{f}E \tag{7.13}$$

This can be written as:

$$v = \mu_{ep}E \tag{7.14}$$

where μ_{ep} is the electrophoretic mobility of that particle. Given that f for a spherical particle (of radius r and η as the viscosity of the buffer solution) is given by:

$$f = 6\pi\eta r \tag{7.15}$$

The mass of the same particle of density ρ is:

$$M = \rho\frac{4}{3}\pi r^3 \tag{7.16}$$

It can be seen that f is proportional to mass, and the electrophoretic mobility is, therefore, proportional to the charge-to-mass ratio:

$$\mu_{eo} \propto \frac{q}{\sqrt[3]{M}} \tag{7.17}$$

In the case of microchannel separations, the mobility μ of the particle is the sum of both the electrophoretic mobility and the electroosmotic mobility (i.e., the motion due to electroosmotic flow). The electroosmotic mobility is dependent on the zeta potential ζ of the surface in addition to the dielectric constant and viscosity of the buffer solution.

The principle of the separation process, then, is to separate two (or more) compounds with different mobilities; the length of the channel needs to be chosen such that one reaches the detector at the end of the separation channel before the other.

The resolution of the system needs to be such that the detector can distinguish between concentration peaks in the compounds of interest when they reach the end of the separation channel. As shown in Figure 7.28, a rectangular plug of a single compound will have spread out by the time it reaches the detector into a peak with variance σ^2. The total variance is a combination of

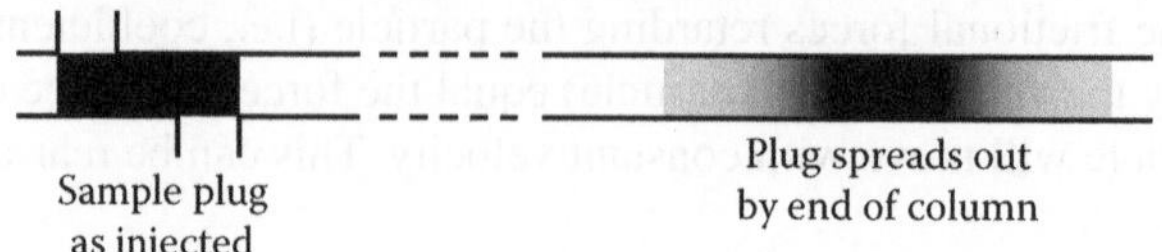

FIGURE 7.28 Rectangular plug spreads out to a peak at the end of the channel, with a variance of σ^2.

that due to the sample injector (no plug will be a perfect rectangle), diffusion during separation, and the detector (even if there was a perfect plug, the detector may not be perfect).

It is common to compare the efficiency of sample separation using the number of theoretical plates, N:

$$N = \frac{L^2}{\sigma^2} \tag{7.18}$$

The theoretical resolution (i.e., the number of different peaks that can be distinguished) of the system can be found:

$$R_T \approx \frac{1}{4}\sqrt{N}\left(\frac{t_b - t_a}{t_a}\right) \tag{7.19}$$

Here t_a and t_b are retention times for the slowest and fastest moving components of the sample, respectively. The retention time is the time spent in the separation channel before reaching the detector, and can be calculated from the channel length and velocity (see Equation 7.14).

Close examination of Equation 7.19 will show that the electric field strength E does not appear to affect the resolution. There are, however, two factors that suggest that a high value of E would be desirable. Firstly, the longer the sample remains in the channel, the more it will diffuse: the variance will increase and, hence, N will decrease. Secondly, a fast separation time is required. This is given by:

$$t_{sep} = \frac{L}{E}\left(\frac{1}{\mu_b} - \frac{1}{\mu_a}\right) \tag{7.20}$$

This implies that a high value of E is desirable. Note that decreasing L will lead to a proportional decrease in resolution (Equation 7.18 and Equation 7.19).

The length of the column is limited by the residence time: the longer the column, the longer it will be before the sample passes the detector, resulting in a slower experiment and greater chance of band broadening due to diffusion effects. This can be compensated for by increasing the electric field strength, the basic limitation of this being the heating caused by the flow of electric current through the buffer in the channel. This has two effects: it will cause band broadening by increasing diffusion and in an extreme situation may denature or damage the biological molecules being separated.

The column can be treated as resistive and the power associated with a resistor (R ohms) is:

$$P = \frac{V^2}{R} \tag{7.21}$$

(See Chapter 11 for additional information.) If the channel has a cross-sectional area of A and the buffer medium has a resistivity of $1/k$ (k being the conductivity), then R is:

$$R = \frac{1}{k}\frac{L}{A} \tag{7.22}$$

and the heat energy developed (power multiplied by time) is:

$$W = k\frac{V^2 A}{L}t \tag{7.23}$$

Further work reveals that the energy per unit volume is directly proportional to the square of the electrical field strength:

$$\frac{W}{volume} = ktE^2 \tag{7.24}$$

This is balanced by the heat dissipated along the length of the channel, which is proportional to the surface area of the channel. Here, miniaturization shows another advantage, i.e., as longitudinal dimensions are reduced, volume will reduce proportionally to the cube and surface area reduces with the square of this reduction. So, as noted before, the surface-area-to-volume ratio increases and the channel is cooled much more efficiently. However, the exact nature of this cooling depends on the geometry of the channel and the materials of construction. Nonetheless, it seems that fields of 200 to 300 kV/m can be achieved in such structures.

Once again, glass is ideal for this because of its excellent dielectric properties and good thermal conductivity. Polymers, although cheaper, have lower thermal conductivities, whereas materials with higher thermal conductivities (such as silicon) suffer from high electrical conductivities or dielectric breakdown of thin insulating layers.

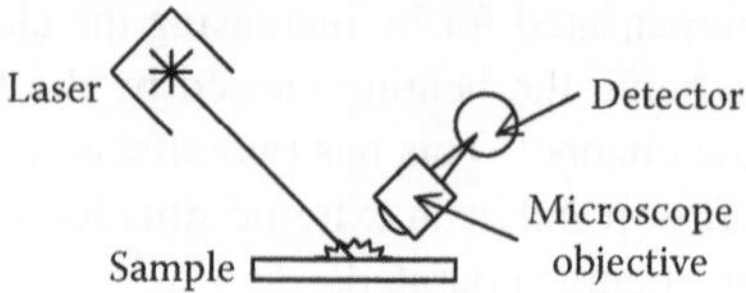

FIGURE 7.29 LIF system. The fluorescence signal is focused by a microscope objective onto the detector; this may be an avalanche photodiode or a photomultiplier tube.

7.4.6 Detection

In order to complete the analysis of a sample, it is necessary to detect the individual components as they leave the separation channel. This section will introduce some of the approaches that have been tried and comment on their advantages and disadvantages.

7.4.6.1 Laser-Induced Fluorescence (LIF)

This is a highly sensitive detection technique, capable of detecting levels down to a few thousand molecules. The approach relies on the compounds of interest fluorescing when exposed to laser irradiation. Some compounds are naturally fluorescent, but the majority will need to be made fluorescent by a derivatization reaction performed either before the sample is introduced into the separation channel (precolumn) or after separation (postcolumn).

The basic system is shown in Figure 7.29. The laser beam is placed at right angles to the microscope objective that is used to collect the induced fluorescence and focus it onto the detector. This may be either an avalanche photodiode or a photomultiplier tube; the latter is bulky and expensive and does not generally lend itself for use in portable equipment, but it is more sensitive and has wider spectral sensitivity.

7.4.6.1.1 Derivatization

In order to make compounds fluoresce, it is necessary to label them with small fluorescent molecules; these are chosen to react with functional groups on the molecules of interest. This will invariably lead to the sample becoming contaminated with unreacted labels; it is necessary to ensure that the peak caused by this at the detector is well separated from those of interest or, if possible, that a label is selected that does not fluoresce in its unreacted state. Some labeling compounds for peptides can be found in Table 7.6 [1].

7.4.6.1.2 Advantages and Disadvantages of LIF Detection

The main advantage of this approach is that it is very sensitive, and furthermore, it can be reliably performed with readily available equipment and reagents. The main problem one is likely to encounter with micromachined devices is background fluorescence of the materials employed (quartz, for example).

If the goal is to produce portable instrumentation, however, this approach does not readily lend itself to miniaturization of the detector. The need to label

TABLE 7.6
Labeling Compounds

Reagent	Comments
Naphthalene dicarboxaldehyde	
o-Pthaldialdehyde	
3-(4-Carboxybenzoyl)-2-quinolinecarboxaldehyde	
Fluorescamine	
Fluoresceine	Can be used in different compounds to label amines
Monobromobimane	Labels sulfhydryl groups
5-Dimethylaminonaphthalene-1-sulfonyl chloride	DANSYL
Fluorescein isothiocyanate	FITC

Source: From Lillard, S.J. and Yeung, E.S., Capillary electrophoresis for the analysis of single cells: laser induced fluorescence detection, in Landers, J.P., Ed., *Handbook of Capillary Electrophoresis*, 2nd ed., CRC Press, Boca Raton, FL, 1997, chap. 18.

complicates the system, and it may also cause problems in some applications: labeled proteins are chemically different from unlabeled, and this may be a particular problem if it is necessary to collect fractions from the column and analyze their activity or composition.

7.4.6.2 Ultraviolet (UV) Absorbance

The quantity of double-stranded DNA in a solution is typically determined by measuring the absorption of UV light at 260-nm wavelength. An optical density (OD) of 0.1 over a path length of 1 cm (that is commonly used in standard equipment) corresponds to a concentration of 5 μg/ml of double-stranded DNA.

The OD can be computed from Equation 7.25:

$$OD = eCl \tag{7.25}$$

where e is the extinction coefficient (20 for 1-mg/ml nucleic acid solutions over a 1-cm path length), C is the concentration of the sample (mg/ml), and l is the optical path length (cm). Proteins also absorb UV light at 260 nm with an extinction coefficient of approximately 0.57.

7.4.6.2.1 Advantages and Disadvantages of UV Absorption

This method requires simpler equipment than LIF, and UV diodes are starting to become available, which may well ease integration problems. It is, however, fairly nonspecific in what it detects; single- and double-stranded nucleic acids can be confused, and RNA and proteins absorb UV light as well as DNA — although by measuring absorption at 280 nm as well as 260 nm, contamination can be detected. Also, the materials used are limited to those that can transmit

UV light — specifically, quartz glass. Nonetheless, it is a well-established macroscale technique.

7.4.6.3 Electrochemical Detection

An electrode positioned in an aqueous chemical solution can supply or remove electrons from compounds in its vicinity depending on its potential relative to that of the aqueous solution. This can be represented symbolically for an arbitrary compound M by the following chemical equation:

$$M^{n+} + ne^- \leftrightarrow M \tag{7.26}$$

The removal of electrons (i.e., going from the left to the right of the equation) is termed *oxidation*, whereas the addition of electrons (going from right to left) is termed *reduction*. Supplying or removing electrons artificially via an electrode will push the equilibrium to the right- or left-hand side of the equation. At some point, the equation will balance, the concentration of both the oxidized and reduced forms being equal. The electrode potential required to achieve this is known as the *redox* potential. This will be different depending on the compounds involved and, therefore, gives a way of detecting not only the presence of a compound but also some indication as to what it may be.

A common arrangement for making redox potential measurements is shown in Figure 7.30. Standard redox potentials can be found tabulated in data books. These are measured with reference to a standard hydrogen electrode (which has been defined as having a redox potential of 0). Most measurements are, however, made with respect to a saturated calomel electrode (potential of +0.244 V). These mercury-based electrodes are chosen because it is relatively easy to create a stable cell on a macroscale. For micromachined devices, carbon thin-film electrodes are popular for making such measurements because they are relatively inert. Chlo-

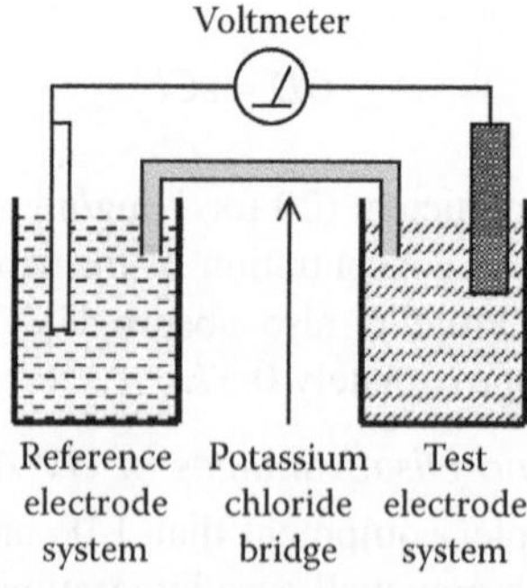

FIGURE 7.30 Apparatus for measuring redox potential. The reference electrode system will be hydrogen or saturated calomel (mercury). The test electrode material is immersed in a solution containing a known concentration of its ions (1 M). The potassium chloride (KCl) bridge that completes the electrical circuit can also be referred to as a salt bridge.

rided silver electrodes can be used as reference electrodes, although they will not necessarily remain stable over long periods of time or great current ranges. Noble metals (gold, iridium, and platinum) do not, on their own, make good electrodes for this application. Although they are inert under most circumstances, they suffer from a baseline drift that is very difficult to control.

7.4.6.3.1 Cyclic Voltammetry

Cyclic voltammetry is a common technique for making electrochemical measurements. It requires a working electrode and a reference electrode, typically connected to the test solution by a salt bridge. Although this can be difficult to arrange formally in a microengineered device, the fact that the flow is laminar can be useful when designing a channel arrangement in which, for example, part of the chip containing a reference solution needs to be electrically connected to another part without the solutions mixing. In a microengineered device, the working electrode will typically be of carbon.

The current passing across the working electrode is monitored. Initially the electrode is held at a potential at which no current flows; in other words, no electrons are being taken from or donated to the compounds in solution. The potential of the working electrode is then increased with a corresponding increase in current as the compounds in solution are oxidized. As the redox point is reached, there is a large increase in the current to a peak. At some predetermined point, the potential is reversed and an inverted pattern can be seen (Figure 7.31).

Only compounds near the electrode can be oxidized or reduced, and the curve obtained depends on local concentrations of the various species involved and how they diffuse toward or away from the electrode. This is particularly true when microelectrodes are used, and this enables some measurements to be made that are related to the diffusion of the species involved toward and away from the electrode. By scanning the voltage at different rates, smaller (slower) or larger (faster) maximum peaks will become apparent, and a set of scan curves can be obtained (note that with the sample plug passing the

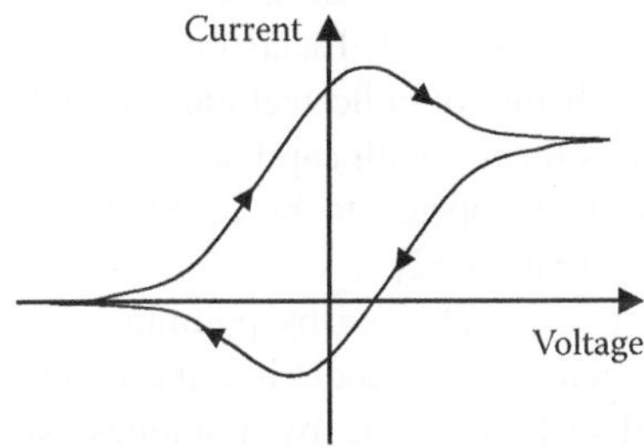

FIGURE 7.31 Cyclic voltammogram for reversible single-electrode reaction with only the reactant present.

electrode quite rapidly, the implementation and calibration of such a system requires some thought).

This assumes that the reaction is completely reversible and diffusion limited. This may not be true if one component becomes involved in a chemical reaction once it has been oxidized or reduced nor will it be true if gas evolves in the system — in fact, gas bubbles are a major problem in any microfluidic system. Obviously, gas will be evolved if the water is electrolyzed into its component parts, hydrogen and oxygen. This will occur at above 0.8 V, so measurements will have to be made below this point.

7.4.6.3.2 Advantages and Disadvantages of Cyclic Voltammetry

Cyclic voltammetry has several advantages; it provides considerable information regarding the composition of a band as it exits the separation column and the procedure is well established on the macroscopic scale. It is not as sensitive as LIF, with detection levels down to about 10^{-10} mol (1 mol is 6.02×10^{23} atoms or molecules). One significant disadvantage, however, is that it can be difficult to perform electrochemical analysis in a high electric field, and device design must take this into account.

7.4.6.4 Radioactive Labeling

As discussed in Subsection 7.3.3, it is possible to label biological molecules of interest by introducing radioactive isotopes. The results can be detected by placing a radiation detector at the outlet of the column or even by dispensing the outlet of the column into wells on a microtiter plate coupled with radiation-sensitive film.

The results of such experiments can be quite illuminating, but the associated radiation hazards usually restrict their use.

7.4.6.5 Mass Spectrometry

The basic principles of MS were introduced in Subsection 7.3.3.5 of this chapter. The MS has four components, an ionizer, an electric field to accelerate the ions, a mass filter, and a (charge) detector. Electrospray ionization appears to be the most compatible one with microfluidic techniques, and has been developed in integrated form as well as for use with capillary electrophoresis systems. One of the mass filter systems that appears to be amenable to miniaturization is the quadrupole mass filter (Figure 7.32).

Electrospray ionization is achieved by pumping the sample through a very narrow (sub-10-μm diameter) glass needle into the vacuum of an MS instrument. An electrode (provided in Figure 7.32 by a stainless steel coupler) applies an appropriate voltage (usually in the kV range) to ionize and accelerate the components of the sample. In a microfluidic system, the sample could be pumped by the electroosmotic flow of the separation channel.

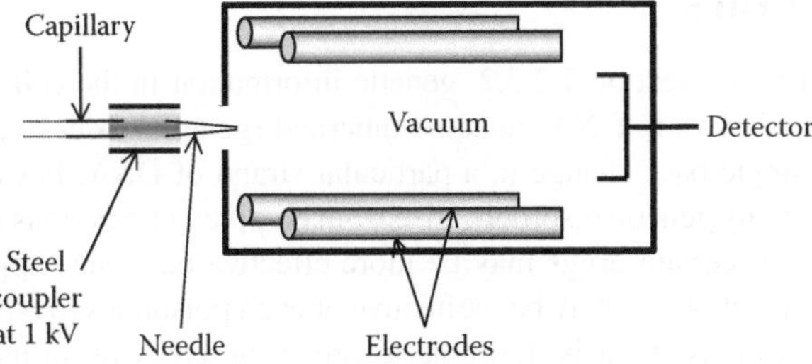

FIGURE 7.32 Electrospray ionization system and quadrupole mass filter coupled to the end of a capillary separation column.

Glass Micropipettes

The glass needles mentioned as electrospray sources can be formed from commercially available glass capillaries. Automated micropipette pullers provide programmed force and temperature control, allowing fine and reproducible control of tip diameters.

The quadrupole mass filter is arranged such that the ionized components of the sample have to traverse a distance surrounded by four electrodes before they arrive at the detector. An alternating voltage of variable amplitude and frequency is applied to these electrodes such that the ions follow a spiral trajectory. In this way, the components of the ionized sample can be separated on the basis of charge and mass.

7.4.6.6 Nuclear Magnetic Resonance

Microcoils designed to act as transceivers for NMR can be integrated onto the microfluidic separation system. To date, the additional paraphernalia required for NMR systems, the magnetic field generators in particular, have limited its use. It does, however, appear that there may be advantages to be gained by miniaturizing NMR systems [2], so further development may be expected.

7.4.6.7 Other Sensors

It should be noted that, in addition to the detection systems discussed earlier, other sensors discussed elsewhere in this volume can be, and have been, successfully employed in microengineered chemical, biochemical, and biological analysis systems. These include simple electrical conductivity sensors, pH sensors, and ISFET sensors. The use of hot-wire sensors or mass-flow-rate sensors in gas chromatography systems should not be neglected (a pulse of gas with different thermal properties from that of the carrier gas will affect the sensor reading as it passes).

7.5 DNA CHIPS

As discussed in Subsection 7.2.3.2, genetic information in the cell is encoded by the sequence of bases in DNA. In some inherited (genetic) diseases, the disease is caused by a single base change in a particular strand of DNA; however, it should be noted that many genetic disorders arise from complex interactions between many genes. Similarly, certain drugs may be more effective on some people owing to a genetic predisposition but may be ineffective or even poisonous to others. For these, and many other reasons, it is desirable to probe the DNA of an individual. Also noted in Subsection 7.2.3.2 was the fact that bases pair up by forming hydrogen bonds, giving rise to the well-known double-helical form of DNA. However, because these hydrogen bonds are relatively weak, double-stranded DNA can be made to break up into the single-stranded form simply by elevating the temperature to above its so-called melting point temperature (98°C). When the DNA solution is cooled, the single-stranded DNA molecules pair up again.

This provides the possibility of probing short strands of DNA to determine their sequence: the DNA of interest is broken up into short fragments, using restriction enzymes (see box). One end of each fragment is labeled with a fluorescent molecule. The DNA is then washed over a chip that has thousands of spots on it, each with a different DNA sequence. Because the sequence of each spot is known, the pattern of fluorescence on the chip can be read off using an automated reader to determine the sequence in the original DNA sample.

Restriction Enzymes

These are catalytic proteins that cut DNA strands at particular points defined by the local sequence of bases, whose number is usually no more than half a dozen. Thus, it is possible to cut DNA up into short lengths and to also know the sequence at the end of each length (because this will depend on the restriction enzyme used).

7.5.1 DNA Chip Fabrication

DNA chips are usually created on a quartz substrate. The surface of quartz, when washed, is covered with OH groups. Silane linker molecules can be attached to this, and large biomolecules (including DNA) can then be attached chemically.

Several approaches have been employed in the fabrication of DNA chips. The simplest is "spotting," which takes various forms. DNA sequences are synthesized and applied by some mechanical means to different areas of the chip. A more sophisticated approach is to use a ink-jet printer type of mechanism to deliver one of the four bases to each spot across the array, and then link it chemically to the previous base before printing on another layer.

Another approach successfully employed by one of the first commercial firms to sell DNA chips is based on photolithography. The linkers on the quartz substrate and the bases used to synthesize the DNA strands are capped with a photocleavable protector. The entire chip is exposed to UV light through a mask, which defines where the protective chemical cap will be removed, a base is added, and the procedure is repeated for each of the other three bases.

7.6 THE POLYMERASE CHAIN REACTION (PCR)

One of the primary functions of the cell is to copy DNA; once the copy has been made, the cell can reproduce by dividing. Copying DNA is performed by a protein machine called DNA polymerase. When given a long single strand of DNA with a short primer hybridized to one end, the polymerase will find where the primer ends and then work its way along the single strand, building a mirror image of the strand (C being matched by G and A by T) so that a length of double-stranded DNA results.

In order to investigate DNA, it is useful to have a large quantity of it. Fortunately, it can be amplified by a procedure known as the PCR, which, because of its usefulness in molecular biology, is also a popular target process for miniaturization. The procedure is as follows:

1. The DNA of interest is purified and broken up by restriction enzymes so that the sequences at the 3 and 5 ends are known.
2. Short primer lengths of DNA are made for the 3 ends. (If the interest is in a section of DNA that starts with a particular sequence, then the primer for this sequence can be made.) These are added in excess of the DNA solution.
3. A polymerase that can survive high temperatures without denaturing (losing its shape and hence its functionality) is added to the solution in excess, as are the phosphorylated forms of the four bases.
4. The double-stranded DNA is melted (heated to 98°C).
5. The sample is allowed to cool to about 60°C, at which point the primers hybridize with the DNA (due to their being in excess), and the polymerase gets to work.
6. The polymerase proceeds from the 3 to 5 end, leaving twice the amount of double-stranded DNA as was originally in the solution.
7. The sequence from item 4 onwards is repeated as often as necessary.

Speed is one of the benefits of miniaturization. In large-scale systems, sample tubes are placed in machined aluminum blocks, which are temperature cycled. The heating and cooling cycle of such large thermal masses limits the speed with which the procedure can be carried out.

7.7 CONDUCTING POLYMERS AND HYDROGELS

These are unusual components of MEMS; brief comments have been included in this chapter because it is useful to present them in a context with chemistry and wet (i.e., fluidic) MEMS.

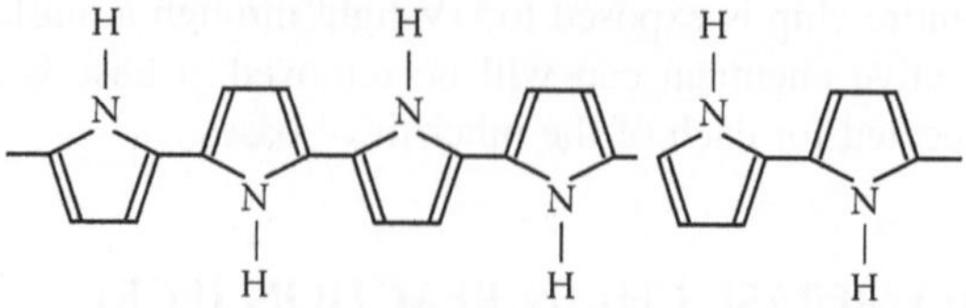

FIGURE 7.33 Section of a polypyrrole chain.

7.7.1 Conducting Polymers

Polypyrroles (Figure 7.33) have been known as electrically conducting polymers for some years now. They are formed of long chains of pentagon-shaped units comprised of four carbon atoms and one nitrogen atom, and are electrically conducting because electrons can drift from one ring to the next along the chain.

Electrically conducting polymers are of interest in themselves, especially because they can be doped to make them semiconducting; plastic integrated circuits would be even cheaper than silicon ones, and flexible circuits would also have many applications. One major application has been in organic light-emitting diodes (OLEDs), which are used to backlight the LCD displays of modern portable devices such as cell phones.

Electrically conducting polymers can be electroplated onto electrodes, and if electroplated from a solution containing other organic molecules, these will become trapped in the electrode matrix. In this way a variety of biosensors can be produced.

They have also been used as microactuators. In ionic solutions, ions can be attracted into the polymer matrix or ejected from it by applying an appropriate electrical bias. This can cause the polymer to contract or expand slightly. This, coupled with selective deposition by electroplating, makes them suitable in some mechanical actuation applications.

7.7.2 Hydrogels

Hydrogels are more effective as actuators than conducting polymers because they exhibit large variations in dimensions in response to various environmental factors, including pH and electric fields. Hydrogels are polymer matrices with functional groups attached to the component monomers such that the polymer would normally be soluble in water. The chains are cross-linked to prevent the matrix from dissolving when immersed in water. Techniques have been developed to photolithographically pattern hydrogels, and a variety of applications are being explored at the time of writing.

REFERENCES

1. Lillard, S.J. and Yeung, E.S., Capillary electrophoresis for the analysis of single cells: laser induced fluorescence detection, in Landers, J.P., Ed., *Handbook of Capillary Electrophoresis*, 2nd ed., CRC Press, Boca Raton, FL, 1997, chap. 18.
2. Olsen, D.L., Lacey, M.E., and Sweedler, J.V., The nanoliter niche, *Anal. Chem. News Features*, 257A–264A, 1, 1998.
3. From notes made by the Author on a presentation by Manz, A., Electrophoresis Microstructures, at a joint meeting of the Microengineering Common Interest Group and the Nanotechnology in the Biosciences Forum on *Microsystems and Chemical Measurement*, National Physical Laboratory, Teddington, UK, 19th March 1996.

8 Integrated Optics

8.1 INTRODUCTION

Modern communications systems rely on fiber optic connections, and optical components can be found in many consumer products (CD and DVD players, digital cameras, etc.). There are many benefits to be gained by the miniaturization of optical components where possible, not only in terms of power reduction and miniaturization. Switching of high-speed optical data channels is still commonly performed by electronic circuits. This means that the signal has to be converted from light in a fiber to an electronic signal, then back, and relaunched down another fiber. Because electronic circuits operate at data rates lower than those possible with optical data transmissions, this represents a communications bottleneck. Similarly, many industrial computer systems and data processing systems are composed of a number of circuit boards plugged into a backplane, a data bus that enables different elements of the system to communicate with each other. It is difficult to maintain the integrity of electronic signals over even relatively short distances (10 cm) as frequencies rise above about 100 MHz. Such buses, therefore, represent a data processing bottleneck.

Movement toward all optical systems, therefore, is a subject of considerable importance. The subject of integrated optics is, consequently, quite large. This section, then, rather than attempting to deal with the entire topic, will be limited to a brief exploration of some applications of MEMS in solving optical problems.

8.2 WAVEGUIDES

In optical communications systems, light is conducted by fiber waveguides. In integrated optical systems, these are usually planar waveguides fabricated on a silicon substrate. Both use the same physical processes to confine and guide light, although the overall structures are different.

8.2.1 Optical Fiber Waveguides

The basic structure of an optical fiber is shown in Figure 8.1. These are cylindrical structures and consist of two parts: core and cladding. The refractive indices of the core and cladding are engineered to ensure that light entering the fiber at one end undergoes total internal reflection so that it arrives at the far end (Figure 8.2), rather than escaping from the fiber at the first opportunity. Fibers are normally made from quartz glass (silicon dioxide), although polymer fibers are available. The core of the fiber is normally between 3 μm and 200 μm in diameter, within a fiber of 140 μm to 400 μm diameter.

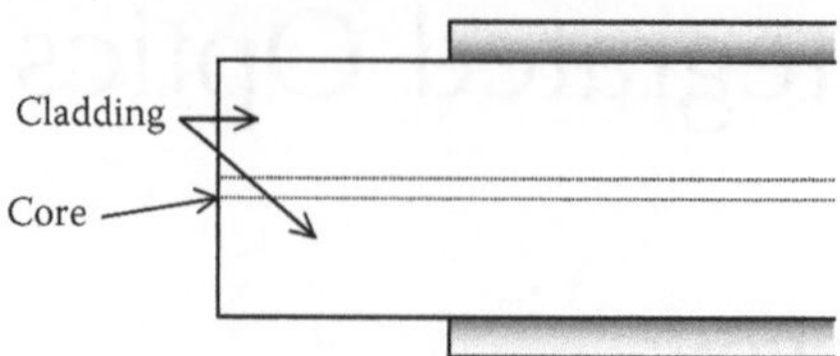

FIGURE 8.1 An optical fiber waveguide (cross section) consists of a core of 3–200 μm and cladding that form a fiber of 140–400 μm diameter. The fiber is usually coated with a protective plastic sheath (indicated).

Fibers are normally provided with additional protective sleeving. To make a connection between two fibers (splice the fibers), the outer protective sheath has to be stripped off, the fibers cleaved cleanly at right angles, and the two cores aligned. Even slight misalignment can cause considerable loss of signal, and this is one area in which microsystems have been deployed: micromanipulation and alignment of optical fibers for splicing.

There are two different kinds of fiber, single mode and multimode. Single-mode fibers have thin cores, usually less than 10 μm in diameter, and light can propagate through them via only one direct path (Figure 8.3a). In multimode fibers, light can propagate by many paths (Figure 8.3b and Figure 8.3c); the difference between the two fiber types is that multimode propagation tends to lead to less signal attenuation (diminution with distance) but more signal broadening (because different parts arrive at the end of the fiber at slightly different times), which results in lower data rates. This can also be a problem if particular properties of the signal (polarization, for instance) are important.

There are a further two classifications to be made for fibers, depending on the profile of the refractive index change between the core and cladding. This can be either a step change, or a graded change. Figure 8.3b and Figure 8.3c show how this affects propagation in multimode fibers.

Optical fibers convey signals with optimal efficiency in the infrared region at wavelengths of about 850 nm, 1300 nm, and 1550 nm.

8.2.1.1 Fabrication of Optical Fibers

The fabrication of glass optical fibers is instructive, because the small dimensions observed are achieved without micromachining. The basic setup is shown in

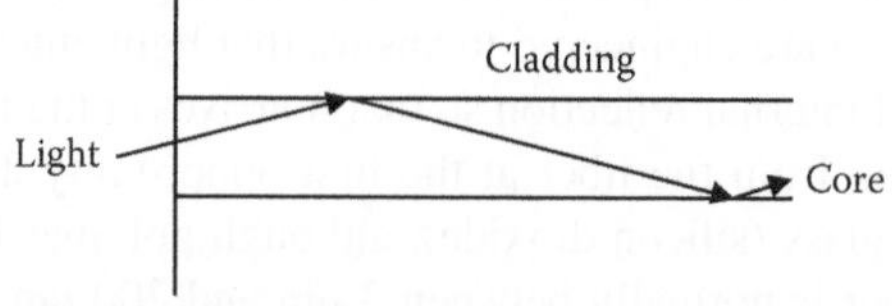

FIGURE 8.2 Total internal reflection. The refractive indices of the core and cladding are controlled such that light entering the core will not escape into the cladding; it will be internally refracted (bent back into the core).

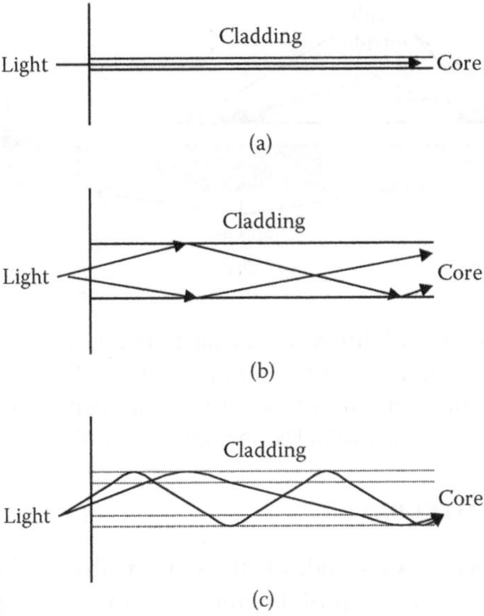

FIGURE 8.3 Propagation in: (a) single-mode fiber, (b) multimode, step index, (c) multi-mode, graded index. Monochromatic light can only take one path along a single-mode fiber, but can take multiple paths along a multimode fiber. It is easier to couple signals into a multimode fiber, but the multiple paths are of different lengths, so a pulse is more spread out when it arrives at the end of the fiber.

Figure 8.4. A quartz tube, a few centimeters in diameter and about 1 m long, is mounted between two rotating chucks. Holes in the chucks allow gas to be passed through the tube, and it is heated to an even temperature by a gas burner.

Refractive index variation is achieved by passing a gas through the tube and doping the inner walls. Once the desired level and depth of impurities have been achieved, the tube is evacuated so that it collapses in on itself. It is then drawn into a fiber on a tower-like structure.

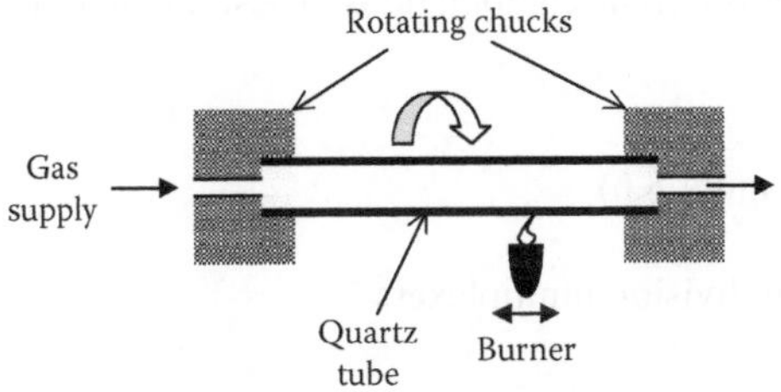

FIGURE 8.4 Doping a glass tube. Once doped, the tube is collapsed by applying a vacuum, and it is then drawn into a fiber (cross section).

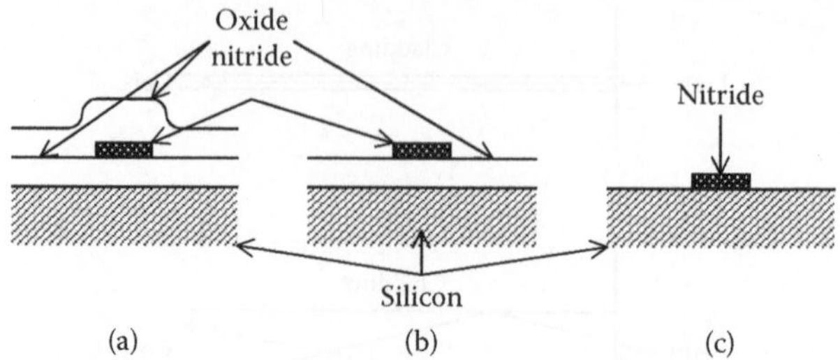

FIGURE 8.5 Cross section of different integrated optic waveguides (note that different glasses may be employed): (a) channel, (b) rib, (c) "strained" silicon. In (a) and (b) light travels through the nitride strip. In (c) the nitride strip induces strain in the underlying silicon, thus changing its refractive index; the light travels through the strained silicon.

8.2.2 Planar Waveguides

The theory behind planar waveguides is the same as that behind fiber waveguides: confining light between two areas of different refractive index. There are three basic approaches that can be used to produce planar waveguides (Figure 8.5).

The first approach (Figure 8.5a) is to duplicate the structure of the fiber waveguide. In this case, the core is of nitride and the cladding of oxide, although these are not the best materials to use, especially, if the films have high levels of hydrogen contamination. The second (Figure 8.5b) is a more basic rib waveguide; again the signal travels through the nitride (or other) core.

Figure 8.5c illustrates a strained silicon waveguide. Here, the change in the refractive index is caused by inducing mechanical strain in the silicon crystal lattice. This is possible because silicon is transparent to infrared light.

All three approaches and variations are under investigation and, in some cases, in use, although most of them involve materials more exotic than oxide and nitride at present.

8.3 INTEGRATED OPTICS COMPONENTS

It is possible to combine planar waveguides with photonic crystals for several applications and in different combinations. These include the production of the following:

- Bends
- Splitters (Figure 8.6)
- Couplers
- Wavelength division multiplexers
- Polarizers
- Optical switches

The optical source used is commonly the laser diode. Although work is in progress to develop laser diodes and photodiodes (for detection) in silicon technology,

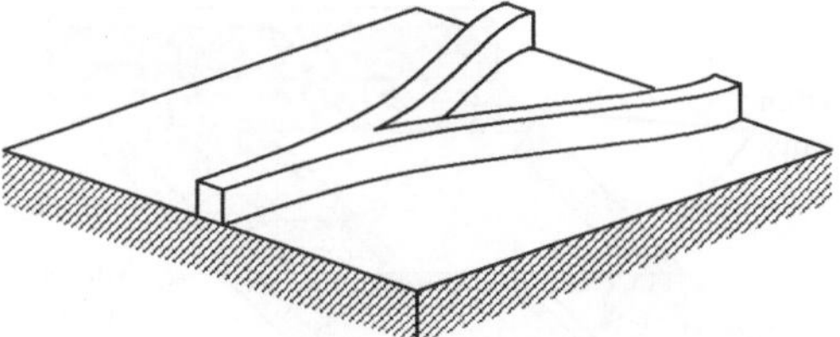

FIGURE 8.6 Shape of standard Y splitter.

these are still generally produced using gallium–arsenide (GaAs) semiconductor technology. For this reason, they normally have to be purchased as separate dies and mounted onto the integrated optics structure. High-power laser diodes also need cooling as they are only about 30% efficient.

8.4 FIBER COUPLING

Because integrated optical systems are frequently intended to be interfaced with fibers, it is necessary to ensure accurate alignment between the fiber and the planar waveguides on the structure. The usual approach to this is the use of V grooves or trenches etched to accommodate the fibers. These can be enhanced by the use of ball lenses, which are readily available optical components. Figure 8.7 shows one structure; note the additional trenches etched for glue overflow during bonding.

8.5 OTHER APPLICATIONS

8.5.1 Lenses

A variety of micromachining technologies can be used to construct an array of lenses for use in optical systems. The most straightforward approach is to deposit and pattern a film of material to leave a number of islands. These are then heated

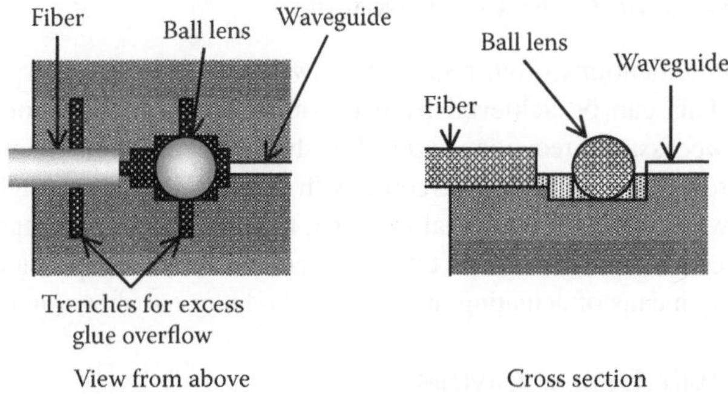

FIGURE 8.7 Trenches and pits can be machined to align fibers with ball lenses and integrated optical components (shown schematically here). Note the additional trenches machined to accommodate excess glue overflow.

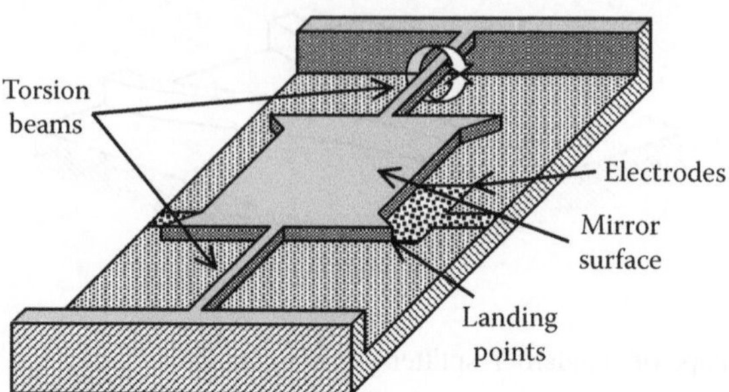

FIGURE 8.8 Principle of electrostatically activated digital mirror device. Two electrodes positioned underneath the mirror tilt it one way or the other. Points on the corners of this structure prevent it from making contact with the electrodes and sticking.

to reflow them, and depending on the nature of the material, these will form balls or, simply, more gently sloping lenses because of surface tension. A new development has been the incorporation of liquids into deformable structures to create lenses that can be focused to different distances.

8.5.2 Displays

Digital mirror displays have been developed and commercially exploited. There are several approaches by which structures similar to that shown in Figure 8.8 (generally using surface micromachining techniques). Aluminum films make very efficient mirrors.

The structure shown in Figure 8.8 can be deflected electrostatically, thus deflecting the path of any light impinging upon it. By combining arrays of such micromirrors, it is possible to create large, bright digital display projectors. The advantage of using mirrors is that high-intensity illumination can be used; this is a limitation for projectors that use more common LCD technology.

8.5.3 Fiber-Optic Cross-Point Switches

Optical communications often require the switching of a signal from one fiber to another. This can be achieved by using microengineered mirror arrays that achieve the necessary precision. Figure 8.9 shows a simplified example of this.

There are challenges to be overcome with this approach. Although the use of mirrors allows potentially less signal loss than the use of integrated optic switches, mechanical considerations have to take into account thermal expansion as well as an appropriate means of actuating the mirrors, the latter not shown in this diagram.

8.5.4 Tunable Optical Cavities

In order to send more data down a single fiber, different wavelengths (colors) of light can be sent down the same fiber, each carrying a different signal (wavelength division multiplexers have been mentioned earlier). This requires

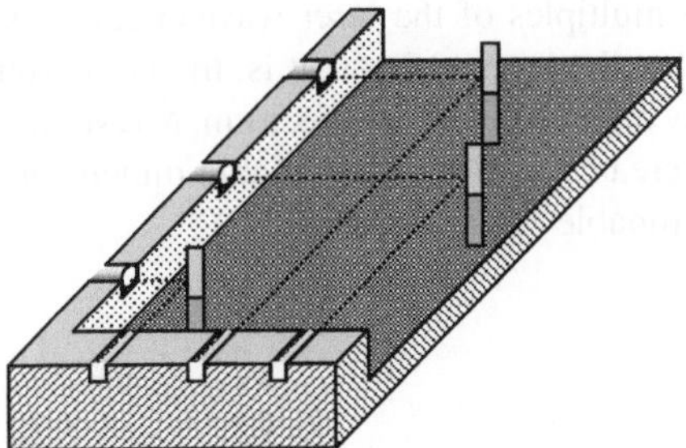

FIGURE 8.9 Optical cross-point switches. These are micromachined fibers designed to align a set of input fibers and a set of output fibers, as indicated. Mirrors are positioned to link a particular input fiber with a particular output fiber. The simplest structures may be LIGA based, with no active parts; mirrors are inserted into slots when the network is set up or reconfigured. More complex devices involve active mirrors that can be reconfigured without human interference. Note the requirement for very precise alignment.

laser diodes of different wavelengths or laser diodes that can be tuned to one of a number of wavelengths.

This is driving the development of a variety of solid-state lasers and associated MEMS. Many laser structures are constructed so that the lasing action takes place within a cavity bounded by two mirrors, one of which is only partially silvered to allow some laser light to escape (Figure 8.10). By constructing cavities with

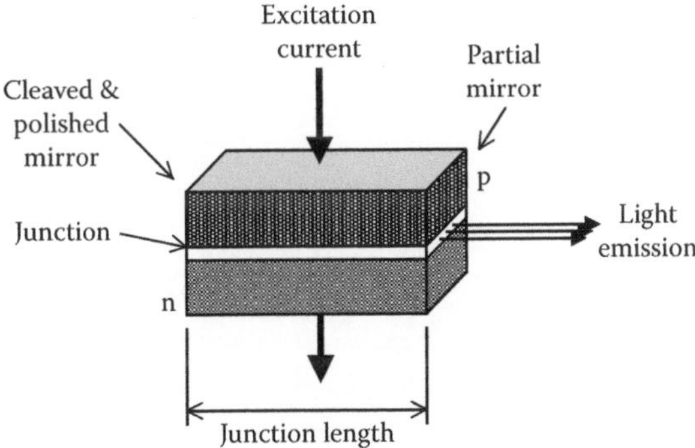

FIGURE 8.10 Semiconductor laser. The junction is highly doped. Holes and electrons recombining cause excess energy to be emitted as light. Photons are reflected back and forth in the cavity, stimulating further recombination, which causes photons to be emitted coherently with the stimulating photon. Light is emitted from the less polished end of the junction. The length of the junction must be an exact multiple (or fraction) of the wavelength of the laser light, which is determined by the semiconductor materials used. It is possible to achieve some degree of tuning by varying the junction length. Other designs have been developed that can lase over relatively wide bandwidths.

dimensions that are multiples of the laser wavelength, it is possible to tune the frequency of light that the laser emits, that is, the laser would normally produce light over a relatively broad band of the spectrum. MEMS techniques are therefore being employed to create cavities with variable dimensions that can be used to create dynamically tunable laser sources.

9 Assembly and Packaging

9.1 INTRODUCTION

Microengineered devices have the potential to be as inexpensive as silicon chips are today. This, however, will only be true when two conditions are met: (1) the fabrication process has a high yield (most of the devices on a wafer function properly and continue to do so after packaging) and (2) batch processing techniques are used for as much of the process as possible (i.e., large numbers of devices per silicon wafer, and a large number of wafers are processed at the same time at each fabrication step).

When developing microengineered devices and complicated microsystems, it is difficult to achieve high yields. However, these must be achieved before putting the device into production, with few exceptions. If the device does something that is very important and cannot be done any other way, then perhaps a low yield and expensive devices can be justified.

Assembling complex devices from many microscopic parts and, in particular, packaging these devices so they can be handled and connected to other components or systems will generally involve handling the devices individually. This can add significantly to the cost of the finished part (tens to hundreds of times the cost of the actual active part of the device depending on the complexity and requirements of packaging). Consequently, the assembly and packaging of devices for commercial manufacturing have to be carefully considered.

9.2 ASSEMBLY

Obviously, if microsystems consisting of many microscopic parts have to be assembled by hand, this can be a costly and time-consuming process. Hand assembly may be acceptable for device development or prototyping. Unfortunately, because the very small parts have to be lined up very accurately (or else they will not go together or will stick), conventional robotic assembly tools are not particularly suited to the task. Consequently, a method for assembling the microsystem or component has to be considered and, ideally, designed at a relatively early stage.

9.2.1 Design for Assembly

The most obvious approach is to design a device that does not need assembling. This is most easily seen in surface-micromachined parts in which the final etch step removes the sacrificial material and releases all the components.

Some microstereolithography processes (Chapter 3, Subsection 3.6.8) also lend themselves to the formation of free structures.

In other cases, the materials required for a particular application may negate such a simple strategy; this is especially true if one wishes to use incompatible fabrication processes (such as bonding laser diodes to integrated optic devices). The following should be considered:

- Can wafers be bonded rather than individual devices?
- Can components be constructed in such a way that they automatically align with one another when brought together (see the next section)?
- What tolerance can be achieved with the alignment tools being used?
- What tolerance can be achieved with the available microactuators or micromanipulators?
- Can components be built into the microsystem to enable or monitor alignment (e.g., test-pad access to optical sensors to facilitate fiber alignment)?

9.2.1.1 Auto- or Self-Alignment and Self-Assembly

Various techniques can be used to automatically align different components of a microsystem. "V" grooves are relatively easy to fabricate in silicon, and these can be used to align optical fibers to waveguides on the chip for integrated optics applications (Figure 9.1).

Owing to the small size of the parts involved, surface tension forces (in liquids such as water) can be used to assemble microengineered devices. For example, surface-micromachined devices can be produced with hinges and latches so that surface tension can be used to draw plates up and latch them into place to form vertical walls.

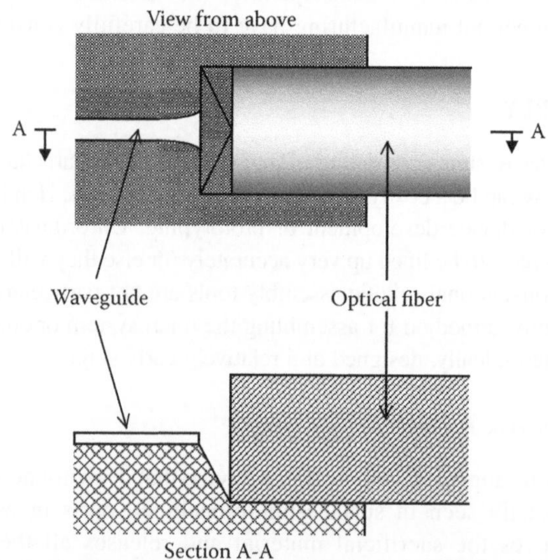

FIGURE 9.1 Use of V groove to align an optical fiber to a strip waveguide.

When parts are soldered or brazed together, careful design may enable the use of surface tension in the molten metal to correctly align the components. This can be readily seen in modern printed circuit boards in which the surface tension in the molten solder aligns small surface-mounted components to their pads, although placement machine errors may have only left them partially overlapping the pad to which they were supposed to bond.

Another possibility that has been proposed is the use of hydrophobic and hydrophilic areas on the surfaces of the parts. When the parts are floated on water, they line up such that the hydrophobic surfaces come together.

9.2.1.2 Future Possibilities

Assembling microparts into microsystems is an area that is receiving more research and development attention as the processes for producing the parts are becoming better developed. One of the areas that received attention under a 10-year micromachines research program sponsored by the Japanese government was the development of a desktop micromachines factory.

9.3 PASSIVATION

Often, parts of micromachined devices have to be exposed to the environment in which they are operating. This means that they have to be protected from mechanical damage and from contamination by dust or liquids that may affect the electronic circuitry. They must also be able to dissipate the heat generated by any active electronic components on chip.

Generally speaking, this has led to devices being coated directly by a thin film of either silicon dioxide or, more commonly, silicon nitride. These are usually deposited at relatively low temperatures using a technique known as plasma-enhanced chemical vapor deposition (PECVD), because high temperatures may affect components already on the device or induce unnecessary mechanical stress.

Silicon nitride is commonly used as it is wear-resistant and provides a good barrier to sodium ions in the environment, which penetrate into oxide layers and destroy their insulating properties. However, under some circumstances a nitride layer alone is not suitable. For instance, in physiological saline solution, under an applied electric field such as may result from active components on a chip, nitride rapidly degrades. Thus, sometimes, multiple layers of oxide and nitride are used: the oxide insulating the nitride from current flow and the nitride protecting the oxide from sodium ions.

More recently, it has become possible to deposit films of diamond. Diamond has excellent resistance to wear, is a good electrical insulator, and a good conductor of heat. However, the deposition process for these films requires a relatively high temperature (around 700 to 900°C); the films are polycrystalline (made of many small crystals) and are relatively difficult to machine.

An alternative to diamond is the so-called diamond-like carbon (DLC, sometimes referred to as amorphous carbon). DLC films can be deposited using

PECVD techniques, and the deposition can be controlled to produce films with different qualities. These, as with diamond, are also biocompatible.

Another film being investigated for various applications is silicon carbide. This, too, is wear- and chemical-resistant. It can also be deposited using PECVD techniques.

9.4 PREPACKAGE TESTING

In Chapter 4, the inclusion of test structures on masks was introduced. As wire bonding and packaging are expensive processes, it is desirable to test devices prior to wafer dicing. Obviously, the degree to which this can be performed depends on the design to some extent. It may be very difficult to test optical or microfluidic components prior to assembly with appropriate input and output ports. Nonetheless, there are some standard techniques that can be employed to ease testing.

The principal inspection tool in any fabrication is the scanning electron microscope (SEM), which is dealt with in Chapter 10. This is coupled with profilometers that can be used to measure step dimensions and optical techniques can be employed to measure certain film thicknesses. Note that these normally only sample one very small area of the wafer.

Beyond these tools, the main test tool is the probe station. This consists of a microscope and a set of tungsten needles mounted on micromanipulators. The wafer is placed on the station and the needles maneuvered onto test pads using the micromanipulators (either automatically or manually). Test signals can be injected, and the results can be measured via other needles or observed using the microscope.

Other optical approaches can be used to make measurements. For instance, interference fringes can be used to monitor membrane deformation.

The test procedure should be considered alongside the initial device design. Requirements will differ considerably for different MEMS devices, but generally include:

- Electronic test structures on wafers incorporating electronic circuitry (ring counters are a standard).
- Mechanical test structures where appropriate, such as:
  - Systems for exciting or deforming structures normally excited or deformed by external forces.
  - Structures for monitoring movement of actuators.
- Additional test pads where possible. When diagnosing a problem, it is desirable to have available as many signals and intervention points as possible.

9.5 PACKAGING

The package that the microsystem or device is finally mounted on has to perform many functions. It will enable the users of the device to handle and incorporate it into their own design. It will allow the attachment of electrical connections, fluid ports, fiber optics, etc., with minimum interference from stray signals or noise in the environment. It may also protect the device in harsh environments, preventing it from mechanical damage, chemical attack, or high temperatures. In many cases, when considering electronic devices, the package must also prevent light falling on the device, because the generation of charge carriers by photoelectric effects will appear as noise. In the case of light sensors, however, the package may be designed to concentrate light at a particular spot.

Owing to the variety of microengineered devices, it is not possible to specify a generic package. However, it is possible to make some general comments. The package must be designed to reduce electrical (or electromagnetic) interference with the device from outside sources, as well as to reduce interference generated by the device itself. Connections to the package must also be capable of delivering the power required by the device, and connections out of the package must have minimal sources of signal disruption (e.g., stray capacitance). The package must be able to dissipate heat generated by the active device to keep it cool. Where necessary, it must also be able to withstand high operating temperatures. It should also be designed to minimize problems due to different coefficients of thermal expansion of the materials used; this is often more important in microengineered sensors and devices than for conventional integrated circuits (ICs). It should also minimize stress on the device because of external loading of the package and be rugged enough to withstand the environment in which the device will be used.

The package also has to have the appropriate fluid feed tubes, optical fibers, etc., attached to it and aligned or attached to the device inside.

9.5.1 Conventional IC Packaging

Conventional IC packages are usually ceramic (for high-reliability applications) or plastic.

With ceramic packages, the die is bonded to a ceramic base, which includes a metal frame and pins for making electrical connections outside the package (Figure 9.2). Wires are bonded between bonding pads on the die and the metal frame (these frames are often manufactured using PCM techniques — see Chapter 3, Section 3.4). The package is usually sealed with a metal lid.

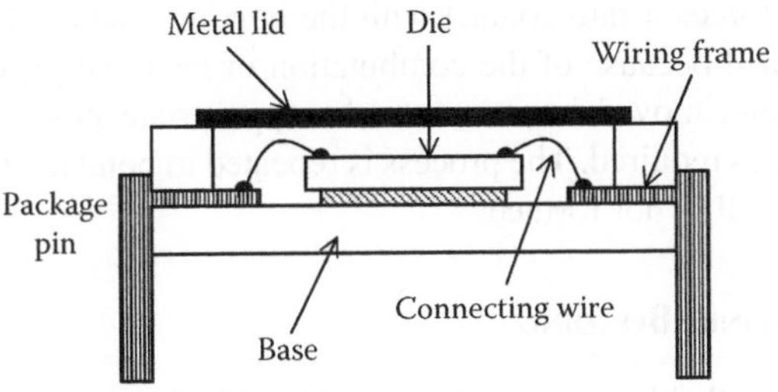

FIGURE 9.2 Conventional IC package. PTFE tape and black wax can be used to protect a wafer during short-term KOH etching (cross section).

With plastic packages, a similar attach-to-base/bond-wires/seal-with-lid process may be used. After wire bonding, however, it is also possible to mold the plastic package around the device. As the substrate of many ICs requires an electrical connection to bias it, the die may be bonded to a metal connector on the base either by thermal methods (melting a suitable metal beneath the die) or by using a conductive epoxy resin.

Epoxy resins are quite often used to attach devices to substrates or to insulate or package them, particularly with prototype micromachined devices. With particularly sensitive devices, however, it is necessary to be aware that some epoxy resins get hot while curing or may shrink slightly, putting mechanical stress on the device.

9.5.2 Multichip Modules

Multichip modules (MCMs) are another aspect of microengineering technology. In the search for ever faster computers and electronic devices, it is desirable to keep the connections between chips as short as possible. This leads to the development of MCMs in which many dies are assembled together into one module. Often thick-film techniques, in which conductors and insulators are screen printed onto ceramic substrates, are used. More exotic techniques include technologies that are being developed to stack up dies one on top of the other.

9.6 WIRE BONDING

There are two conventional ways of bonding wires to chips. These are thermocompression bonding and ultrasonic bonding. Commonly, fine (25-μm diameter) aluminum wire is used, but gold wire is also used quite often. For high-current applications (e.g., to drive magnetic coils or for heaters), consider larger-diameter wires or multiple connections.

9.6.1 Thermocompression Bonding

In thermocompression bonding, the die and the wire are heated to a high temperature (around 250°C). The tip of the wire is heated to form a ball; the tool holding it then forces it into contact with the bonding pad on the chip. The wire adheres to the pad because of the combination of heat and pressure. The tool is then lifted up and moved in an arc to the appropriate position on the frame, dispensing wire as required. The process is repeated to bond the wire to the frame, but this time a ball is not formed.

9.6.2 Ultrasonic Bonding

This is used when the device cannot or should not be heated. In this case, the wire and bonding surface (pad or frame) are forced together by the tool, and ultrasonic vibration is used to compress the surfaces together to achieve the desired bond.

This can be combined with the previous technique to achieve a temperature lower than that of thermocompression bonding.

9.6.3 Flip-Chip Bonding

Flip-chip bonding is another new technique associated with microengineering. Here, small beads of solder are formed on bonding pads on the die. The die is then mounted facedown on the base and heated until the solder melts and forms a contact between metal tracks on the base and the bonding pads.

There are several methods by which solder bumps can be formed on the die or wafer. Two common commercial methods are vapor deposition or electroplating. Some companies have developed screen-printing techniques or printing through dry film resists as alternatives. It is possible to achieve pitches down to 200 μm, but it is of greater interest to the microengineer to note that this technique provides some degree of self-alignment during assembly because of the surface tension of the molten solder. Note that when using this approach, or soldering generally, the device has to be able to withstand application of heat at least for a short period of time. When the final assembly incorporates structures that act as heat sinks, soldering ovens may need to be set to relatively high temperatures.

9.7 MATERIALS FOR PROTOTYPE ASSEMBLY AND PACKAGING

Although some of these materials may be used for production devices, they are listed here because most of them are commonly found in MEMS laboratories in various applications.

- Black wax
- Dental wax
- Polyethylene glycol
- Silver-loaded epoxy
- Epoxy adhesives
- UV-curing adhesives
- Cyanoacrylate adhesives
- Photoresists
- PTFE tape

Black wax, commonly known as "apiezon" is useful for temporarily holding parts together and is used during low-temperature wet-etching processes to protect part of the wafer. Being opaque, it also protects electronic components from optical interference. It can be applied in liquid form when dissolved, commonly in toluene, but it is often difficult to remove. Solvents include toluene (often used hot) and xylene, and ultrasonic cleaning baths have also been employed to help remove the wax. It is insoluble in the two common clean room cleaning solvents: acetone and isopropyl alcohol.

Dental wax is also available for holding small parts temporarily. It comes with a variety of melting temperatures. Another material, not a wax, that has

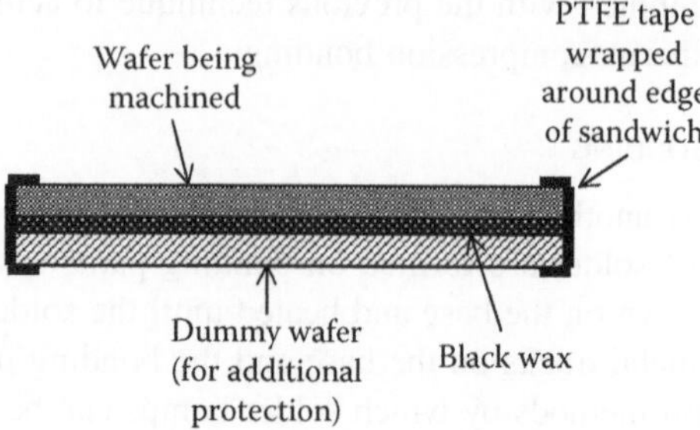

FIGURE 9.3 PTFE tape and black wax can be used to protect a wafer during short-term KOH etching (cross section).

unusual properties but is not commonly found in clean rooms is polyethylene glycol. This is available in various molecular weights and, hence, different melting points. It is also soluble in water.

Electrical contacts can be made using silver-loaded epoxy. This is commonly used to ensure the electrical bonding of the back of a die to the substrate to enable biasing, but it can also be used to attach wires to bonding pads. If this is done by hand, it is a good idea to have relatively large bonding pads (400 μm^2 with the same distance between them) at the edge of the die.

There are a variety of epoxy glues available. These form strong bonds and are chemical-resistant. You should be aware that some will shrink on curing, some require heat to cure, and some take a long time to cure during which individual parts have to be clamped together.

UV-curing adhesives are becoming very popular. One problem with epoxies is the need to hold parts together while the adhesive is curing, which can be problematic. UV-curing epoxies have the advantage that curing is much more rapid and occurs only when the adhesive is exposed to UV light. This offers some advantages to prototype MEMS assemblies. It also makes these epoxies useful for mass production, because they can be dispensed and cured as required without having to worry about postmixing curing time or heating.

Cyanoacrylate adhesives, commonly known as "super glues," offer instantaneous adhesion of parts. They do not offer great resistance to shear forces and cannot be used if the device is to be exposed to water or humid environments.

Photoresists provide a form of temporary adhesive or passivation layer. There is not much more to be said about this application, except that generally they are not very good in either role (with the exception of epoxy-based resists such as SU-8).

Polytetrafluoroethylene (PTFE) tape is commonly found for temporarily taping wafers together. It is not, in itself, an adhesive (it is better known for its application to nonstick cookware), but it is water- and chemical-resistant and amalgamates with itself to some extent, so it can be used for very short-term fixes. One common use is in conjunction with black wax to protect wafers during etching (Figure 9.3).

10 Nanotechnology

10.1 INTRODUCTION

The boundary between micro- and nanotechnology is blurred in much the same way as that between microtechnology and miniature technology. The popular perspective is that nanotechnology is some sort of futuristic technology that involves building little machines out of atoms, but if we develop our definition of microengineering (working with tolerances in the order of micrometers) in relation to nanotechnology (i.e., working with tolerances in the order of nanometers), then even at the time of writing nanotechnology has already arrived on the scene. Integrated circuits are being manufactured with 90-nm feature sizes, and many powders and solgels (such as the spin-on ceramics referred to in Chapter 1 of this book) have particle sizes within the nanometer range.

Photolithography and, more notably, e-beam lithography can be used to produce nanometer structures. Micromachining techniques can be refined, and additional machining techniques implemented, to produce nanoelectromechanical systems (NEMS). The popular image of robots built out of atoms is being developed, but this is a small part of molecular nanotechnology, which also has to compete with bionanotechnology, where parts are taken from nature's nanomachines (cells, bacteria, and viruses) and assembled into new configurations.

This chapter is a brief review of many aspects of nanotechnology, but will start with an exploration of scanning probe microscopes (SPMs), after touching briefly on the MEMS workhorse microscope, the scanning electron microscope (SEM). It was the use of SPMs to write the initials of a company on the atomic scale with a few atoms that sparked off much of the popularity that nanotechnology now enjoys.

10.2 THE SCANNING ELECTRON MICROSCOPE

Although optical microscopes are frequently found and used in MEMS work, it is the SEM that is the true workhorse. In Chapter 1, where e-beam and UV lithography were compared, the much smaller wavelength and, hence, superior resolution of the electron became apparent. The same is true in the world of microscopy, and basic SEMs can be obtained at a relatively low cost.

The basic form of a SEM is illustrated in Figure 10.1. The sample is placed on a stage in a sample chamber. This can normally be rotated, tilted, or moved in x, y, or z directions. The microscope itself is housed in a column that sits atop the sample chamber. Generally speaking, the sample chamber and column need to be evacuated, but some of the more expensive SEMs will operate under low

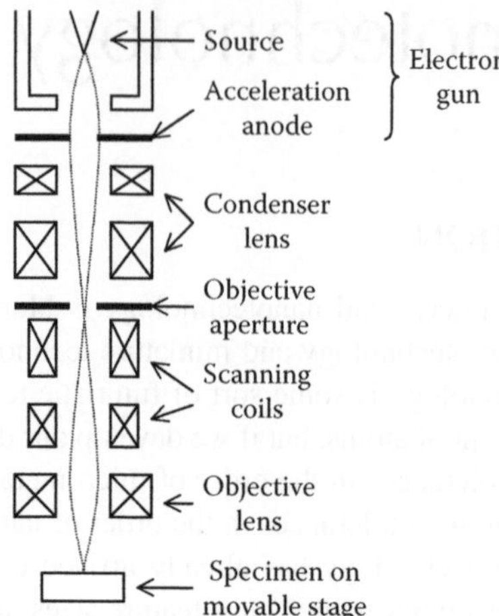

FIGURE 10.1 Outline of scanning electron microscope operation. The electron gun is provided with mechanical adjustments to align it with the column. The entire unit is evacuated.

pressures, allowing live biological specimens (insects such as mites, etc., that can survive harsh environments) to be imaged.

Electrons are produced at the top of the column from an electron gun or point source such as a sharpened tungsten needle. The electrons are accelerated through electrodes toward the target, and the beam is focused to a spot by electrostatic and magnetic lenses. These are lenses produced using either concentric electrodes or electromagnetic coils, the exact arrangement depending on the manufacturer of the instrument.

The beam (spot) is then scanned across a tiny portion of the surface of the object by the "deflection coils," the magnification being a combination of spot size and the area over which it is scanned. At the same time that the beam is scanned across the target, a second beam is scanned across a cathode ray tube. The signal from the electron beam is amplified and reproduced on the CRT to form an image.

One thing to note about the SEM is that because the sample is being continuously bombarded with electrons, it will soon get charged. A charged sample will cause artifacts to appear in the image. One solution to this is to reduce the intensity of the beam, but it is normally better to ensure that the sample can conduct any charge to ground by evaporating a thin layer of gold on to it. This does make the sample unusable from a practical point of view, but if the aim is to measure dimensions or verify the final result of the process, then it saves a lot of time used for adjusting beam parameters. This is especially true if the structure has a good cover of insulating films.

Alternative instruments for process measurement include optical techniques for measuring steps and film thicknesses (assuming the film is transparent), usually based on interference fringes, and mechanical profilers. Talysurf is a well-known brand name of one such device, and the atomic force microscope (AFM), which will be discussed in the following text, is another mechanical profiler.

10.3 SCANNING PROBE MICROSCOPY

The principles of scanning probe microscopy are illustrated in Figure 10.2. The sample is mounted on a moveable and brought up to a probe. The tip of the probe is sharpened to a fine point and is designed to interact with the sample in some manner. The stage is operated to scan the specimen beneath the probe, and a feedback control system keeps the distance between the probe tip and the sample constant. This enables images with atomic-scale resolution to be built up. This section will consider three different forms of SPM: the scanning tunneling electron microscope (STEM), atomic force microscope (AFM), and the scanning near-field optical microscope (SNOM). The principal difference between these can be found in the means of interaction between the probe and the sample, which is further reflected in the probe tip itself and the means of controlling it.

10.3.1 Scanning Tunneling Electron Microscope

The STEM is an electron microscope that operates on the phenomenon of tunneling. When two atoms are brought close together, an electron in the outer orbital of one atom can sometimes disappear and reappear in the outer orbital of the adjacent atom. It, in effect, "tunnels" through the energy barrier that separates the two atoms. The phenomenon is termed "tunneling" because the electron never appears between the two atoms; it is either associated with one or the other.

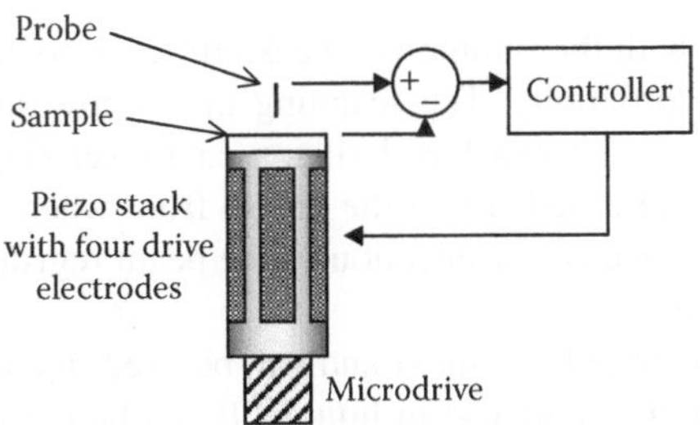

FIGURE 10.2 Basic principles of scanning probe microscopy. The sample is placed on the top of a long stack (10 cm) of piezoelectric disks. A microdrive provides coarse movement to bring the sample close to the probe, but all fine movement is achieved by the piezo stack. A feedback control unit keeps the distance between the probe tip and the sample constant; the amount of drive needed by the electrodes to achieve this is directly related to the topology of the surface. Scanning is achieved by bending the piezo stack; because it is so long and the distances scanned are very short, it looks like a horizontal translation.

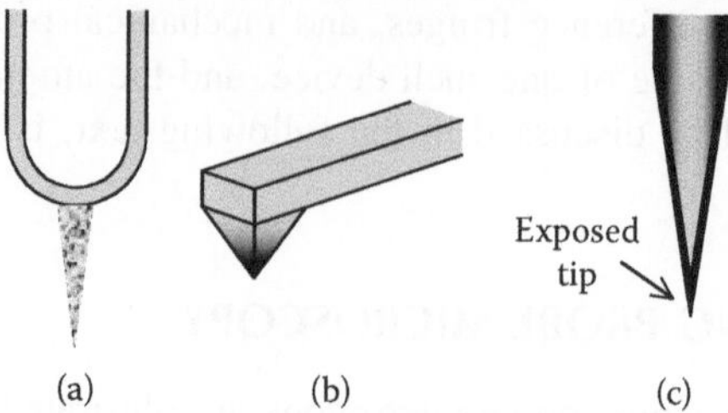

FIGURE 10.3 (a) STEM probe tip. This is formed from a U-shaped tungsten wire, which is dipped and withdrawn from electroplating solution to build up the point, (b) AFM probe tip; a pyramidal structure on a silicon cantilever beam, (c) SNOM probe tip; the core of an optical fiber is shaped by dipping and withdrawing from an etch solution; this is then coated with metal by evaporation and a small hole is formed at the tip by spark discharge.

The frequency of tunneling depends upon the distance between the probe tip and sample and also upon the electrical potential bias applied between the probe and sample; this has the effect of increasing or decreasing the height of the energy barrier for a given distance. The tunneling current between the probe and sample is monitored: as the sample is scanned beneath the probe tip, the height is adjusted to maintain a constant tunneling current and, hence, a constant distance between the probe and sample. Note that the sample has to be electrically conductive for this to work. STEM is normally carried out in a vacuum. Highest resolution is achieved with cryogenically cooled stages.

The STEM tip is traditionally an electrolytically etched tungsten needle (Figure 10.3a). The tip is formed by repeatedly dipping and withdrawing a tungsten wire into an etch solution. A bias current assists in the etching process.

10.3.2 Atomic Force Microscope

The AFM interacts with the sample via weak attractive forces that exist between two atoms in close proximity. The scanning tip is, now, normally a sharpened point at the end of a micromachined silicon cantilever (Figure 10.3b). A laser beam is directed toward the back of the probe, from where it is reflected onto a quadrant photodiode detector. Deflection of the beam registers as an asymmetric response from the detector.

The AFM is remarkably resilient and can be used in air, in (relatively) dirty laboratory environments, and also in liquids. It can be used to image biological material with minimal preparation. For optimal operation, the spring constant of the probe needs to be calibrated (rather than relying on the manufacturer's data). Various other modalities, such as magnetic resonance and temperature measurements, are currently being combined on AFM tips to provide additional information about the sample.

Note that at the highest resolution, the AFM affects the sample mechanically. Just as the probe tip is attracted down toward the sample, atoms in the sample are attracted toward the probe tip.

10.3.3 Scanning Near-Field Optical Microscope

The tip of an SNOM is shown schematically in Figure 10.3c. The probe consists of an optical fiber that is etched (by dipping and withdrawing) down to a fine point. A metal film is then evaporated on this and a small opening made at the tip, by spark discharge.

The fiber is illuminated by laser light, but the opening at the tip is narrower than the wavelength of the light employed. In optical waveguides, a small portion of the optical wave projects outside the waveguide. This is the "evanescent" field and can be used to excite fluorophores when sufficiently close. Similarly, disturbances to the evanescent field in other situations (e.g., planar waveguides) can be used in sensing applications.

The SNOM functions, in the first place, as an AFM. Usually, the fiber is made to vibrate, and changes in the frequency of oscillation indicate interaction with the sample. AFM probe tips with appropriate optics are also being developed. The advantage of the SNOM is that it can be made to interact optically with the sample. This makes it particularly useful in biology, where fluorescent markers can be excited by the SNOM and used to link chemistry with mechanical structure.

10.3.4 Scanning Probe Microscope: Control of the Stage

Large movement of the SEM stage is achieved using small geared electric motors. However, the main control element is a piezoelectric stack. This is a cylinder several centimeters (4–5 cm) in length and has three or four electrodes at the outer edges of the cylinder. By activating all the electrodes simultaneously, the stack can be made to lengthen, bringing the specimen closer to the probe, or shorten, moving it further away. By activating the electrodes asymmetrically, it is possible to cause the stack to bend. Because of the long radius of curvature compared to the horizontal motion, this is used to scan the sample in the x and y directions beneath the probe.

10.3.5 Artifacts and Calibration

Despite the efforts made to sharpen probe tips, they are still relatively blunt on the atomic scale. As a consequence, sharp steps and narrow trenches in the sample are often smoothed over (Figure 10.4).

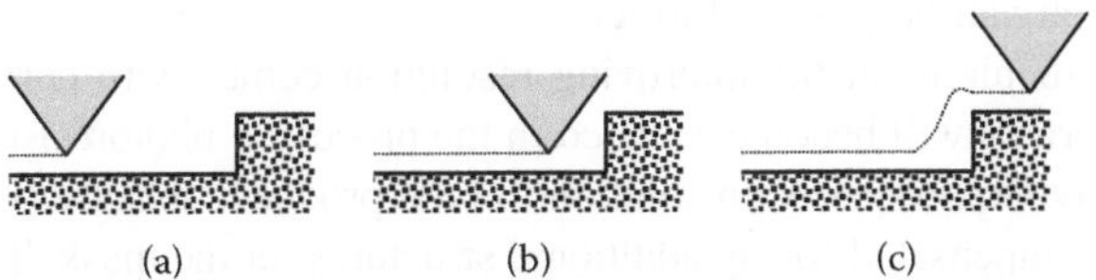

FIGURE 10.4 AFM tip approaching a sharp step. At (b) the side of the probe tip starts to interact with the corner. This leads to a curved scan path, (c) dotted line, so the step is incorrectly reproduced.

There are two other limitations of SPMs of which to be aware. The first is that the area examined is a tiny fraction of the surface and may not be typical. The second is particularly applicable when using AFMs. These often come with software that will allow the rms surface roughness to be calculated at the push of a button. At AFM scales, this is not necessarily a very useful absolute measurement, because the rms roughness is fractal in nature (as you increase magnification, you are measuring a different rms roughness).

The spring constant of AFM probes needs to be calibrated. This is normally performed by measuring the fundamental frequency at which the probe oscillates.

When working to atomic-scale resolution, it is possible to calibrate the microscope using an atomically flat surface with a known interatomic spacing. The arrangement of carbon atoms in graphite sheets provides one possibility. Mica provides another. Adhesive tape applied to the sample and pulled away reveals a clean, atomically flat surface.

10.4 NANOELECTROMECHANICAL SYSTEMS

As suggested in the introduction, micro- and nanotechnologies are converging. One strand of this convergence is the demand placed on lithography systems by the integrated circuit industry. Another avenue is NEMS. Lithography techniques can be adapted to produce nanostructures, but there are also some micromachining techniques, covered in Part I, that can be adapted to produce nanostructures.

10.4.1 NANOLITHOGRAPHY

The principal tool for nanolithography was covered in Chapter 1: direct-write e-beam lithography. This can be used to pattern structures down to 10-nm minimum feature size, but there are several limits to this. Highest resolutions are achieved with thin resist films, therefore the processes employed to form structures have to take this into account (harsh etches, for instance, have to be used with care).

10.4.1.1 UV Photolithography for Nanostructures

The use of UV photolithography for high-resolution printing applications was alluded to in Chapter 1. In summary, highest resolutions are achieved with short wavelength illumination (i-line), projection printing, and thin-resist films. However, resolution can be pushed further.

The first requirement for improving resolution comes with corners on mask structures. Corners will become rounded in the processed photoresist film because they ultimately require perfect resolution to be reproduced exactly (Figure 10.5a). This can be compensated for by additional structures on the mask. These will not be reproduced in the patterned photoresist but act to bulk out the corner to improve its reproduction (Figure 10.5b). Similarly, by artificially enlarging very small structures on the mask they can be reproduced in the resist film (Figure 10.5c).

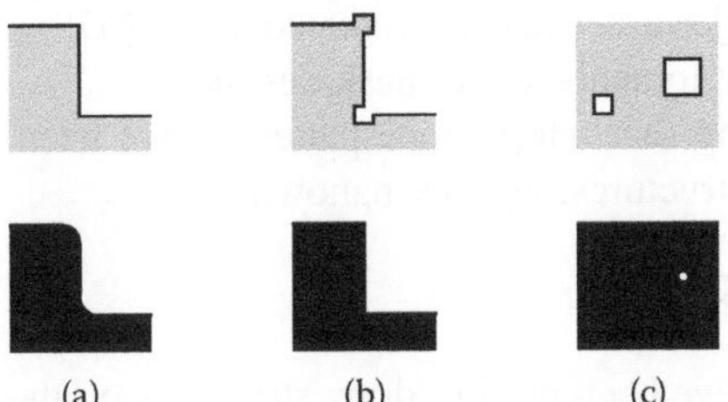

FIGURE 10.5 (a) Corners on the mask (gray) become rounded in the resist profile unless compensation is employed, (b) compensated corner, (c) small features can be produced in the resist by designing larger features that will be reduced down by optical effects.

This approach requires quite a sophisticated analysis of the optical system, but this can be added to industrial CAD systems. It is generally known as optical proximity correction and may also be applied for structures that are close together on the mask.

10.4.1.1.1 Phase-Shift Masks

Another method by which high resolutions can be achieved is to employ a phase-shift mask. This approach employs a mask that causes the phase of the UV illumination to shift (by 180°) around the edges of the structures on the mask. This creates a controlled interference pattern, which enables higher resolutions to be achieved.

Two approaches are available for this. One is the use of molybdenum silicide instead of chrome on a quartz mask. This transmits a small percentage of the incident UV radiation, which is phase shifted. The other approach employs a machined chrome-on-quartz mask (Figure 10.6). Where the light enters the etched areas of the mask, it will be phase shifted. This can be used to create an interference pattern to print lines smaller than the wavelength of the light employed; a similar effect can be achieved by depositing and patterning rather than etching films on the mask, thickening it. Note that a similar effect can be created with x-ray masks by varying the thickness of the attenuating material.

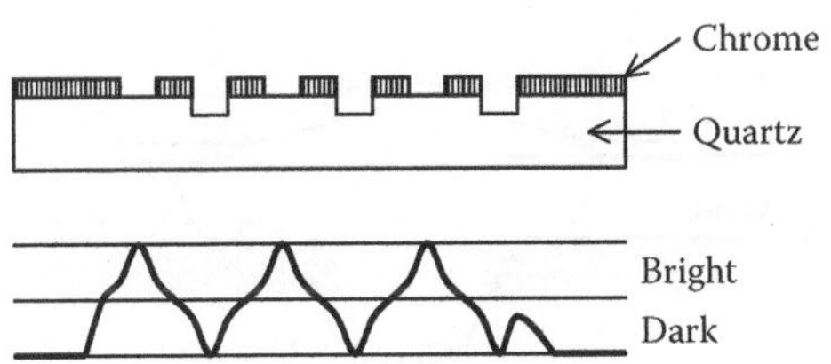

FIGURE 10.6 Machined chrome-on-quartz phase-shift mask. The machining causes the incident light to emerge from the mask 180° out of phase with the light passing through unmachined areas.

Phase-shift masks are usually produced by CAD systems equipped with appropriate software to analyze the mask design.

It is worth noting that interference patterns have been productively used to make various nanostructures, notably nanowires.

10.4.1.2 SPM "Pens"

The tips of SPMs have been used to draw structures on the surface of an appropriate material. The most basic of these approaches is to use the probe tip of an AFM to scratch a design on the surface. This can also be employed to create nanostructures by scratching away layers that have been created by the Langmuir–Blodgett (LB) technique, explained in the following subsection. A series of monolayers of different compounds can be deposited and then cut through to reveal the interior structure.

A further approach is to dip the tip of the AFM probe into a liquid and write on a surface, as with an ink pen. This has been reported to produce 30-nm wide lines, writing with alkanethiols on a gold surface [1]. The use of the STEM with electron-sensitive photoresists is also possible.

10.4.2 Silicon Micromachining and Nanostructures

Whereas basic photolithography systems cannot be used to produce nanostructures, it is possible to adapt silicon micromachining techniques to produce structures that bridge the micro–nano division. The simplest approach is to use timed over-etching to etch microstructures down to nanostructures. This approach has been used to form AFM probe tips. In Figure 10.7, for example, an oxide pillar is etched down by immersion in a slow-timed wet etch.

Oxide, nitride, and metal films can be deposited with submicron thickness. This means that it is possible to produce structures with submicron vertical feature sizes, such as steps with 100-nm heights, without having to resort to special techniques. Horizontal dimensions will, however, still be on the order of microns if standard photolithographic techniques are used. Thin-beam structures can be implemented in silicon using concentration-dependent etching, or electrochemical etching, but with very shallow diffusion or implantation of the impurities.

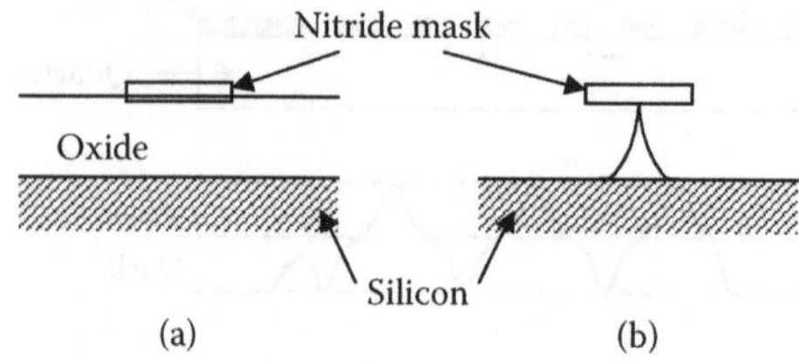

FIGURE 10.7 Forming a fine point by wet etching of oxide. Compare with Figure 2.20 in Chapter 2.

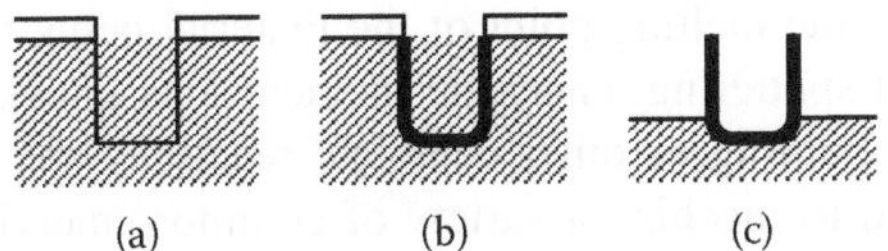

FIGURE 10.8 Nanostructures formed using thermal oxide: (a) trench etched through nitride mask, (b) thermal oxide, (c) nitride stripped and silicon etched (e.g., in TMAH).

Thermal oxidation has proved to be fertile ground when it comes to the production of nanostructures. High-quality oxide film can be carefully controlled because it grows slowly, and it also grows on all exposed silicon regardless of orientation. Figure 10.8 shows how this can be used to produce freestanding walls of submicrometer thickness. A nitride mask is employed, and a vertical trench is etched in the silicon wafer (Figure 10.8a). Thermal oxide is grown on the walls of the trench (Figure 10.8b). After this, the nitride is stripped in phosphoric acid. The wafer is then etched in a silicon etch that has a high selectivity over oxide (e.g., TMAH), leaving the freestanding structures (Figure 10.8c).

Thermal oxidation can also be used to close up microstructures. Figure 10.9 outlines an approach that has been used to create membranes with pore dimensions of less than 100 nm. In Figure 10.9, a silicon membrane has been prepared with a pyramidal pore etched through it using KOH (Figure 10.9a). An SOI wafer can be used for this, for example. The smallest dimensions of the pyramidal pore will depend on many variables, such as control of the KOH etch process, thickness and thickness variations of the membrane, tolerances of the photolithography process, etc. However, quite large pores can be closed up by thermal oxidation, as in Figure 10.9b.

Finally, it is worth noting that when setting up deposition equipment, one is frequently faced with a number of artifacts or defects in the films produced: islands, pinholes, ears, etc. These, typically, have submicron dimensions, and if it is possible to produce them in a controlled manner, they can be used as nanostructures. Some quantum dots were first produced as defects during a deposition process.

10.4.3 Ion Beam Milling

Ion beam milling was introduced in Chapter 2. This was divided into showered-ion-beam milling (SIBM) and focused-ion-beam milling (FIBM). The former thins out samples simultaneously over a large area, and the latter only at a focal point.

FIGURE 10.9 (a) Pyramidal pore in membrane, (b) closed up by thermal oxidation.

Both operate below the melting point of the material being machined, using the process of physical sputtering: energetic incident ions knock atoms off the substrate material, no burning or other chemical reactions are involved. Ion beam milling can be used to machine a variety of common materials, including most of those found in semiconductor manufacturing, diamond, ceramics, etc.

SIBM is capable of machining at rates of microns per minute over areas of tens of square centimeters. The equipment, usually, consists of an evacuated column with an ion source at the top and the sample at the bottom. The ion source is often a gaseous plasma, and ions are extracted from this and accelerated in a beam of several centimeters diameter toward the target. SIBM can be used to polish or control the profile of microstructures down to features of tens of nanometers.

The rate of material removal of SIBM depends on the material being machined, ion type, the energy of the ions, and the angle of incidence (with a maximum in the 40 to 60° range). Ion beam milling also causes subsurface damage, which is also dependent on the material involved and the ion energy.

FIBM normally uses a liquid-metal ion source, normally gallium, which produces a beam of metal ions. This is focused to a spot of less than 10 nm in a manner similar to an SEM. The beam is directed to particular parts of the structure to be machined, and is capable of cutting trenches with sub-100-nm width and trimming structures to the order of 10 nm. Unlike SIBM, however, it only mills a small area at one time, making it slow for use in batch production.

FIBM can also be used to deposit materials in a localized vapor deposition process. The vapor phase of an organic or organometallic precursor is delivered to the chamber in the region of the incident beam (Figure 10.10), where it is decomposed by the beam. Furthermore, FIBM systems with gallium ion sources can implant gallium ions into titanium (this is an unwanted side effect in many FIBM processes). However, in titanium, gallium impurities act as an etch stop (above $\sim 1 \times 10^{15}$ cm^{-3}) when etching is performed with SF_6 in a plasma etcher. Structures with a width of 250 nm can be produced using this approach.

The similarity of the FIBM and SEM has already been noted. It is also possible to monitor the process of machining or even image the substrate by

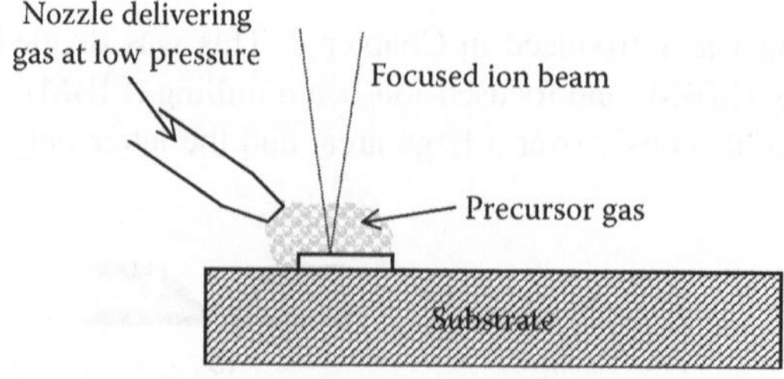

FIGURE 10.10 Use of a focused ion beam to deposit material; the nozzle is normally formed from a glass capillary drawn down to a fine opening.

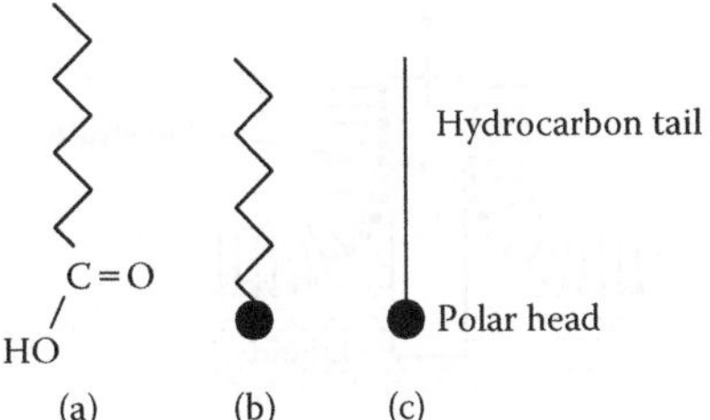

FIGURE 10.11 Schematic ways of representing an amphiphilic molecule: (a) explicit drawing of head, schematic drawing of hydrocarbon tail, (b) head drawn as circle, (c) head shown as circle and tail indicated by line.

monitoring secondary ions or electrons coming from the target, amplifying the signal, and displaying it, or by monitoring the current from the substrate.

10.5 LANGMUIR–BLODGETT FILMS

The Langmuir–Blodgett (LB) technique allows films of one-molecule thickness to be built up on solid substrate materials. The materials that can be used to form these layers are water-insoluble amphiphilic organic molecules such as fatty acids. These consist of a hydrophilic polar head (such as a carboxyl, amine, alcohol, or carboxylic group) that dissolves in water, and a long hydrophobic hydrocarbon tail, that does not; Figure 10.11 shows a typical schematic depiction (Chapter 7 contains an introduction to organic chemistry). These are dissolved in a volatile organic solvent that does not dissolve in water. A small drop of this solution is deposited onto the surface of a tank of water, and when the solvent evaporates the amphiphilic molecules remain with the hydrophilic head dissolved in the surface of the water and the tail projecting from it.

The coating process takes place in a tank with a balance to measure the surface pressure of the film, and computer-controlled barriers confine the area over which the film can spread. If a few molecules are scattered over a large surface area, they will interact very rarely and form a fairly random film. As the surface area is reduced, they will be compressed into a highly ordered state, the solid phase (Figure 10.12). The changes between the disorganized gaseous phase and organized solid phase are observable by changes in the surface pressure.

LB films are applied by dipping the substrate into the tank and withdrawing it (Figure 10.13) using feedback control of the barriers to maintain the surface

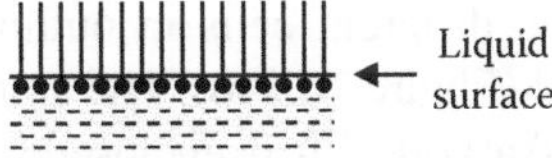

FIGURE 10.12 Amphiphilic molecules compressed into the ordered solid phase on the surface of a bath of water.

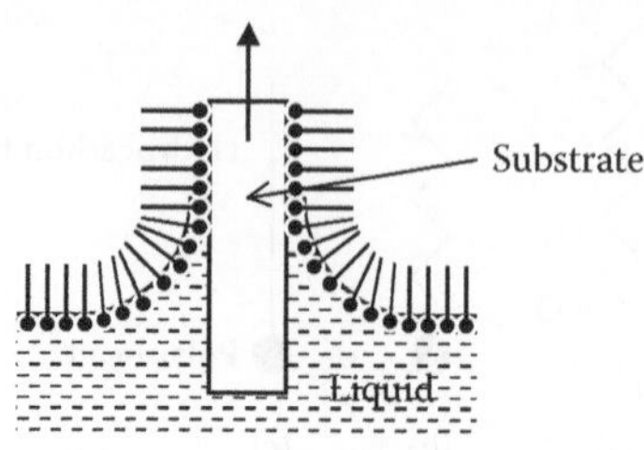

FIGURE 10.13 Forming a Langmuir Blodgett film on a hydrophilic substrate.

pressure indicative of the solid phase. The orientation of the molecules in the film thus produced depends on the molecules themselves, the substrate, and the precise technique employed. Generally, a hydrophilic substrate will have the orientation illustrated in Figure 10.13, with the tails pointing away from the substrate, whereas a hydrophobic substrate will have the opposite orientation. Deposition can occur on dip, withdrawal, or both, and it is possible to build up multiple layers. Additionally, it is possible to build up alternating coats of different amphiphilic molecules.

10.6 BIONANOTECHNOLOGY

One could say that nanotechnology is now in an analogous position to the early days of aviation. We know what is possible with bionanotechnology because it is visible all around us in nature; we only have to work out how it is done. Rather than trying to design our own molecular machines from scratch (molecular nanotechnology) or adapting comparatively well-understood micromachining processes (NEMS), bionanotechnology borrows components found in nature, usually the living cell, and puts them together in new configurations.

Another source of inspiration is from viruses. Many viruses survive outside the cell by being enclosed in sophisticated capsules formed of proteins, which can be considered as basic nanostructures (Chapter 7 includes a brief introduction to the molecules mentioned in this section). One of the basic uses of such structures is as formers for nonbiological structures; metal films can be evaporated over these, and some of the biological materials mentioned in the following text, for instance.

Most of the nanostructures employed in bionanotechnology will either be manufactured by genetic engineering of bacteria (usually *Escheria coli*) or extracted from cells by a complex separation process.

Figure 10.14 shows two different cells in outline: the animal cell (Figure 10.14a) and a bacterium cell (Figure 10.14b). The animal cell has a clear nucleus and is therefore termed "eukaryotic." It is enclosed in a lipid bilayer membrane formed of amphiphilic molecules with polar heads on the inner and outer surfaces of the membrane and tails in the center. Various proteins project through or are

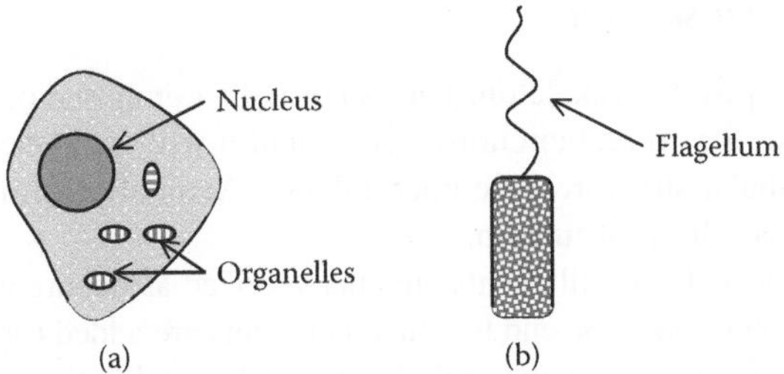

FIGURE 10.14 (a) Eukaryotic cell, (b) prokaryotic cell with flagellum; procaryotic cells are generally one or two orders of magnitude smaller than animal cells.

dissolved in this membrane. It also contains a number of small membrane-bound structures called "*organelles*," which have specific functions. The bacterial cell is also highly organized, but does not contain membrane-bound organelles. Normally, it will have a tough protein coat, and this one has been depicted with a flagellum, which helps it to move around in a liquid environment. It should be borne in mind that these descriptions are very general.

The basic fuel for the cell is adenosine triphosphate (ATP), which is converted (hydrolyzed) to adenosine diphosphate (ADP) or adenosine monophosphate (AMP) when any work is done. ATP is generated either through photosynthesis or by breaking down fuel molecules (to ethanol in anaerobic conditions or carbon dioxide and water in aerobic conditions, i.e., when oxygen is available).

In the following discussion, it is worth remembering that many processes in the cell require a number of different components to bind together, break apart, or change shape at different stages. Similarly, most molecular machines in the cell, even relatively simple ones, are composed of several components. These may be identical macromolecules that join together, such as machines formed by two copies of the same protein, or they may be composed of entirely different types of macromolecules: proteins and RNAs, for example.

10.6.1 Cell Membranes

Many different processes take place in the cell membrane. The cell needs to maintain a particular chemical composition within itself in order to function. This is achieved by pumps embedded in the cell membrane. These are complex proteins or multiprotein structures, which, on encountering the item that they are required to pump (e.g., a sodium ion), change shape (this usually involves the hydrolysis of ATP) to move it from one side of the membrane to the other. In other circumstances, ions moving down the concentration gradient can be used to drive processes without the involvement of ATP.

Artificial lipid bilayer membranes can be constructed in the form of closed vesicles, and genetically engineered proteins can be embedded in them.

10.6.2 The Cytoskeleton

Eukaryotic cells possess a skeleton. This is normally composed of tubulin or actin filaments. Under the correct circumstances, tubulin will spontaneously polymerize into long tubular structures, the microtubules. Actin also forms filaments that are less rigid than those of tubulin.

Microtubules and actin filaments are characterized as having a plus end and a minus end; the plus end is the end to which new units are added (or removed from) when the structure grows in the cell. Microtubules and actin filaments are the highways along which one type of molecular motor runs. Normally, these move things about in the cell, but they can be adapted by engineers to do other tasks.

10.6.3 Molecular Motors

The molecular motors that travel along microtubules are known as kinesins and dyenins. Kinesins "walk" along the microtubule towards the positive end, and dyenins are negative-end directed. A molecular motor that encounters a microtubule will walk along it until it reaches the end, where it will fall off.

Several different kinds of kinesins and dyenins are encountered in the cell, but they progress along the microtubule in a similar manner. These motors possess two "feet" that interact with the microtubule (see Figure 10.15 for schematic examples). It is thought that they progress along the microtubule by detaching one foot, deforming the molecule, reattaching the foot, and then detaching the other foot and moving it up to the first one. In this way they remain permanently attached to the microtubule.

Myosins walk along actin filaments. Myosin I has one foot, and myosin II has two. The action of myosin II is different from two-footed kinesins in that it probably acts to draw two actin filaments past each other (Figure 10.16b).

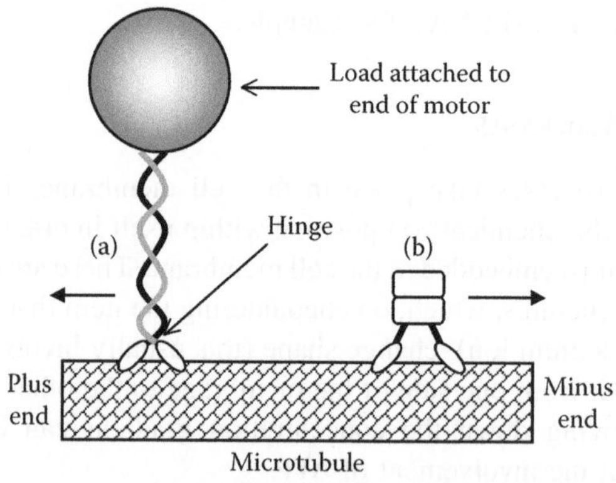

FIGURE 10.15 (a) Kinesin, carrying a load, (b) dyenin.

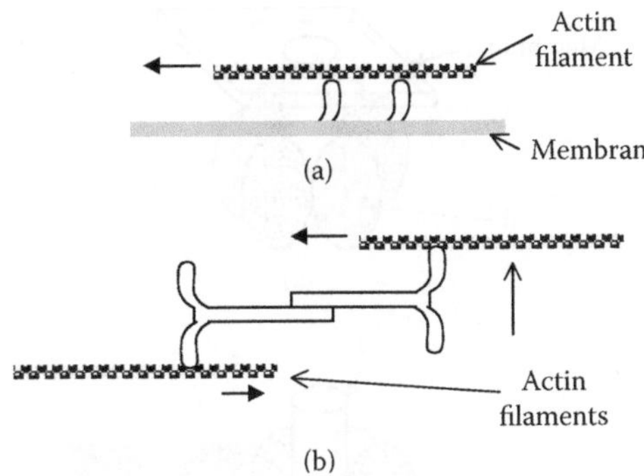

FIGURE 10.16 Possible roles of: (a) myosin I, (b) myosin II.

Myosins are positive-end directed. Figure 10.16 shows some roles for myosin in the cell.

Linear molecular motors can be made to do useful work for the nanoengineer by bonding the motor to one component and forming mictotubules on a second one. In an aqueous environment, with sufficient ATP, the motors will drive one component over the other.

The flagella motor is used by bacteria to move around in aqueous environments. The flagellum itself is composed of flagellin protein subunits, but the motor that drives it is a sophisticated rotary motor. This is shown schematically in Figure 10.17.

The smallest rotary motor so far discovered is ATPase: the molecular assembly that synthesizes ADT from ADP and phosphorous. It is driven by a pH gradient across a cell membrane; if there is no pH gradient, then the reaction reverses (ATP is converted to ADP and hydrogen ions are pumped across the membrane). The assembly consists of one γ subunit, which is inserted in the membrane, and three α and

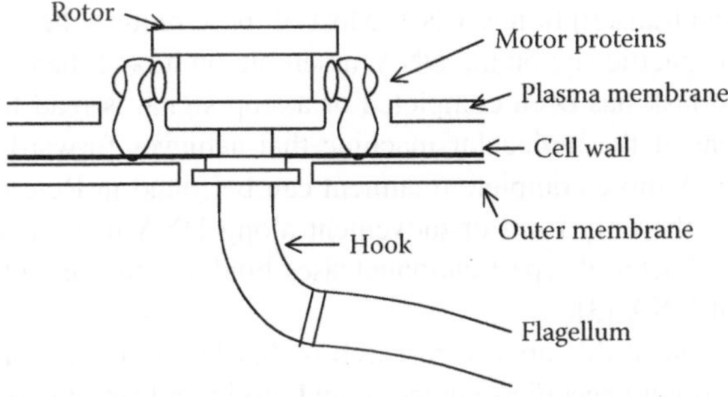

FIGURE 10.17 Flagellum motor. The motor is composed of a number of macromolecular assemblies that perform the roles that one may expect to find in a rotary motor — bearings, rotor, stator, etc. These are shown schematically.

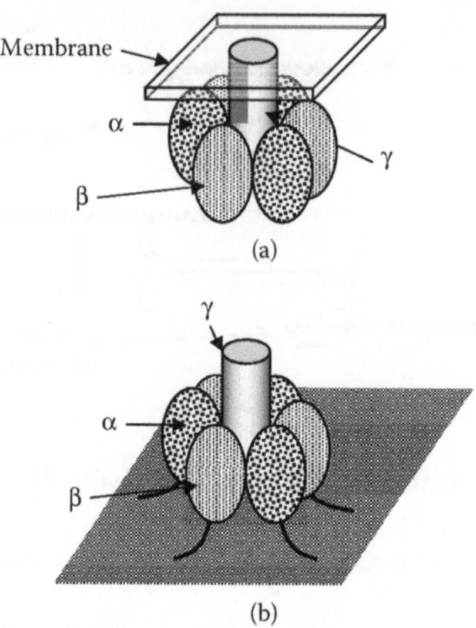

FIGURE 10.18 ATPase: (a) in cell membrane, (b) bonded to solid substrate; the α and β subunits normally revolve around the γ subunit (in (b) the α and β units are fixed, so the γ unit has to revolve).

three β subunits, which cluster around the γ subunit alternately (Figure 10.18a). When ATPase is operating, the α and β subunits spin around the γ subunit. This can be turned into an ATP-fueled rotary motor by bonding the α and β subunits to a solid substrate and allowing the γ subunit to rotate (Figure 10.18b).

10.6.4 DNA-Associated Molecular Machines

Many operations are carried out on the DNA double helix in cells, such as repair, duplication, and transcription to RNA. Most of these require the protein machine to recognize a specific site on the DNA, assemble on it, and then travel along the DNA until the task has been completed or a stop signal is read from the DNA. The component of the molecular machine that achieves forward motion is the DNA helicase. A more complete treatment can be found in Reference 2.

This is not the only form of movement along DNA that can be found. For example, some bacterial type I endonucleases bind at one site and then reel in a large length of DNA [3].

The attraction of this area of research is that DNA strands can be designed and synthesized with specific sequences, and biotin and streptavidin can be used to bind the ends to specific points. Unlike kinesins and dyenins, DNA sequences can be defined so that the binding machines assemble at specific points and then travel to other specific points where they disassemble. At the time of writing, however, this area of research is yet to be explored in detail.

10.6.5 Protein and DNA Engineering

Given that proteins with sophisticated functions are produced in nature and that the means to manufacture these through genetic engineering exist, it would appear to be worthwhile investigating how proteins could be designed to perform specific tasks. There are a number of problems associated with this.

Although proteins only have two basic structural features, α helices and β sheets, their function is dependent on how they fold into their tertiary structure and what parts are exposed or hidden (e.g., the overall pattern of surface charge). In the cell, folding into the correct shapes is often assisted by specialized proteins. At present, it is impossible to use computer analysis to predict how proteins with more than three to five amino acids will fold.

For this reason, protein engineering is limited to selecting a likely candidate, analyzing its tertiary structure (if known), identifying likely points at which to make changes, and trying these out. One is then faced with having to produce the new proteins in bacteria, which is still something of a "black art."

DNA and RNA engineering is slightly easier, because it is possible to predict which base will pair with which. It is, for example, possible to engineer single strands of DNA that close and open like tweezers because of base pairing or melting, depending on the temperature.

10.7 MOLECULAR NANOTECHNOLOGY

There are several disadvantages to assembling molecular machines from biological components found in nature, aside from the fact that they are still poorly understood as engineering materials. One problem is that they normally have to operate in chemically complex, and quite specific, aqueous environments, which slightly limits their application. A further problem is that many life-forms have evolved to look on these components as food.

Molecular nanotechnology is commonly envisioned as the diamandoid structures popularized by K. Eric Drexler [4]. This approach proposes the creation of nanomachines using carbon chemistry, but instead of the long-chain approach found in nature, the bodies of these structures are composed by carbon atoms bonded to each other in a diamond-like manner (tetrahedral bonds as seen in silicon; Chapter 2). Surface chemistry and surface charge are still important and have to be designed to provide sufficient attraction to hold the machines together and sufficient repulsion to float bearings apart. The overall appearance of these designs is similar to macroscopic and MEMS machines: bearings, gears, etc.

Although some progress has been made in this area, these designs still remain mainly as computer models. The chemistry to create them is complex, and a variety of approaches have been proposed. One popular suggestion has been the use of an AFM or STEM to enhance the chemistry by holding atoms in position. These SPMs can be used to manipulate individual atoms: a small electrical pulse applied to the probe can detach an atom from the surface and attach it to the probe tip, and reversing the polarity of the pulse places it back again. This approach has been used to write slogans with atoms and may prove useful for data storage, but the nanoassembler still seems a long way off.

10.7.1 Buckminsterfullerene

Relatively recently, two new forms of carbon were discovered. One of these was the C_{60} molecule: sixty carbon atoms bonded together in a soccer-ball-like structure. This was followed by carbon nanotubes: carbon sheets rolled up into tubular structures. Carbon nanotubes can be single walled or multiple walled (one inside another). These structures are produced by arc decomposition of carbon rods under controlled conditions.

"Bucky balls" and carbon nanotubes are still being explored in terms of electrical, optical, and mechanical properties. They have been proposed for a variety of applications. Nanotubes have been investigated as reinforcement for composite materials, elements of quantum transistors, and even AFP probe tips, for instance.

10.7.2 Dendrimers

Dendrimers are a bridge between bionanotechnology and diamondoid molecular nanotechnology. They are highly branched spherical molecules, built up from a core molecule by successive reactions of acrylic acid (Figure 10.19a) and a diamine (Figure 10.19b). Each layer of acrylic acid and diamine is referred to as a "generation." By the fifth generation the molecule has developed a fairly organized spherical structure. Each generation leaves the surface of the molecule with amine terminations and each half generation with carboxylic acid terminations (Figure 10.20 shows how a dendrimer structure builds up). This leaves considerable scope for modification of the surface chemistry, and a wide choice of diamines and cores provide structural flexibility.

Assemblies of different dendrimers are being explored for biological and medical applications, as are their optical and electronic properties.

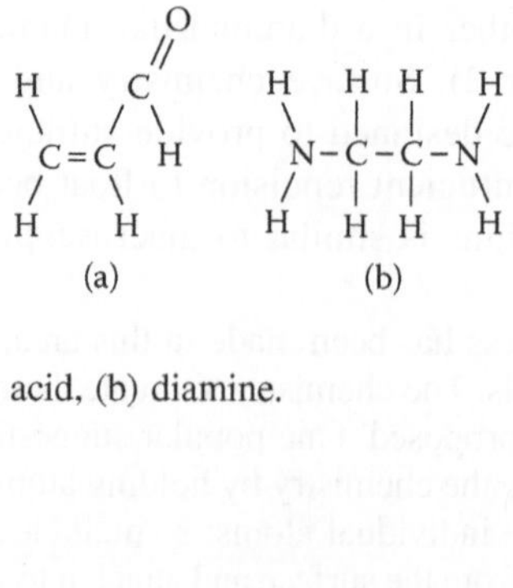

```
           O
          //
  H      C       H   H H   H
   \    / \       \  | |  /
    C=C    H       N-C-C-N
   /   \          /  | |  \
  H     H        H   H H   H
     (a)             (b)
```

FIGURE 10.19 (a) Acrylic acid, (b) diamine.

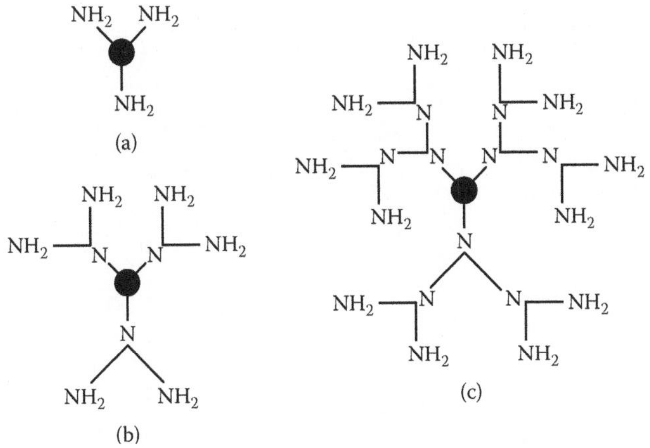

FIGURE 10.20 Synthesis of a dendrimer structure: (a) triamine core, (b) first generation, (c) second generation.

REFERENCES

1. Piner, R.D., Zhu, J., Xu, F., Hong, S., Mirkin, C.A., Dip-pen nanolithography, *Science*, 283, 661–663, 1999.
2. Alberts, B., Bray, D., Lewis, J., Raff, M., Roberts, K., Watson, J.D., *Molecular Biology of the Cell*, 3rd ed., Garland, New York, 1994.
3. Dryden, D.T.F., Reeling in the bases, *Nat. Structural Molecular Biol.*, 11, 804–806, 2004.
4. Drexler, K.E., *Nanosystems*, John Wiley & Sons, New York, 1992.

Part III

Interfacing

III.1 INTRODUCTION

Almost all MEMS devices will eventually need to be interfaced to a control system, and this will invariably be some form of electronic control system. In the laboratory, a PC with an appropriate interface card can be employed, but this will still require some additional interfacing circuitry between the card and the MEMS device. In a finished product, it is quite possible that a microcontroller will be employed, which will also require signal-conditioning circuits.

MEMS devices are often quite unusual in terms of electronic requirements. They are frequently on the high side of current and voltage requirements (hundreds of milliamperes and tens of volts) compared to modern electronics. On the other hand, this is still on the low side compared to what is termed "power electronics."

This section attempts to introduce some basic, and quite generic, electronic designs that will help when planning how individual MEMS devices will be used in practice, and will facilitate test circuit design and construction.

The treatment focuses mainly on analog electronics design, although the transistor is introduced as a switch in Chapter 11. It is, therefore, worth noting some aspects of digital control at this point.

Digital logic devices come in a number of different varieties, operated at a number of different voltage levels. However, from the interfacing point of view, both PC I/O cards and microcontrollers are relatively robust in digital terms. They normally operate with 5 V logic levels and can generally source or sink 10 mA. When CMOS logic compatibility is specified, this means that the output range is split into three parts: the lower third from 0 V upward represents a logic 0, the upper third a logic 1, and anything between these is undefined. TTL compatible logic levels are: 0 to 0.7 V logic 0, greater than 1.8 V is logic 1. Finally, if pressed for a solution, it is worth remembering that 4000 series CMOS devices are available and operate at up to 15 V, unlike most modern logic devices.

The subsequent chapters are set out as follows. Chapter 11 attempts to provide as comprehensive an introduction to basic electronics as possible within this format, ending up with operational amplifiers, the backbone device for analog electronics, and op-amp applications and filtering. Chapter 12 is entitled

"Computer Interfacing," but focuses mainly on analog-to-digital and digital-to-analog conversion. Many of the approaches employed in digital interfacing — transistor switches, relays, and optoisolators — can be found in Chapter 13, which also deals with using transistors to boost the output of operational amplifiers.

The material is presented in a manner that is intended to give a good basic grounding in electronic design and show the pathways along which solutions lie. At some point, however, it is possible that more advanced material will be required, for which specialized electronic textbooks will be needed [1,2,3].

REFERENCES

1. Horowitz, P., Hill, W., *The Art of Electronics*, 2nd ed., Cambridge University Press, Cambridge, New York, Melbourne, 1989.
2. Bogart, T.F., Beasley, J.S., Rico, G., *Electronic Devices and Circuits*, 5th ed., Pearson, 2000.
3. Lander, C.W., *Power Electronics*, 3rd ed., McGraw-Hill, London, 1993.

11 Amplifiers and Filtering

11.1 INTRODUCTION

Although there are many readily available data acquisition boards and units that can be readily interfaced to a PC, providing highly flexible analog and digital inputs and outputs, it is still quite probable that it will be necessary to use additional analog-signal-conditioning circuitry to either drive a MEMS device or to amplify and filter the signal received from a sensor. If a self-contained unit is to be created, then a suitable microcontroller can be selected, hopefully one with the required analog-input/output (I/O)-interfacing capabilities. Even so, it may still be necessary to implement filters for antialiasing and noise reduction purposes. Many of these requirements can be addressed using operational amplifiers (op-amps), and the concepts gleaned in studying these can also be employed when considering other problems.

Before introducing this topic, it may be useful for the uninitiated to consider the following very brief introduction to fundamental electronic components (or elements).

11.1.1 Quick Introduction to Electronics

For the purposes of an introduction and a very brief analysis, we consider idealized electronic components. The symbols employed for these components are shown in Figure 11.1. For brevity, the ideal component will be described along with its time and frequency response characteristics. The frequency response assumes an applied sinusoidal signal with frequency of ω rad/sec (1 Hz is 2π rad/sec). Note that this is only for sinusoidal signals.

The lowercase letter j is used to represent the square root of 1. This is standard practice in electronics, where the letter i, which is used for the same purpose in mathematics and physics, refers instead to electric current.

11.1.1.1 Voltage and Current Conventions

In practice, voltage is measured with respect to one point (in the circuit), i.e., the ground (GND) or 0-V line. When considering individual components, the concept of a voltage drop is introduced. The voltage drop is the voltage (or potential, giving rise to the more ambiguously named "potential difference") measured at one end of the component (usually marked with a – symbol) subtracted from the voltage measured at the other end of the component (usually marked with a + symbol); see Figure 11.2 for examples of both positive and negative voltage drops. The voltage drop between two arbitrary points in a circuit (A and B) is written as V_{AB}.

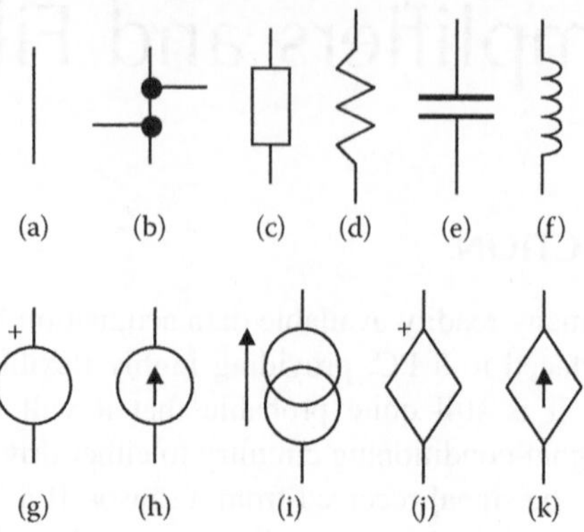

FIGURE 11.1 Component symbols: (a) wire, (b) wire with connections, (c) resistor, (d) alternative resistor symbol, (e) capacitor, (f) inductor, (g) voltage source, (h) current source, (i) alternative current source symbol, (j) controlled voltage source (may be replaced by a voltage source with an equation written next to it), (k) controlled current source (may be replaced by a constant-current source symbol with an equation written next to it).

Less commonly, the voltage drop across a particular component is written with the name of the component as the subscript; for instance, the voltage dropped across R1 is V_{R1}. Note that this does not indicate the polarity of the voltage drop.

Voltage can be considered as the electrical driving force, driving current around the circuit.

Units

- Voltage is measured in volts — V
- Current is measured in amperes (amps) — A

Conventional current flows from the most positive part of the circuit (highest voltage) to the most negative (lowest voltage), and is indicated by the letter I.

The direction in which current flow is assumed is indicated by an arrow on the circuit diagram. A negative value for a current, obtained from a series of calculations,

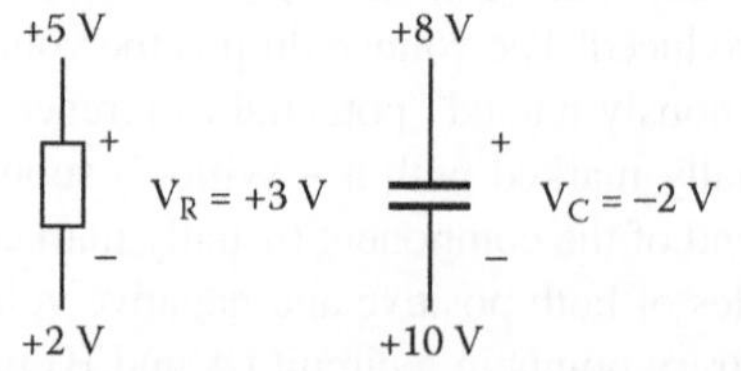

+5 V
+
$V_R = +3$ V
−
+2 V
+8 V
+
$V_C = -2$ V
−
+10 V

FIGURE 11.2 A voltage drop of +3 V across a resistor and 2 V across a capacitor.

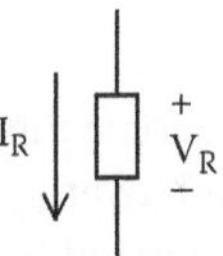

FIGURE 11.3 Voltage and current marking for a resistor.

means that actual current flow will be in the direction opposite to the marked arrow. Figure 11.3 shows conventional marking for voltage drop and current flow across a resistor element. (Note that it is not absolutely necessary for the arrow to be marked from + to –, as long as any departures are reflected in the equations used to analyze the circuit. Students unfamiliar with this topic are recommended to follow the conventions.)

> **DC and AC**
>
> The distinction is often made between a direct current and an alternating one. This is a misnomer, as DC is usually used to refer to a constant voltage — one that does not vary over time. AC is usually applied to refer to a voltage that varies in a sinusoidal manner over time.
>
> - DC values are normally written using capital letters, V and I to represent voltage and current.
> - AC values and transient signals are normally written using lower case letters, v and i to represent voltage and current.

11.1.1.2 The Ideal Conductor and Insulator

Figure 11.1a and Figure11.1b show the ideal wire that is used to connect individual circuit elements in circuit diagrams. It allows electrical current to flow unimpeded, so any amount of current can flow and yet no voltage drop will be developed across the wires at any time (this condition may also be referred to as a "short circuit").

One may consider the converse, the ideal insulator: any value of voltage may be placed across it and yet no current will flow (this may also be termed an "open circuit"). It is useful to visualize these conditions, as a good grasp of them can assist in the understanding of various problems.

11.1.1.3 The Ideal Resistor

The ideal resistor element is shown in Figure 11.1c and Figure 11.d. The current flowing through it is related by Ohm's law to the voltage dropped across it. Both the time and frequency domain equations are identical:

$$v = iR \quad \text{time and frequency domain} \tag{11.1}$$

11.1.1.4 The Ideal Capacitor

The ideal capacitor (see Figure 11.1e for symbol) is a component that stores charge. Its basic physical form is two parallel, overlapping, conducting plates separated by a narrow insulating gap. It is characterized by the following equations:

$$v = \frac{1}{C}\int i dt \quad \text{time domain} \tag{11.2}$$

$$v = i\frac{1}{j\varpi C} \quad \text{frequency domain} \tag{11.3}$$

Notice that Equation 11.3 resembles Equation 11.1, but with the R replaced by $1/j\omega C$. When performing frequency domain analysis, it is often convenient to transform the entire circuit into the frequency domain by replacing C with $1/j\omega C$ and then essentially applying Ohm's law to derive the equations.

Impedance, Resistance, and Reactance

Complex numbers in the frequency domain are referred to as *impedances*. An impedance (complex number) is made up of a real (or *resistive*) part and an imaginary (or *reactive*) part. It has already been seen that the reactive component may be a capacitance (i.e., a capacitive reactance), an inductance (see next section), or a combination of both.

From Equation 11.2 it can be seen that an infinite current is required to cause an instantaneous change in voltage. This is impossible in practice, so it will always take time to charge or discharge capacitors. (Note that this illustrates one of their applications: smoothing out unwanted oscillations in voltage supply lines.)

From Equation 11.3 it can be seen that at high frequencies (large ω), the capacitor appears to be a short circuit. For low frequencies (DC conditions), the capacitor appears to be an open circuit. (Note that the frequency domain analysis works only for sinusoidal signals.)

11.1.1.5 The Ideal Inductor

Inductors (Figures 11.1f) store energy in a magnetic field. They are commonly wound coils. The ideal inductor can be described by Equation 11.4 and Equation 11.5:

$$v = L\frac{di}{dt} \quad \text{time domain} \tag{11.4}$$

$$v = ij\omega L \quad \text{frequency domain} \tag{11.5}$$

Notice that an instantaneous change in current requires an infinite voltage drop across the inductor. If an attempt is made to switch an inductive load (a circuit that is being electrically powered and contains an inductive element) off, the current has to go somewhere. In the absence of any provision made for this, it will break down the insulation of the switch, causing a spark (arc).

Also note that the inductor is a short circuit to DC and an open circuit at high frequencies.

More Units

- Resistance is measured in ohms — Ω
- Capacitance is measured in farads — F
- Inductance is measured in henries — H

11.1.1.6 The Ideal Voltage Source

The ideal constant-voltage source (Figure 11.1g) will maintain a constant voltage drop across its terminals, no matter what. This means current may flow into or out of either terminal, and it can even supply the practically impossible infinite current necessary to charge a capacitor instantaneously. (This makes it somewhat similar to an ideal wire with a constant voltage drop placed across it.)

11.1.1.7 The Ideal Current Source

The ideal constant-current source (Figure 11.1h and Figure 11.1i) will maintain a constant current flow in the direction indicated by the arrow, no matter what voltage drop appears across it. (This makes it somewhat similar to an ideal open circuit with a constant current flowing through it.)

The previous descriptions may be a little counterintuitive (current flowing through an open circuit?), but they are useful. Deeper consideration of the concepts introduced in this section will reveal other useful qualities or analogies, some of which will become apparent later (see the book by Bogart for a complete introduction to the subject [1]).

Note on Strange-Seeming Analogies

In the case of the ideal current source, it can be found in Subsection 11.1.1.2 that no current will flow through an open circuit irrespective of the voltage dropped across it. Turning the analogy around, one could argue that the open circuit is a special case of the ideal current source, one with a 0-A current rating.

11.1.1.8 Controlled Sources

Often, the voltage or current in one part of a circuit is mathematically related to a voltage or current in another. This is represented in the circuit diagram by the controlled voltage and current sources (Figure 11.1j and Figure 11.1k). In this case, the voltage appearing across the terminals (or current flowing through the component) is mathematically related to some other parameter in the circuit — the relationship normally being noted next to the source. Otherwise, these ideal components exhibit the same properties as the constant sources mentioned before. Controlled sources appear in models for amplifying components, where the output is some multiple of the input, for example.

11.1.1.9 Power Calculations

Power can be computed by multiplying the voltage drop across a circuit element by the current passing though it:

$$p = iv \tag{11.6}$$

In the case of a resistor (or resistive load), this equation can be readily employed to yield the power dissipated by the component. Ohm's law can be employed to transform Equation 11.6, in which case it will be seen that the power dissipated is equal to the square of the current multiplied by the value of the resistor; one may conclude that reducing the current flowing through a resistive element is most effective in reducing the power dissipated. In the case of reactive loads (capacitive or inductive loads), the same equation may be employed, but the power will depend on the waveform being applied.

If Equation 11.6 is applied to voltage or current sources, it will normally yield a negative result — the voltage drop across the circuit element being in opposition to current flow. This indicates that the source is supplying energy to the circuit, not dissipating it. As capacitors and inductors both store energy, they extract energy from the circuit (positive value) during one part of the cycle and deliver it (negative value) during another part of the cycle (the signs noted in parentheses being correct if normal sign conventions have been followed).

11.1.1.9.1 Switching Losses

The ideal switch acts either as a short circuit (on state) or open circuit (off state). In either state, no power would be dissipated by the switch. The nonideal switch (in this case the transistor can be considered when employed in a switching capacity) would appear as a very-low-value resistor when in the on state and a very high value resistor when in the off state. It would therefore dissipate some power when in either (static) state. During the act of switching, however, the current rises (or falls) as the voltage dropped across the switch falls (or rises). At high switching frequencies, this dynamic power dissipation (power dissipated during a change of state) is typically greater than the static power dissipation. This is particularly true when highly capacitive loads are involved because the time taken to charge and discharge the capacitance through the (ideal and non-ideal) resistances in the circuit and, hence, the time taken for the switch to open or close will be relatively long.

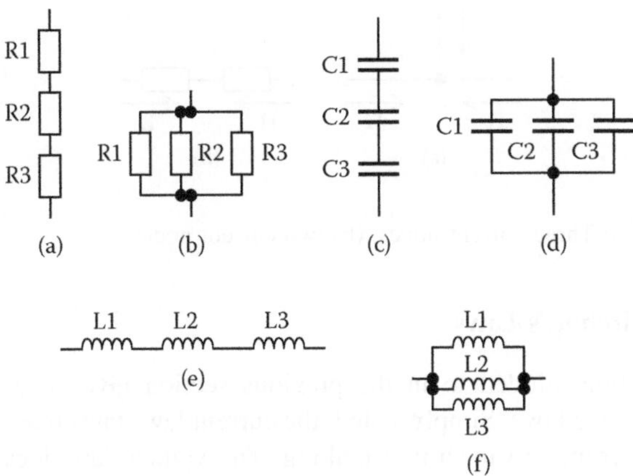

FIGURE 11.4 (a) Resistors in series, (b) resistors in parallel, (c) capacitors in series, (d) capacitors in parallel, (e) inductors in series, (f) inductors in parallel.

11.1.1.10 Components in Series and Parallel

Component elements may be placed in series (Figure 11.4a, Figure 11.4c, and Figure 11.4e) or parallel (Figure 11.4b, Figure 11.4d, and Figure 11.4f). Consideration of Equation 11.1 to Equation 11.5 and Figure 11.4 will show that several instances of the same component connected together in such a network will have the same combined effect as a single instance of that component of a value computed as follows:

$$R_{TOTAL} = R1 + R2 + ... + Rn \quad \text{resistors in series} \tag{11.7}$$

$$R_{TOTAL} = \frac{1}{\frac{1}{R1} + \frac{1}{R2} + ... + \frac{1}{Rn}} \quad \text{resistors in parallel} \tag{11.8}$$

$$C_{TOTAL} = \frac{1}{\frac{1}{C1} + \frac{1}{C2} + ... + \frac{1}{Cn}} \quad \text{capacitors in series} \tag{11.9}$$

$$C_{TOTAL} = C1 + C2 + ... + Cn \quad \text{capacitors in parallel} \tag{11.10}$$

$$L_{TOTAL} = L1 + L2 + ... + Ln \quad \text{inductors in series} \tag{11.11}$$

$$L_{TOTAL} = \frac{1}{\frac{1}{L1} + \frac{1}{L2} + ... + \frac{1}{Ln}} \quad \text{inductors in parallel} \tag{11.12}$$

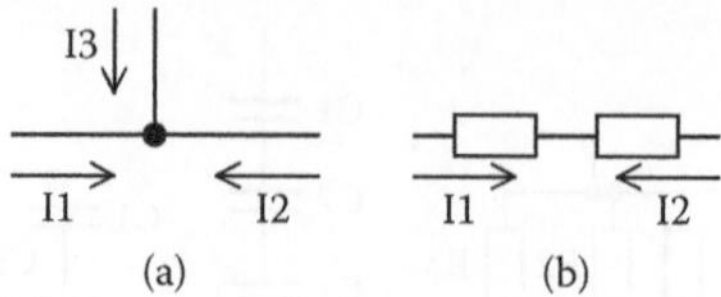

FIGURE 11.5 (a) Three current nodes, (b) two current nodes.

11.1.1.11 Kirchoff's Laws

The consideration alluded to in the previous section gives rise to Kirchoff's current and voltage laws. Simply stated, the current law states that current cannot simply appear from or vanish into nothing. The voltage law does the same for voltages.

Kirchoff's current law, more formally, states that the sum of currents flowing into a node (a junction between components) must be zero. In Figure 11.5a:

$$I1 + I2 + I3 = 0 \tag{11.13}$$

It should be recalled that when analyzing a circuit and computing the actual value of a current, a positive result means that the current actually flows in the direction indicated, i.e., into the node, and a negative result will indicate that the current in fact flows in the direction opposite to that indicated, i.e., out of the node. Figure 11.5b shows the junction between just two components, two resistors, marked up for analysis such that:

$$I1 + I2 = 0 \tag{11.14}$$

Clearly, at any instant that the circuit is in operation the current will flow into the node through the one resistor and the same current will flow out through the other (unless, of course, no current at all flows). This is the same as saying that the same current flows through both resistors, and Kirchoff's current law formalizes this for multiple currents (components).

Kirchoff's voltage law states that the sum of voltage drops around a closed path in a circuit must be equal to zero. This is explained in Figure 11.6. Considering Figure 11.6a, it is apparent that the voltage dropped across the circuit enclosed by the box, V_{AB}, must be equal to the source voltage (V_{SOURCE}). Starting at node A in the circuit and working clockwise, applying Kirchoff's voltage law, it can be seen that:

$$V_{AB} - V_{SOURCE} = 0 \tag{11.15}$$

Notice that the source voltage appears with a negative sign in the equation. This is because in working clockwise around the circuit the voltage dropped

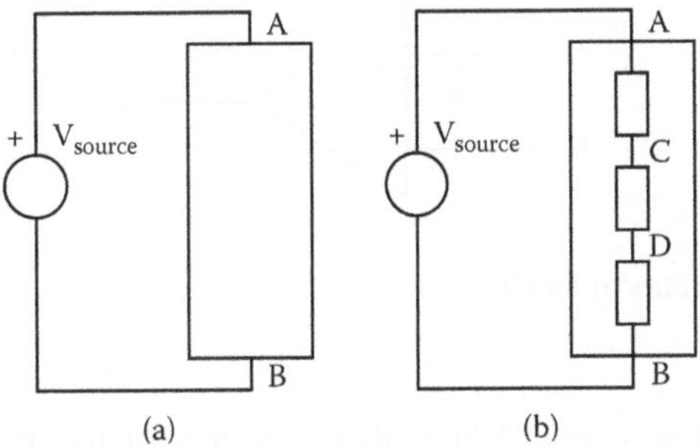

FIGURE 11.6 (a) Voltage source and box, nodes A and B, (b) box expanded to show series of resistors.

across the source is being considered from node B to node A, which is in the opposite sense to the polarity marked for the source. Rearrangement of Equation 11.15 will show what was suspected all along, that $V_{AB} = V_{SOURCE}$.

The box in Figure 11.6a in fact contains three resistors in series (Figure 11.6b). It now makes more sense to apply Kirchoff's voltage law, giving:

$$V_{AC} + V_{CD} + V_{DB} - V_{SOURCE} = 0 \tag{11.16}$$

or

$$V_{R1} + V_{R2} + V_{R3} - V_{SOURCE} = 0 \tag{11.17}$$

Kirchoff's Laws

- The sum of currents flowing into a node is zero.
- The sum of voltage drops around a closed path is zero.

The statement of Kirchoff's laws within this section is not the formal definition but a form that the author has found to be of most practical use in circuit analysis.

11.2 OP-AMP

The op-amp is the most convenient circuit element to employ in many analog circuits. The symbol for the op-amps is shown in Figure 11.7. It has three terminals, a noninverting input (marked with a plus sign), an inverting input (marked with a minus sign), and an output. The component itself will have two power supply connections; these are usually to the symmetrical positive and negative supply rails (+15 V and −15 V, for instance). Some have additional

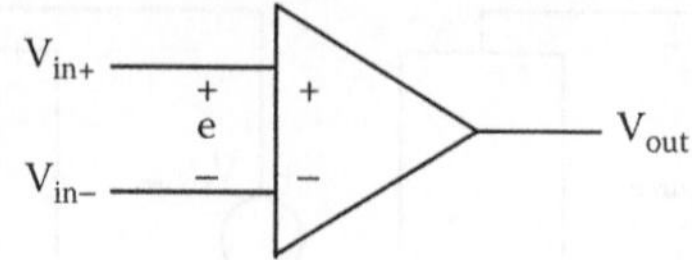

FIGURE 11.7 Operational amplifier.

connections to allow certain (internal) parameters of the amplifier itself to be adjusted — an offset null being common.

The difference between the voltage applied to the noninverting input (V_{IN+}) and the inverting input (V_{IN}) is commonly termed the *error voltage*, marked as e in Figure 11.7. The op-amp has a very high open-loop gain, A_{OL}, and the output voltage is related to the input error voltage by the formula:

$$V_{OUT} = e \times A_{OL} \tag{11.18}$$

Open-loop gains in the region of 200,000 or greater are not unusual.

Gain, Transconductance, and Transimpedance

The gain of a circuit is the ratio of the output parameter of interest to the input parameter of interest. Normally this will be either the voltage gain (output voltage/input voltage) or current gain (output current/input current); as a consequence it is represented by a dimensionless number. Sometimes it will be the ratio of the output current to the input voltage or the output voltage to the input current. In the former of these two cases, it is commonly referred to as the transconductance and normally measured in siemens, the unit of conductance (the reciprocal of resistance). In the latter, it is normally referred to as transimpedance and measured in ohms. Examination of Equation 11.1 will make it apparent as to why these terms and units are relevant.

11.2.1 The Ideal Op-Amp

The ideal op-amp is principally characterized as having:

- infinite open-loop gain
- infinite input impedance
- zero output impedance

Additional points can be appended to this list. Suffice it to say that the two inputs do nothing more than sense the voltages appearing at them, and the output acts as a perfect voltage controlled voltage source.

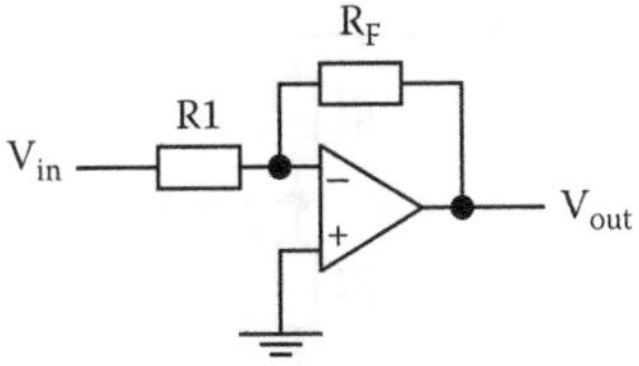

FIGURE 11.8 Inverting amplifier.

The issue of infinite open-loop gain appears nonsensical, but, in practice, it means that for a noninfinite V_{OUT}:

$$e \to 0 \quad \text{as} \quad A_{OL} \to \infty \tag{11.19}$$

This enables an amplifier with an arbitrary closed-loop gain to be constructed using an op-amp and two external resistors. Figure 11.8 shows an op-amp in inverting configuration with negative feedback (resistor R_F being the feedback resistor).

If an ideal amplifier is assumed, no current flows into or out of the inputs (infinite input impedance). Thus, by applying Kirchoff's current law it can be seen that the current flowing through R_F also flows through R1. Furthermore, given that the ideal op-amp has infinite open-loop gain, from Equation 11.19 it can be seen that the voltages at both input terminals of the amplifier are the same. As one of these (the noninverting input) is connected to 0 V (or ground, symbolized by the three diminishing lines; the symbol is not normally labeled), it follows that the voltage at the inverting input must also be 0 V. Applying Kirchoff's current law then gives:

$$\frac{V_{OUT} - 0}{R_F} + \frac{V_{IN} - 0}{R1} = 0 \tag{11.20}$$

This can be rearranged to give the classic equation for the closed-loop voltage gain (A_{CL}) of an ideal op-amp in inverting configuration (Figure 11.8):

$$A_{CL} = \frac{V_{OUT}}{V_{IN}} = -\frac{R_F}{R1} \tag{11.21}$$

The gain of the resulting amplifier is set by the ratio of the two resistors, and it is called an *inverting amplifier* because of the negative gain.

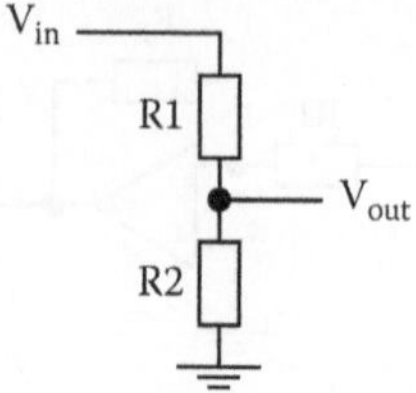

FIGURE 11.9 Potential divider.

The Potential Divider

Although gains of less than unity can be achieved with an inverting amplifier, a signal that is too large can be more economically attenuated (reduced in amplitude) by a potential (or voltage) divider. This consists of two resistors in series (Figure 11.9), characterized by the equation: $V_{OUT} = V_{IN} \frac{R2}{R1+R2}$

A similar analysis can be applied to the noninverting configuration of Figure 11.10, yielding:

$$\frac{V_{OUT}}{V_{IN}} = 1 + \frac{R_F}{R1} \tag{11.22}$$

Because the input signal is connected directly to the very-high-impedance noninverting input of the op-amp in the noninverting configuration and the gain will be at least unity, it is common to see it used as a unity-gain buffer amplifier, the inverting input being connected directly to the output with no resistors involved.

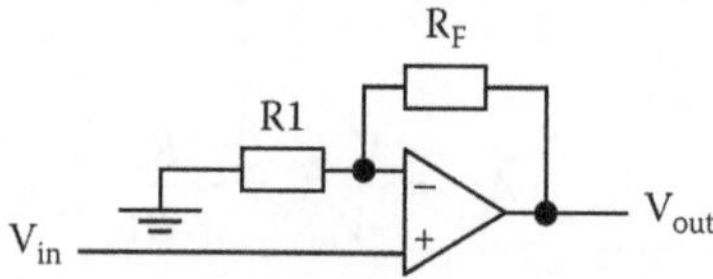

FIGURE 11.10 Noninverting amplifier.

Decibels

Gains are often expressed in decibel units. These are formally defined in terms of power:

$$\text{gain in dB} = 10\log_{10}\left(\frac{P_{out}}{P_{in}}\right)$$

It is, however, more common to find voltage gains expressed in dB:

$$\text{gain} = 20\log_{10}\left(\frac{V_{out}}{V_{in}}\right)$$

Filter cutoff points (Section 11.3 of this chapter) are taken to be the frequencies at which attenuation is by a factor of 3 dB. Beyond this, attenuation may well increase by a factor that is a multiple of –20 dB per decade (i.e., the signal is attenuated by an additional 20 dB, 40 dB, 60 dB, etc., for every factor-of-ten increase in frequency, depending on the design of the filter).

Consideration of the concepts presented in the previous sections of this chapter will show that if the object is to measure a voltage that arises from a source that is not an ideal voltage source (i.e., it could be modeled as an ideal voltage source — the signal that is to be measured — within a network incorporating at least one series resistor and possibly capacitive and inductive elements as well), then connecting this source to an input with a finite impedance will degrade the signal; exactly what will happen will depend on the nature of the source and the input impedance of the measuring circuit.

Power Transfer

In many cases, it is required to transfer maximum power from the source to the input of the measuring circuit. In this case, the input impedance of the measuring circuit should be the same as the output impedance of the source.

11.2.1.1 Nonideal Sources, Inverting, and Noninverting Op-Amp Configurations

Op-amps are used to condition signals from sensor elements. For the most part, these are not going to be ideal voltage or current sources. Most nonideal voltage sources can be modeled as an ideal voltage source, Vs, in series with a source

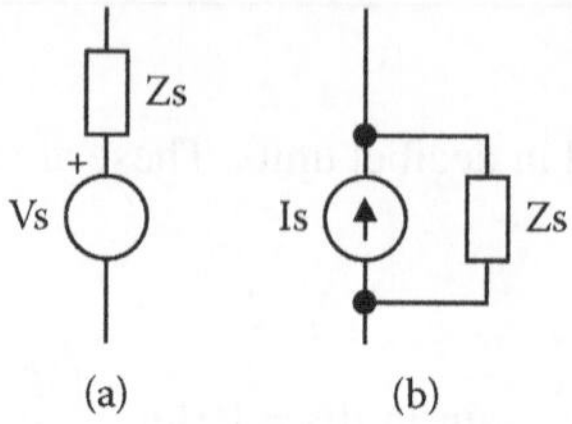

FIGURE 11.11 (a) Nonideal voltage source, (b) nonideal current source.

impedance, Zs, as illustrated in Figure 11.11a. The nonideal current source is illustrated in Figure 11.11b. For simplicity, only a nonideal voltage source with a resistive source impedance (source resistance, Rs) will be considered, as illustrated in Figure 11.12. This is a very common model for a signal source.

Returning to Figure 11.10, it can be seen that the input signal is connected directly to the high-impedance input of the op-amp. When the nonideal source of Figure 11.12 is connected to this configuration (Figure 11.13a), it is apparent that no current, ideally, flows through Rs. It follows that there is no voltage drop across Rs, and therefore V_{IN+} = Vs. In other words, the ideal circuit amplifies the signal of interest, Vs, according to Equation 11.22.

Turning to Figure 11.8, it is apparent in the inverting amplifier configuration, the source "sees" one end of R1. It has already been noted that in this case, V_{IN} will be the same as V_{IN+}, which in this case is tied to 0 V. The source sees the resistor, R1, connected to ground. In Figure 11.13b, it can be seen that Rs and R1 form a voltage divider (see box titled "The Potential Divider") that attenuates the signal of interest before it can be amplified. This has noise implications, as will be discussed later, as well as design implications.

The overall gain of the circuit will, therefore, be dependent on Rs. As this is beyond the designer's control, the noninverting configuration is preferred. The noninverting configuration is also used to buffer voltage dividers and filters constructed using passive components (capacitors, inductors, and resistors) before the signals enter the next stage of a circuit.

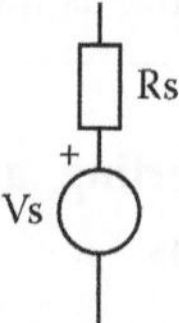

FIGURE 11.12 Nonideal voltage source with the source impedance, Zs, replaced by a resistance, Rs.

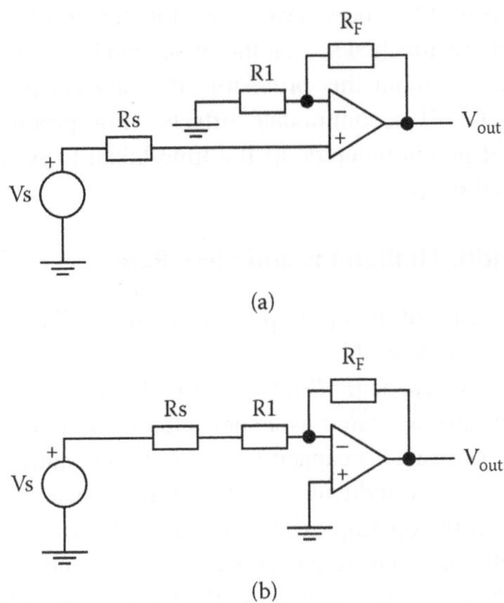

FIGURE 11.13 Nonideal voltage source (Figure 11.12) connected to: (a) noninverting amplifier, (b) inverting amplifier.

11.2.2 Nonideal Op-Amps

Op-amps must be supplied with appropriate power rails. One obvious problem with this is that the output voltage cannot swing beyond those of the rails. The circuit must be designed and supplied with appropriate power rails so that under normal (expected) operating conditions the outputs do not saturate — that is, the output does not drive to go beyond the power rails. Common situations in which this may occur are:

- At the input stage, when a signal is buried in noise that will be filtered out in later stages. If the gain is such that the amplified noise causes the output to saturate, then the input signal will be lost.
- Where a small AC signal is superimposed on a large DC offset, in such a case an AC-coupled amplifier may be employed (see following text).

A further limit on the output stage of the op-amp is that it can only supply (source or sink — source refers to currents flowing out of the device and sink to currents flowing in) limited current (±10 mA, for example). This limitation will be dealt with in Chapter 13.

Op-amps also exhibit an offset voltage, i.e., a nonzero output when the output is expected to be 0 V. This is due to asymmetries within the circuitry itself, as it is never possible to achieve a perfect match between two components. Some op-amps are designed to have tolerably low offsets, others may be laser-trimmed to minimize offsets, and still others have provision for an external potentiometer (variable resistor) to be attached so that the offset can be adjusted once the circuit has been built. Even without this provision, it is always possible to design an external circuit that will accommodate offsets. (For production versions, the design should avoid potentiometers, as the time taken to set them up adds considerably to the final costs).

11.2.2.1 Bandwidth Limitations and Slew Rate

The incredibly high gain of the op-amp leaves it susceptible to instability, unless something is done to prevent this.

When discussing the capacitor (Subsection 11.1.1.4), it was noticed that at high frequencies the capacitor appeared as a short circuit. In an electrical circuit there are lots of parasitic (unwanted) capacitors where two conductors run near to one another. There may also be inductive loops that allow one signal to couple magnetically to another. If the op-amp exhibited its incredibly high gain at all frequencies, it is quite likely that high-frequency signals at the output would couple back to the input through these routes and cause the whole system to oscillate.

Therefore, op-amps normally only operate at their maximum open-loop gain up to a few hertz (1 Hz to 10 Hz is typical). Thereafter, the open-loop gain of the amplifier drops off at, usually, about 20 dB per decade. The frequency at which the gain becomes 1 (i.e., unity) can be found from the gain–bandwidth product listed in the amplifier's data sheet. Typically, this will be in the 1 MHz to 10 MHz range.

The op-amps chosen must have sufficient open-loop gain at the frequency of interest for it to operate. As the open-loop gain approaches the closed-loop gain, the op-amp will cease to approximate an ideal one and the design equations discussed previously will fail. Most modern op-amps will operate at moderate closed-loop gains up to frequencies in the 100-kHz range or above, and amplifiers with hundreds of megahertz bandwidth are available, although special care has to be taken when designing circuits that involve such high frequencies.

It may sometimes be found that although an op-amp with sufficient gain–bandwidth product has been selected, the circuit still fails to function as expected, with the output waveform being distorted in shape — particularly for square and other non-sinusoidal waves. This is due to another limiting parameter of the nonideal op-amp, the slew rate. This is the maximum rate at which the output of the amplifier can change and is usually expressed in volts per microsecond. It can be thought of as the op-amp's ability to track a fast-moving input signal.

Gain–Bandwidth Product and Slew Rate

- The gain–bandwidth product (in Hz) gives the frequency at which the open-loop gain of the op-amp drops to unity.
- The slew rate (in V/μsec) is the maximum rate of change that the op-amp output is capable of.

Slew rate problems are commonly encountered when the op-amp is being asked to drive a rapidly changing signal over a significant part of its output voltage range. The following example illustrates the difference between slew rate and gain–bandwidth product.

Consider a circuit with a closed-loop gain of 100, supplied by ±12-V power rails. The op-amp chosen has a gain–bandwidth product (also known as f_t) of 10 MHz and a slew rate of 1 V/μsec. It is intended to amplify a 100-kHz sinusoidal signal.

Working backwards from f_t at 10 MHz, assuming a 20 dB per decade change in open-loop gain (up to the maximum rated open loop gain), it can be seen that the open-loop gain will be 10 at 1 MHz and 100 at 100 kHz, ample for this application. If the input signal is 0.1 V in amplitude (0.2 V peak to peak), then the output signal will be 1 V in amplitude, and it is found that this is the case in practice.

If, instead, a 1-V amplitude signal is applied to the circuit, then it would be logical to expect a 10-V amplitude signal to appear on the output — the gain–bandwidth product is sufficient, as are the power rails. In practice, however, this is not so due to the slew rate limitation.

With a 0.1-V amplitude input signal, the output swings from 0 to 1 V in one quarter of a cycle, and it changes by 1 V every 2.5 μsec (or 0.4 V/μsec). This is within the maximum 1 V/μsec slew rate rating. With the 1-V amplitude signal applied to the input, however, the output is forced to swing from 0 to 10 V every quarter cycle (i.e., 10 V every 2.5 μsec or 4 V/μsec). The output stage of the amplifier simply cannot keep up, so although there is sufficient open-loop gain available, the slew rate limits the output performance.

11.2.2.2 Input Impedance and Bias Currents

Although op-amps do have finite input impedances, these are usually incredibly high, and circuit design can boost these further. Some op-amps have rated input resistances that approach, or are in excess of, the resistance between the two pins of the socket into which the chip may be inserted. Where input impedance is of concern, it will attenuate signals in a potential divider effect. The input capacitance of op-amps may be of greater concern when considering high frequencies, or even high-impedance or high-capacitive sources. Data sheets of precision op-amps (e.g., the OPA111) usually provide design guidelines to maximize the performance of the integrated circuit (IC) in various circumstances.

Another factor at the input stage that may be of concern in bioMEMS and sensing applications, but is not often dealt with in textbooks on electronics, is the question of input bias currents. These are seen in electronics more as an annoyance, part of the nonideal aspect of the circuit, i.e., something that has to be compensated for but not something that could damage a system. In some biochemical situations, the passage of unknown currents, no matter how small, could not only cause erroneous results but could also degrade (micro)electrodes through unpredictable electrochemical activity. They, therefore, need to be understood and dealt with.

TABLE 11.1
BJT, JFET, and MOSFET Input Stages Compared

Input	Noise	Bias Currents
BJT	Low noise	High bias currents
JFET	Medium noise	Medium- to high-bias currents
MOSFET	Highest noise	Very low bias currents (fA) possible

Input bias currents arise because the transistors in the input stage of the amplifier need small currents in order to operate correctly. There are three types of input stages: bipolar junction transistor (BJT), junction field effect transistor (JFET), and metal-oxide-semiconductor field effect transistor (MOSFET). Their selection is summarized in Table 11.1.

The best op-amps for most sensing applications will be precision BJT amplifiers with either JFET or MOSFET input stages, depending on the input impedance and bias current requirements. Once past the initial stage, however, the requirements can usually be relaxed and cheaper op-amps can be used (the 741 op-amp, for example, is an old workhorse and very well known, often, the performance of other op-amps are summarized by the way they relate to the 741).

11.2.2.3 Common-Mode Rejection Ratio and Power Supply Rejection Ratio

Two further parameters that are often quoted with respect to amplifier performance are the common-mode rejection ratio (CMRR) and power supply rejection ratio (PSRR).

Op-amps and some other forms of amplifier have both inverting and noninverting inputs. If the same signal were to be applied simultaneously to both inputs, then the output expected would be zero. This is important in sensing applications in which remote signals (radio signals or the main power supply, for example, the latter appearing as an interfering signal at 60 Hz, or 50 Hz in Europe) couple into the leads attaching the sensor to the amplifier and interfere with the signal of interest. If the two leads follow exactly the same path, then the same signal should be induced in both leads and rejected by the amplifier. The CMRR of the amplifier is a measure of its ability to perform this. Instrumentation amplifiers, dealt with in a following section, are designed to have high CMRRs.

The power rails supplying an amplifier should ideally be absolutely smooth and without any variations from their specified voltages. This is practically impossible to achieve, as the switching or drawing of large currents in one part of the circuit will be reflected by glitches and variations in the power rails in other parts of the circuit. The PSRR is a measure of the amplifier's ability to ignore such variations.

These and many other nonideal aspects of amplifiers are dealt with in textbooks on electronics in greater depth. Particularly recommended to the reader is *The Art of Electronics* by Horowitz and Hill, an excellent and comprehensive textbook [2].

11.2.3 Noise

There are many different sources of noise in electronic circuits that corrupt the signals being measured or the accurate output of driving signals. Many noise sources are characterized as *white noise*. This is essentially a random signal, and is characterized by having equal signal power at all frequencies. Notice that this means that a true white noise signal would have infinite power, so such a thing does not exist. Nevertheless, it is a convenient model to use.

The amplitude of a white noise source is given as its root mean square (rms) amplitude, called the noise density.

Root Mean Square

This is the value used in power calculations for AC signals (e.g., $P = \frac{V_{rms}^2}{R}$). It is calculated as the name indicates. For any signal, square the required parameter (multiply it by itself), compute the mean (the average level over one period or cycle), and then calculate the square root of this value.

11.2.3.1 Combining White Noise Sources

When working out the overall performance of a system, it is necessary to combine several noise sources. Assuming that the signals are uncorrelated, that is to say that they have no mathematical relationship to one another, and as they are represented by rms amplitudes, they cannot be combined by simple addition. To combine two or more noise sources, the noise densities are squared, and the square root of the sum is taken to give the noise density of the combined source. Consider, for example, the two voltage noise sources in Figure 11.14. Here, the equivalent total noise density is given as:

$$V_{NTOTAL} = \sqrt{V_{N1}^{\,2} + V_{N2}^2} \tag{11.23}$$

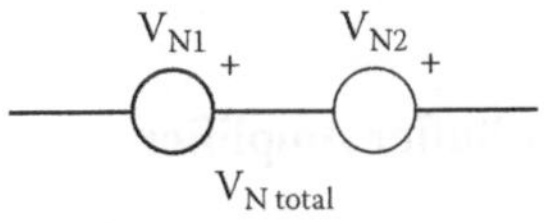

V_{N1} + V_{N2} +

$V_{N\,total}$

FIGURE 11.14 Two white (voltage) noise sources combined, note that the polarity (+) marks make no sense with noise sources; they are included here to indicate voltage noise sources.

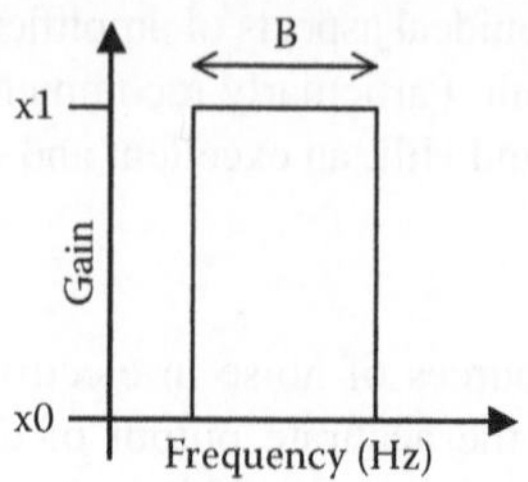

FIGURE 11.15 Ideal filter response for noise bandwidth calculations.

11.2.3.2 Thermal Noise

The white noise (or Johnson noise) exhibited across the terminals of an open circuit resistor is given by:

$$V_n^2 = 4kTBR \tag{11.24}$$

where V_n is the noise voltage; k Boltzmann's constant (1.38×10^{23} J K^{1}); T, the absolute temperature in Kelvin; R, the resistance; and B, the equivalent noise bandwidth. The noise source will be bandwidth limited, and the noise bandwidth is that of a perfect bandpass filter through which the noise will theoretically, be measured (Figure 11.15 shows the response of an ideal filter). For a 1-k resistor at room temperature (293 K), the noise voltage is given as:

$$V_n = \sqrt{4 \times 1.38 \times 10^{-23} \times 293 \times B \times 1000}\ \text{V}_{\text{rms}} \tag{11.25}$$

or

$$\frac{V_n}{\sqrt{B}} \approx 4.02\ \text{nV}/\sqrt{\text{Hz}} \tag{11.26}$$

As the bandwidth is not known until the component is placed into a circuit, noise performance is often quoted on data sheets and elsewhere in units of V/√Hz.

11.2.4 Op-Amp Applications

A variety of different op-amp circuits are illustrated in Figure 11.16. Of these, the inverting configuration and noninverting configurations have been dealt with in Subsection 11.2.1.

11.2.4.1 The Unity-Gain Buffer Amplifier

From Equation 11.22, it can be seen that a noninverting amplifier configuration must have a closed-loop gain of at least 1. Moreover, the input impedance is the very high input impedance of the op-amp itself (in the case of the inverting configuration, it is equivalent to the input resistor). The circuit of Figure 11.16c

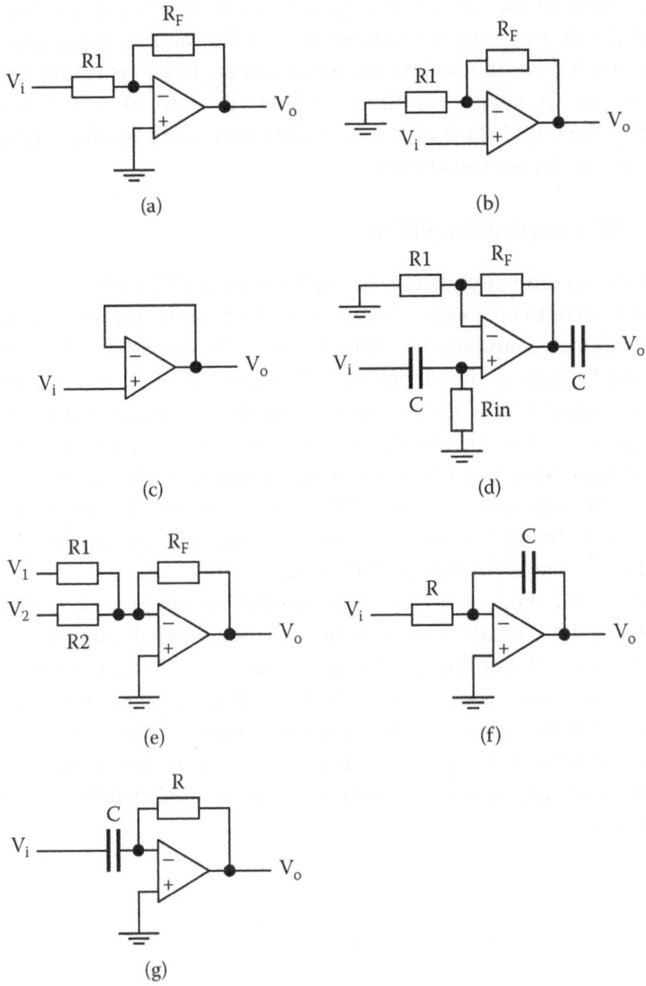

FIGURE 11.16 (a) Inverting amplifier, (b) noninverting amplifier, (c) buffer amplifier, (d) AC-coupled amplifier, (e) summing amplifier, (f) integrator, (g) differentiator.

is often used to "buffer" a signal from a nonideal source, before it is fed through further signal processing circuitry (filters, etc.) — the output stage from the op-amp is normally expected to be much better than the source itself.

Notice from the foregoing discussion on noise, if a noninverting amplifier with any gain above unity can be used and any filtering technique can be employed, then this would be even better as the first stage in a signal conditioning circuit. A simple unity-gain buffer (Figure 11.16c) would only be employed if high-amplitude noise was expected to swamp the signal prior to filtering. In such a situation, amplification at this initial stage may cause the output of the amplifier to saturate (drive to the positive or negative power rail) because of the amplified noise, and the desired signal would be lost each time this happened. A typical example of such noise in sensitive measurement applications is interference from the main electricity supply (at 60Hz, or 50Hz in Europe), which may well be picked up by the initial stage of sensitive instruments (try waving an oscilloscope probe around in the air sometime).

11.2.4.2 AC-Coupled Amplifiers

So far, all the op-amp circuits considered have been DC-coupled amplifiers. If a fixed voltage were to be coupled to the input of either the inverting, noninverting, or buffer amplifier configurations (Figure 11.16a to Figure 11.16c), then a fixed voltage would appear at the output. In Subsection 11.1.1.4, it was noted that the capacitor can appear to be a short circuit at high frequencies and an open circuit at low frequencies. This means that it can be used to block DC signals and only pass AC signals. This is particularly useful when a small alternating signal of interest is superimposed on a large DC offset, perhaps due to electrochemical action, or maybe the AC signal is being used to measure a parameter that is being controlled by a (lower-frequency) DC signal.

Figure 11.16d shows a noninverting amplifier configuration with capacitors added to block DC signals. This is termed an AC-coupled amplifier.

The AC-coupled amplifier in Figure 11.16d has a finite input impedance at high frequencies (the capacitor represents an infinite resistance at low or DC frequencies). The magnitude of this impedance should approximate to *Rin* at the frequencies of interest, the capacitor being selected so that at these frequencies the magnitude of the capacitive reactance is much smaller than the value of the resistor, that is:

$$\frac{1}{\omega C} << Rin \tag{11.27}$$

Obviously, there will be a signal frequency below which the capacitor will represent a significant impedance (i.e., will block the signal) and above which it will pass the signal. This occurs at:

$$\omega = \frac{1}{CRin} \tag{11.28}$$

Notice that frequency is given in radians per second (conventionally symbolized using ω). To obtain the frequency in hertz (i.e., cycles per second, normally symbolized by *f*), divide by a factor of 2π. The component values should be selected so that this cutoff frequency is at least an order of magnitude below the frequencies of interest. This is, in fact, a rudimentary filter — the design of filters is dealt with in the following sections.

The question arises as to why one would want to place *Rin* in the circuit in any case. The op-amp in a noninverting configuration has a very high input impedance. If *Rin* were to be taken out of the circuit, it would be reasonable to assume that an AC-coupled amplifier with very high input impedance and a very low cutoff frequency would result. *Rin* takes into account other nonideal qualities of real op-amps, specifically, input bias currents. These are DC currents that flow through the inputs of the amplifier, and are normally insignificantly small (BJT input amplifiers have the highest bias currents, JFET next, and MOSFET the lowest). Without the presence of *Rin*, there would be nowhere for these currents to flow, and the input would float to one rail or the other, saturating the output (imagine a very small current flowing through a very high resistance, in this case, the parasitic resistance of the capacitor — no component is perfect). Even with *Rin*, these currents will still flow, although they will flow to 0 V via the resistor and only a very small offset voltage will appear at the input. The amplifier and *Rin* should be chosen so that this offset is maintained within reasonable limits (i.e., will not cause the amplifier to saturate). An input resistor is even recommended with MOSFET input op-amps. Although the bias currents are infinitesimally small, it is better that they flow to ground through a known path rather than an uncontrolled path across the circuit board.

The final point to note is that the bandwidth (cutoff frequency) of the output will be limited by the input resistance (or impedance) of the next stage.

11.2.4.3 Summing Amplifiers

It is possible to design op-amp circuits that perform a variety of mathematical functions. The basic circuits, of course, perform multiplication by a constant. Figure 11.16e shows a simple summing amplifier configuration. It will be recalled from Subsection 11.2.1 that the inverting input of the op-amp must be more or less at 0 V (this is sometimes called a *virtual earth* or *virtual ground*). Following an analysis similar to that carried out in Subsection 11.2.1 for the inverting amplifier configuration, it can be shown that

$$-V_o = V_1 \frac{R_F}{R1} + V_2 \frac{R_F}{R2} \tag{11.29}$$

Clearly, if all the resistors are of the same value, then V_o will be the negative sum of the two input voltages. Further study will show that any number of signals can potentially be combined in this manner.

11.2.4.4 Integrators and Differentiators

Figure 11.16f shows a theoretical integrator circuit, and Figure 11.16g shows a theoretical differentiator. These circuits are, for various reasons, not suitable for practical implementation, but they can be readily analyzed by assuming an ideal op-amp. In Figure 11.16f, the current that flows through the feedback capacitor must be the same as that flowing through the resistor, and the inverting input will be a virtual earth. Equation 11.2 can be rearranged to give the current flowing through a capacitor in terms of the time domain derivative of the voltage across it. The analysis will yield:

$$v_o = -\frac{1}{RC}\int v_i dt \tag{11.30}$$

Thus, the circuit effectively integrates the input (and scales it). A similar analysis can be carried out on Figure 11.16g with the result that

$$v_o = -RC\frac{dv_i}{dt} \tag{11.31}$$

As they stand, these circuits are not practical. Bias currents and offset voltages in the integrator will cause the output to drift over time, as will the charge leakage from the capacitor. Also, when performing a mathematical integration, an initial condition must normally be met, i.e., the capacitor must be charged to some initial value. It is common practice to use a switch to reset (discharge) the capacitor, but for accurate work, the circuit has to be carefully designed and offsets and biases minimized. The integrator circuit, and before that, electromechanical integrators, used to be the key element in analog computers created to solve complex differential equations. Owing to the practical problems of implementing these solutions, however, digital computers are now almost exclusively used to solve such problems.

The differentiator was never a popular circuit element, principally because it tends to be unstable. Remembering that a capacitor appears to be a short circuit to high-frequency signals and drawing an analogy between Figure 11.16g and Figure 11.16a, it is apparent that the differentiator will have very high gain at high frequencies. As these are easily picked up from the environment (for example, radio signals, other electronic equipment, electrostatic discharges, from humans), it is normal to add components to limit the bandwidth of the differentiator.

Although these circuits are little used for signal conditioning, they are still of interest as very rudimentary filters, as can be seen by considering the nature of the capacitor at high and low frequencies and drawing parallels with the inverting amplifier circuit: the integrator is a low-pass filter (allowing low frequencies through and attenuating high frequencies) and the differentiator is a high-pass filter. Similarly, it can be seen that filters have some element of integration or differentiation in their function. Filters are treated in some detail in Section 11.5.

11.2.4.5 Other Functions

It is clear from the foregoing discussion that various mathematical functions can be implemented using op-amps, provided that a component with an appropriate response can be placed in the feedback circuit. One additional function that used to be fairly common was the logarithm (and antilog) circuit, which made use of the logarithmic response of the BJT. These would facilitate the multiplication and division of two signals. However, with the advent of powerful digital signal processors and computing techniques, coupled with the difficulty of designing stable and accurate analog circuits and their inflexible nature, most signal processing has now migrated to the digital domain. This is dealt with in Chapter 12.

There are times when it may be desirable to compare one signal with a threshold level; this function is accomplished using a comparator circuit. Dedicated comparator ICs are usually faster or have specialized outputs with lower open-loop gains than op-amps. These, or op-amp comparators, are usually operated with a little positive feedback so that noisy signals do not cause rapid repeated switching of the output as the signal approaches and passes the threshold; i.e., some hysteresis is introduced. This configuration is known as a *Schmitt trigger*.

The op-amp is the fundamental building block of hundreds of useful circuits: oscillators, level detectors, filters, etc. Only some of these are dealt with here, but many ideas can be found in Horowitz and Hill [2] or the many books on hobby electronics that are available.

11.3 INSTRUMENTATION AMPLIFIERS

The concept of CMRR was introduced in Subsection 11.2.2.3 for op-amps; it describes how good the circuit is at removing a signal that appears simultaneously on both inputs.

Many sensors are particularly bad signal sources; they often produce very small signals with very high source impedances, subject to drift, and, worse still, they are often connected to the nearest amplifier by a long cable. Cables are particularly susceptible to picking up stray signals radiated from a variety of sources. The worst offenders are electronic equipment, especially computers, televisions, cell phones, and the main power supply. Heavy industrial equipment and electric motors are even worse. Often these signals can be filtered out, but this assumes first that the interfering signal is smaller than the signal of interest (otherwise, the first stage amplifiers may saturate) and second that it is well outside the frequency range of interest. Even so, there are situations in which a differential measurement is preferable to an absolute measurement.

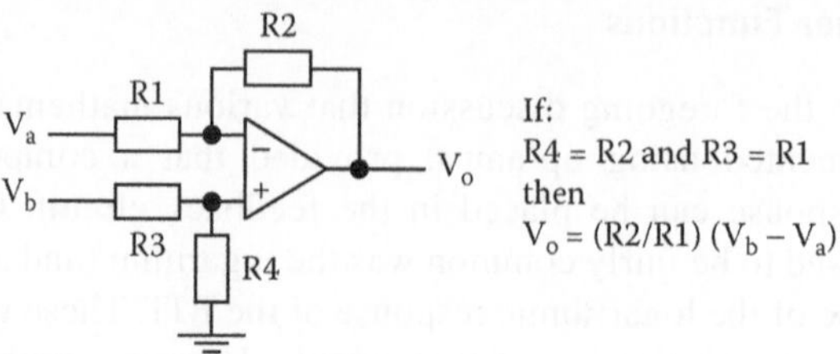

FIGURE 11.17 Basic differential amplifier.

> **Differential and Absolute Values**
>
> An absolute value is measured with respect to a specific reference point, usually the equipment's 0 V or ground line. Two wires will connect the sensor to the input of the instrumentation: the signal wire and ground.
>
> A differential value is the difference between two measured values (e.g., voltage A minus voltage B). So if the absolute value of signal A with respect to 0 V is 1000 V and of signal B is 1001 V, the differential value will be 1 V. Measurements are normally made with three wires — one to carry each signal and one reference signal (often, but not always, 0 V). Without a fixed reference, the circuit will make up its own paths for stray currents, which causes many problems.

The basic single-op-amp differential amplifier circuit is shown in Figure 11.17. Strictly speaking, this is a differential input amplifier — the output is single ended (i.e., not differential). Truly differential amplifiers do exist, but they are relatively uncommon. The disadvantage of this circuit for use in sensor applications is immediately apparent: it has a low input impedance. For this reason, the three-op-amp circuit, or instrumentation amplifier, is used (Figure 11.18).

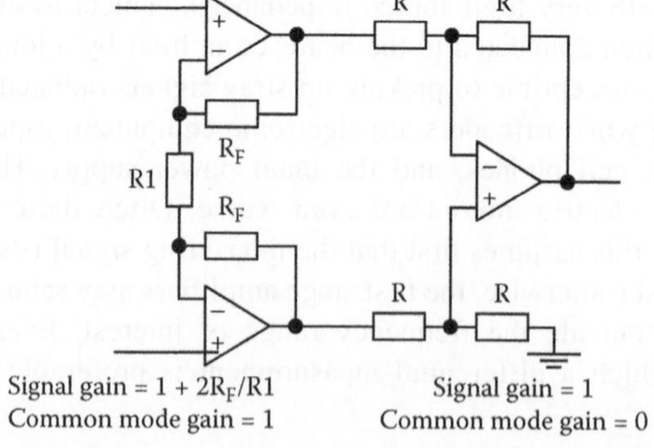

FIGURE 11.18 Instrumentation amplifier; the gain of each of the two stages is indicated.

Notice that the common-mode voltage may be tapped from the center of R1 (by making R1 two resistors of equal value). This can then be fed back as the reference signal through the third connection (the shield connection of the cable) to increase the input impedance of the circuit.

Cabling

If it is not possible to mount the sensor close to the signal-conditioning equipment and digitize the signal as soon as possible, then the wrong choice of cabling can result in interference problems. The basic choice is:

- Twisted pair — two wires twisted together; good for fairly strong signals in benign environments.
- Screened cable — one or more signal wires with one or more braided wire or foil screens; the screen will carry 0 V (ground) or the reference signal; better immunity in noisy environments for multiple signals.
- Coaxial cable — one central core separated from a screen by an insulating layer; both screen and the core share the same longitudinal axis (hence the name); used for high frequency signals (hundreds of kHz up).
- Fiber-optics — uses light and not electrical current and is therefore almost completely immune to electrical interference; drawbacks include cost and difficulty of installing the system.

Note that the coaxial cable has a characteristic impedance (usually 50 ohms) this needs to be matched by the driving circuit (50 ohms output impedance) and receiving circuit (50 ohms input impedance), otherwise it will suffer from signal distortions due to reflections, etc.

Both the simple differential amplifier and instrumentation amplifier can be purchased as single ICs. The gain is normally fixed or adjusted by using an external link or resistor. The internal circuitry should be laser-trimmed to provide a very accurate circuit and eliminate the problems that may otherwise be experienced when creating a circuit from individual op-amps, such as component tolerances and offsets.

11.4 WHEATSTONE BRIDGE

In many cases, small variations in the parameter to be sensed cause variations in the impedance or resistance of a sensor element. These can be detected most sensitively by use of a bridge configuration, commonly called the *Wheatstone Bridge* (Figure 11.19).

In practice, the voltmeter would be replaced by an instrumentation amplifier. The bridge consists of a measurement arm and a reference arm, and is initially

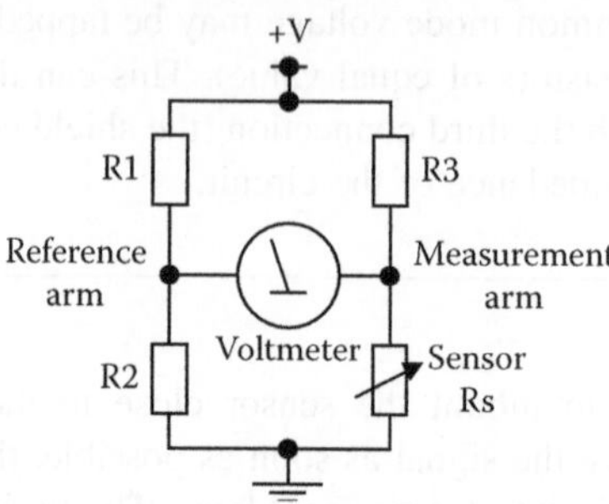

FIGURE 11.19 Wheatstone bridge; when the ratio of Rs to R3 is the same as R2:R1, then the voltmeter will read 0 V and the bridge is said to be balanced; normally, a small variable resistor (trimmer) is included in the bridge to balance it initially; then, any change in Rs will result in a nonzero reading on the voltmeter.

adjusted so that it is balanced, i.e., the meter reads 0 V. Thereafter, any changes in the resistance of the sensing element will cause the bridge to become unbalanced, and the magnitude of the change can be deduced and related to the change in the parameter being sensed.

The bridge configuration has a number of advantages, the most notable is that it can be used to reduce the magnitude of interfering signals caused by changes in parameters other than those being sensed. One good example of this is where implanted piezoresistors are used to sense the bending of silicon beams (e.g., those supporting seismic masses in accelerometers). These are also sensitive to changes in temperature, and it is common to fabricate an additional dummy beam upon which a reference resistor is implanted (Figure 11.20).

In some cases, closed-loop control is used to effect the sensing and extend the range over which the sensor will operate. Figure 11.21 shows how this operates for a thermal mass-flow-rate sensor. Here, a heating element is cooled by the passage of fluid or gas. The rate of cooling is related to the mass-flow rate of the substance over a heated temperature-sensing element, commonly a platinum resistor. The bridge is supplied with sufficient current to ensure that the sensing element is heated to its working temperature and balanced under zero-flow conditions. As the sensing element is cooled, the current flowing through the bridge is increased to rebalance the bridge, the current flowing through the bridge being related to the rate of cooling of the heating element and, hence, the mass-flow rate. A larger reference sensor is also included in the bridge to adjust for changes in the ambient temperature. This must be much larger then the platinum-sensing element as it should not be heated by the current flow through the bridge.

11.4.1 The Capacitor Bridge

It is quite easy to microengineer sensors that transduce mechanical changes into changes in capacitance. Although these sensors are normally relatively precise, the circuits necessary to measure changes in capacitance are more complex than

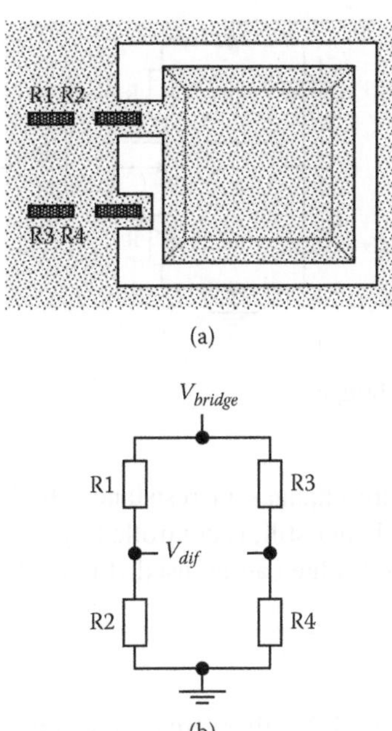

FIGURE 11.20 Bridge configuration of piezoresistors in beams supporting a seismic mass (see also Chapter 5): (a) piezoresistors (dark rectangles) are implanted into the beam suspending an accelerometer mass, seen from above, and a reference pair is implanted on a dummy beam; (b) the four resistors are connected in a bridge circuit; if all resistances match, then V_{diff} will be 0 V; if the beam bends as a result of acceleration, then R2 will change its resistance and the bridge circuit will no longer be balanced; to make measurements over a range, an actuator would be included (e.g., an electrostatic actuator created by mounting an electrode below the mass and using the mass as the opposite pole); then, under acceleration, the actuator is activated to force the mass back to its original position and zero the bridge; acceleration is then measured by the drive level to the actuator.

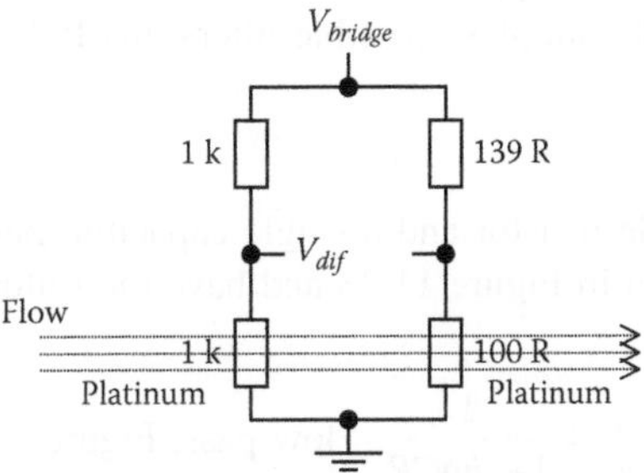

FIGURE 11.21 Thermal air mass-flow-rate sensor; the 100R (R is often used as an alternative to Ω) platinum resistor is the main sensing element and the 1 k platinum element adjusts for ambient temperature; the bridge will balance when the 100R platinum element has been heated to 100°C (by current flowing through the bridge).

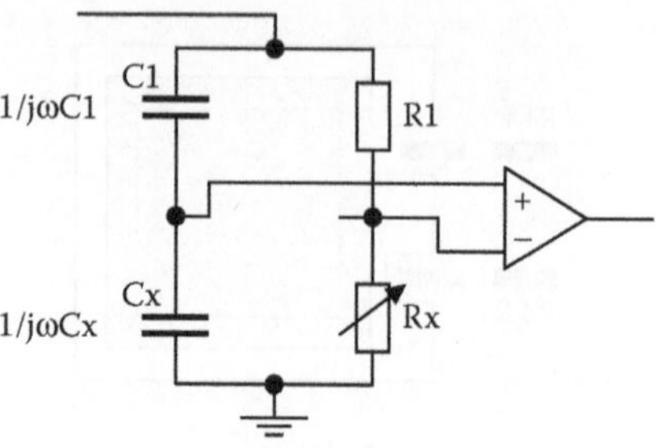

FIGURE 11.22 Capacitor bridge.

those required to measure changes in resistance. If the expected value of the capacitance is known or, better still, is controlled by feedback to remain at a fixed value, then a capacitance bridge can be used (Figure 11.22).

11.5 FILTERING

As discussed in Subsection 11.2.3, there are a large number of sources of electronic and electrical noise that can interfere and obscure a weak signal that one is trying to measure. Some of these will be white noise sources and some will be at a fixed frequency. The main power line interference at 60 Hz (in the U.S., or 50Hz in Europe) is one example of the latter case that often causes problems. Furthermore, Equation 11.24 suggests that by reducing the bandwidth of the system to the minimum necessary, interference from white noise sources can be reduced.

Finally, it may be that the signal of interest is superimposed on another much larger signal of a different frequency. One example would be where the aim is to use the same set of electrodes to control the position of an electrostatic actuator — a small AC measurement signal would be imposed upon a much larger DC actuation signal. In this case, an AC-coupled amplifier of the type discussed in Subsection 11.2.4.2 may be appropriate. The AC-coupled amplifier in Figure 11.16d incorporates one of the simplest possible filters, the RC filter.

11.5.1 RC FILTERS

These employ a single resistor and a single capacitor, hence the term RC filter. Both forms are shown in Figure 11.23 and have the following characteristics:

$$\frac{v_o}{v_i} = \frac{1}{1 + j\omega CR} \quad \text{low-pass, Figure 11.23a} \tag{11.32}$$

$$\frac{v_o}{v_i} = \frac{j\omega CR}{1 + j\omega CR} \quad \text{high-pass, Figure 11.23b} \tag{11.33}$$

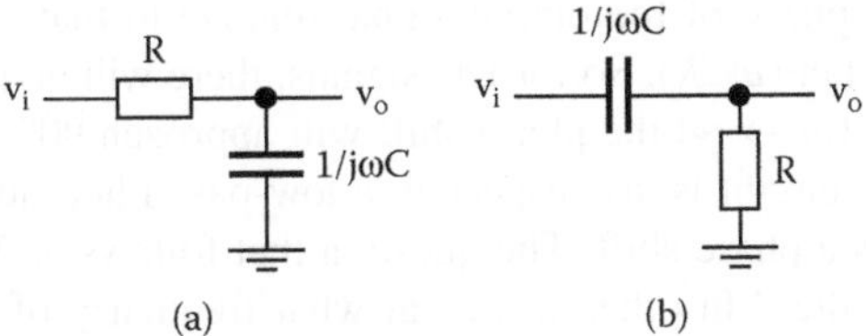

FIGURE 11.23 (a) Low-pass RC, (b) high-pass RC — both with frequency domain labels.

The functions of the circuits in Figure 11.23 should be clear by simply examining them and substituting the capacitor with a short circuit at high frequencies and an open circuit at low frequencies. They can also be explored by examining the equations, but first it is helpful to convert these into real and imaginary parts by multiplying the numerator and denominator (by 1 jωCR), and rearranging the terms:

From Equation 11.32:

$$= \frac{1}{1+\omega^2C^2R^2} - j\frac{\omega CR}{1+\omega^2C^2R^2} \tag{11.34}$$

and from Equation 11.33:

$$= \frac{\omega^2C^2R^2}{1+\omega^2C^2R^2} + j\frac{\omega CR}{1+\omega^2C^2R^2} \tag{11.35}$$

And finally converting these to magnitude and angle form:

From Equation 11.34:

$$= \sqrt{\frac{1}{1+\omega^2C^2R^2}}\angle\tan^{-1}(-\omega CR) \tag{11.36}$$

and from Equation 11.35

$$= \sqrt{\frac{\omega^2C^2R^2}{1+\omega^2C^2R^2}}\angle\tan^{-1}\left(\frac{1}{\omega CR}\right) \tag{11.37}$$

Equation 11.36 and Equation 11.37 give a handle on the functioning of the circuit in case of sinusoidal signals (this does not work for nonsinusoidal signals). All the following discussion applies to Equation 11.36, but a similar working can be applied to Equation 11.37. First, note that the magnitude of the resulting signal will be related to the input signal by the factor $\sqrt{(1/(1 + \omega^2C^2R^2))}$. For DC signals, ($\omega = 0$), then the output will be the same as the input signal. For high frequency signals, ($\omega \to \infty$), then the output will approach 0.

Similarly, the phase of the output signal relative to that of the input signal will be shifted by tan[1](ωCR). So for DC signals, there will be no phase shift, but at high frequency ($\omega \to \infty$) the phase shift will approach 90°.

Therefore, the circuit is, as suspected, a low-pass filter, and it is possible to say that it exhibits a phase shift. The question that follows is: What is the cutoff frequency of the filter? In other words, at what frequency of input signal does the output signal start to become significantly attenuated? Referring back to the box entitled "Decibels," it was suggested that a reduction in amplitude by 3 dB was a useful measure of the filter cutoff point. By referring to this box, it will be seen that this is equivalent to a loss of half the power of the signal, or an attenuation of the signal amplitude by a factor of $1/\sqrt{2}$ (approximately 0.707). From Equation 11.36, it can be seen that this occurs when:

$$\omega^2 C^2 R^2 = 1 \tag{11.38}$$

Notice also that at the cutoff frequency, the phase shift is 45°. By letting the cutoff frequency be ω_0, it is possible to go on to say that at one decade below ω_0 (i.e., $\omega_0/10$) the amplitude of v_o is approximately equal to the amplitude of v_i and the phase shift is close to 0. Similarly, for each decade (factor of 10) increase of frequency above ω_0, the amplitude of v_o drops further by 20 dB, and at $10\omega_0$ the phase shift stabilizes at approximately 90°. This information is conveyed by a "Bode plot" (Figure 11.24a). The same information is plotted for the high-pass filter (Equation 11.37) in Figure 11.24b. Note that both these Bode plots have been normalized by setting $R = C = 1$, giving a 3 dB point at $\omega = 1$. The cutoff frequency of both filters can, therefore, be set by selecting appropriate values of R and C:

$$\omega_0 = \frac{1}{CR} \tag{11.39}$$

Again, do not forget that ω_0 is in radians per second. Divide this by 2π to get the frequency in hertz. Also, note that if the next circuit element has a noninfinite input impedance (i.e., is not an op-amp in the noninverting configuration), it will load the filter circuit and the resulting response will not be as predicted. Component tolerances will also have some effect on the cutoff frequency.

Equation 11.29, then, allows one to design a simple RC filter with an arbitrary cutoff frequency. This may be a high-pass or low-pass filter, depending on the circuit architecture (Figure 11.23a or Figure 11.23b), which will have unity gain in the pass band and attenuate the signal by 20 dB per decade in the stop band. For a low-pass filter, the phase shift will be 0° in the pass band, increasing to 45° at the 3-dB point and to 90° in the stop band. For a high-pass filter, it will be +90° in the stop band, falling to 45° at the 3-dB point and 0° in the pass band.

Of course, a simple low- or high-pass filter may be insufficient for the application in question. There are two other filter configurations that may be of use, the band-pass filter and band-stop filter. These can be created by cascading RC low-pass and high-pass filters with buffer amplifiers (see Figure 11.25a and Figure 11.25c). Their characteristics can be quite easily determined by using the

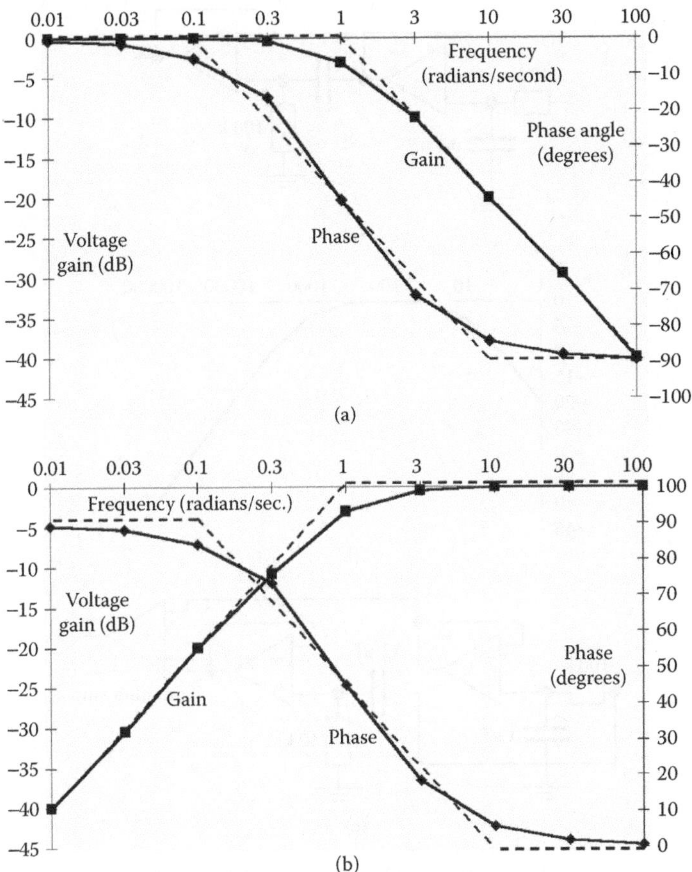

FIGURE 11.24 Bode plots; thick lines represent computed values, dashed lines show linear approximations. (a) Low-pass RC filter, (b) high-pass RC filter; normalized for R = C = 1.

Bode plot: draw the Bode plot for each separate element and then add them together to give the combined filter Bode plot (Figure 11.25b and Figure 11.25d).

Notice that RC filters do not make particularly good filters, especially if the signal of interest is at, e.g., 1 kHz and there is interference at 100 kHz that is 100 times larger. In such a situation, one would desire the interfering signal to be attenuated by a factor of 1000 (60 dB) so that its effect on the signal of interest becomes negligible. Also, if the 3-dB point is chosen to be 1 kHz in order to get maximum attenuation at 100 kHz, then:

1. The signal of interest will be attenuated by 3 dB.
2. The signal of interest will be distorted by a 45° phase shift.
3. The signal of interest can only be a sinusoidal signal, or close it will be even more distorted.

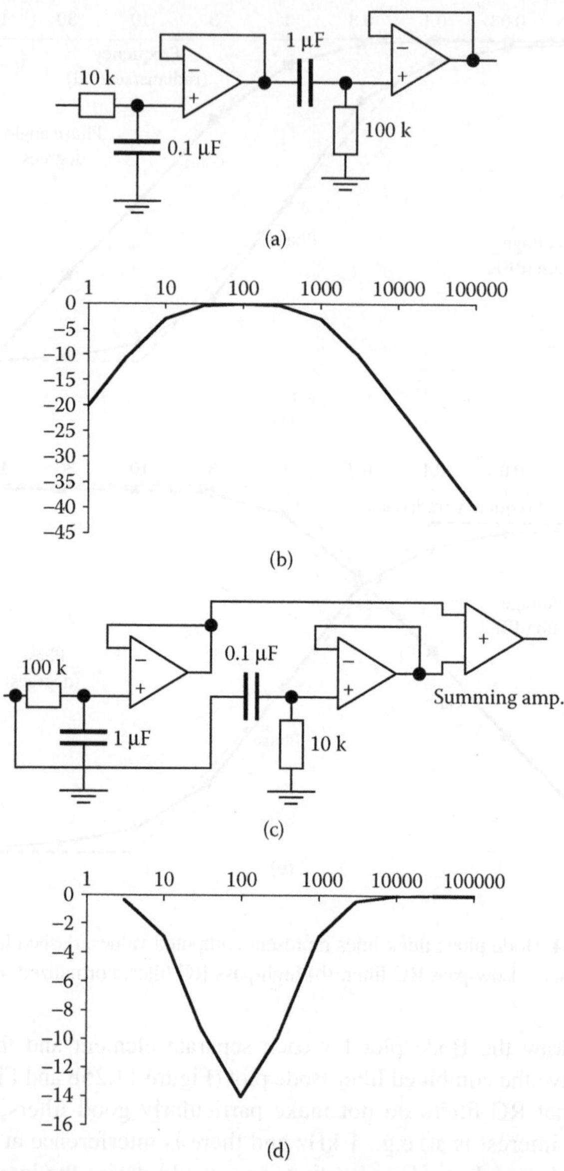

FIGURE 11.25 (a) Band-pass filter using simple RC filters, (b) Bode plot of this (gain in dB vs. frequency in radians per second), (c) band-stop filter with RC elements, (d) Bode plot of this (gain in dB vs. frequency in radians/second).

To design a filter, it is necessary to ensure that the frequency of the signal of interest lies well within the pass band and that the nearest interfering frequency is sufficiently attenuated. This can be done by employing more complex filter designs, as discussed in the following section.

Nonsinusoidal Signals

The frequency domain analysis of filters is only applicable to sinusoidal signals. Fortunately, however, most other signals of interest can be created by adding up a number (infinite number) of sinusoidal signals of different amplitudes and phases. Determining the effect of the filter on nonsinusoidal signals, therefore, involves determining how it affects each of the sinusoidal components in terms of attenuation and phase shift (changing the relative phase of the components will change the shape of the resulting signal, thus it is important to consider phase effects as these can extend a long way from the 3 dB point). Working out the sinusoidal components of signals is performed by using Fourier analysis, or Fourier transforms. These are not dealt with in this volume. Generally speaking, given a signal with a fundamental frequency of ω_0 [i.e., 2π/(the time before the signal starts to repeat itself; the period, in other words)], then signals that are similar in shape to a sinusoidal wave will not have significant components at very high frequencies, whereas a rapidly changing signal such as a square wave will have significant components at two, four, six, and eight times the fundamental frequency. To avoid distortion, then, the 3 dB point of a low-pass filter would need to be at least two decades beyond the fundamental frequency, preferably more. Also, if this is an interfering signal, then a high-pass filter would need to be chosen that is two or three decades (preferably more) above ω_0, so that high-frequency components of this signal do not appear in the pass band.

11.5.2 Butterworth Filters

The RC filters dealt with in Subsection 11.5.1 are termed passive filters because only passive components (a resistor and a capacitor) are required to implement them. By incorporating inductors, it is possible to build very complex passive filters. It is easier, however, to incorporate op-amps into the circuit to form active filters; this became apparent in Subsection 11.5.1 when the buffer amplifier was introduced so that RC filters could be joined together without the first section being loaded by the following section.

There are three different types of continuous-time filters (as opposed to switched-capacitor filters or digital filters, which sample the signal at discrete points) in common use. These are the Butterworth, Bessel, and Chebyshev filters.

They have the following characteristics:

- Butterworth: maximal gain in the pass band, reasonable cutoff.
- Bessel: minimal phase distortion, poor cutoff.
- Chebyshev: controllable ripple in pass-band gain; this is traded off against a steep initial cutoff (the more ripple that can be tolerated, the sharper the initial cutoff).

These comments assume filters of equivalent complexity. Of these filters, the Butterworth is relatively easy to synthesize, and is therefore introduced here. There are a number of continuous–time-filter ICs on the market that use different approaches to implement the different filter types. Given modern computing power, however, for many transducer applications, complex continuous-time filters are not required because digital filtering and digital signal processing techniques have taken their place.

The transfer function of a Butterworth filter takes the form:

$$H(s) = \frac{A_{vo}}{B_n(s)} \tag{11.40}$$

The transfer function is a formal way of writing v_{out}/v_{in}. B_n is the appropriate Butterworth polynomial selected from Table 11.2. The subscript n gives the *number of poles* of the filter, a technical term also referred to as the *order* of the filter. Here, it is only necessary to know that for each increase in n, the cutoff rate increases by a further 20 dB per decade. For a single-pole filter, the gain drops off at 20 dB per decade, for a two-pole filter it is 40 dB per decade, three poles give 60 dB per decade, and so on. The fact that the equations are given as a function of s will also be neglected for now. Note, however, that the frequency response can be found by replacing s with $j\omega$.

Butterworth filters are relatively easy to deal with because only a first-order filter circuit and a second-order filter circuit are required. Higher-order filters are made up by cascading combinations of first- and second-order circuits.

TABLE 11.2
Butterworth Polynomials

Poles	Polynomial	Cutoff (dB per decade)
1	$(s + 1)$	20
2	$(s^2 + 1.414s + 1)$	40
3	$(s + 1)(s^2 + s + 1)$	60
4	$(s^2 + 1.848s + 1)(s^2 + 0.7654s + 1)$	80
5	$(s + 1)(s^2 + 1.618s + 1)(s^2 + 0.6180s + 1)$	100
6	$(s^2 + 1.932s + 1)(s^2 + 1.414s + 1)(s^2 + 0.5176s + 1)$	120

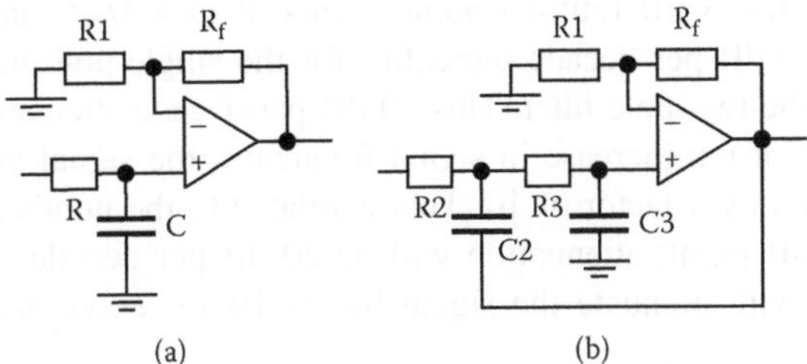

FIGURE 11.26 Low-pass filter circuits: (a) first-order filter circuit, (b) second-order filter circuit. These can be used to synthesize Butterworth filters; high-pass filters are created by changing R2 with C2 and R3 with C3 (R with C for the first-order circuit).

Examples, based on the *Sallen and Key* filter, are shown in Figure 11.26. These are low-pass filters; to make high-pass filters simply switch the Rs and Cs around.

Figure 11.27 shows the typical frequency response of one- and two-pole low-pass Butterworth filters. Straight line approximations have been drawn to show how the analytical responses deviate from these.

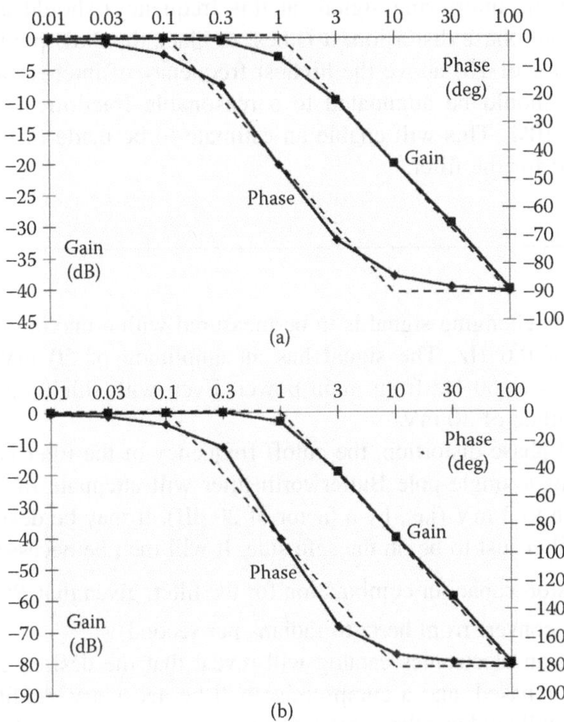

FIGURE 11.27 Normalized frequency response of Butterworth low-pass filters: (a) one-pole, (b) two-pole (frequency axis is in radians per second); straight line approximations are shown as dashed lines.

Notice that the 3 dB (cutoff) point occurs at $\omega = 1/RC$ and the signal is attenuated by 20 dB per decade thereafter for the single-pole filter, and 40 dB per decade for the two-pole filter. This 20-dB-per decade attenuation means that for every factor of ten increase in signal frequency, the signal amplitude at the output is reduced by a factor of 10. This is related to the number of filter poles — after the 3 dB point, attenuation will be 20 dB per decade per pole (i.e., a four-pole filter will attenuate the signal by 80 dB for every decade above the 3 dB point).

Furthermore, the phase shift of the filter is 45° at the 3 dB point for the single-pole filter and 90° at the 3 dB point for the double-pole filter. The phase shift will distort the signal for one decade on either side of the 3 dB frequency.

11.5.2.1 Synthesizing Butterworth Active Filters

The procedure described here is for synthesizing a low-pass filter. The procedure for synthesizing high-pass filters is the same, but discussion will focus on the lowest frequency of interest, as opposed to the highest.

The first step is to decide the order of the filter. The maximum frequency of interest and the lowest interfering frequency should be determined. The magnitude of the interfering signal (at this frequency) should also be determined. To avoid phase distortion, it is best to place the 3 dB point at least one decade (a factor of 10) above the highest frequency of interest, and the interfering signal should be attenuated to a reasonable fraction of the signal of interest, e.g., 10%. This will enable an estimate to be made of the number of poles required for the filter.

Example

A very slowly changing signal is to be measured with a maximum frequency of interest of 0.6 Hz. The signal has an amplitude of 50 mV. The main interference is at 60 Hz from main power. Even with shielding, this signal has an amplitude of 30 mV.

To avoid phase distortion, the cutoff frequency of the filter is selected to be 6 Hz, thus a single-pole Butterworth filter will attenuate the main interference down to 3 mV (i.e., by a factor of 20 dB). It may be desirable to use a two-pole filter just to be on the safe side. It will then be necessary to find a suitable resistor–capacitor combination for the filter, given that $RC = \frac{1}{2\pi f}$ (it is necessary to convert from hertz to radians per second).

Studying an electronics catalog will reveal that the desired combination cannot be purchased, and a compromise will be necessary. Additionally, the components will not have the exact resistance or capacitance marked; they will vary by a given tolerance (±1%, ±5%, etc.) about the given value.

At this point the cutoff frequency, ω_0 (in radians per second), has been determined. From this it is possible to find R and C from Equation 11.41. (A likely value for one of the components is chosen, and the equation is rearranged to give the value of the second).

$$\omega_0 = \frac{1}{RC} \tag{11.41}$$

At this point, the gain of the circuit needs to be determined. From Table 11.2 it can be seen that all orders of Butterworth filter can be obtained by cascading first- and second-order filter circuits. The polynomials in Table 11.2 have been given for normalized filters, i.e., $\omega = 1$ and gain = 1. It is possible to write a general version of the first- and second-order equations as

$$H_1(s) = \frac{A_v}{\dfrac{s}{\omega_0} + 1} \tag{11.42}$$

and

$$H_2(s) = \frac{A_v}{\dfrac{s^2}{\omega_0^2} + 2\zeta\dfrac{s}{\omega_0} + 1} \tag{11.43}$$

respectively, where A_v is the voltage gain of the stage. For a single-pole Butterworth filter stage, the gain is equal to unity. For a two-pole stage, the gain of the stage is given by:

$$A_v = 3 - 2\zeta \tag{11.44}$$

The value of 2ζ can be readily extracted from Table 11.2. Thus, a four-pole Butterworth filter (of any cutoff frequency) can be made up of a couple of two-pole filter stages, one with a gain of 2.235 (3 – 0.765) and the other with a gain of 1.152.

Appropriate values can then be selected for the gain-setting resistors. In the Sallen and Key filter (Figure 11.26), these act like a noninverting amplifier:

$$A_v = 1 + \frac{R_f}{R_1} \tag{11.45}$$

11.5.2.2 Approximating the Frequency Response of a Butterworth Filter

The frequency response of a filter is normally drawn on a log–log plot or a Bode plot. This requires log–linear graph paper because the vertical scale, gain, is given in decibels, which is a logarithmic unit (see box in Subsection 11.2.1). Phase shift in degrees is plotted against logarithmic frequency.

First, the filter is examined to determine if it is high pass or low pass. Then the pass-band gain is determined. This is converted to decibels and marked temporarily on the graph paper.

The cutoff frequency is found, normally in hertz, from $f = \frac{1}{2\pi RC}$ (or Equation 11.41 for radians per second). This is marked as a point on the graph at the pass-band gain level.

The order (number of poles) of the filter is determined, giving the cutoff rate (20 dB per decade per pole). If the filter is a low-pass filter, then a line is drawn from the 3 dB point, sloping down to the right (increasing frequency) at this rate. For a high-pass filter, the line slopes down to the left. The pass-band gain is then drawn in as a horizontal line.

Finally, the phase part of the Bode plot is constructed. The cutoff frequency is located again and a point marked beneath this on the graph paper. For a high-pass filter this will be at 45° times the number of poles and for a low-pass filter -45° multiplied by the number of poles. The frequencies one decade above and one decade below this point are located. For a high-pass filter, the phase shift one decade or more above this point is 0° and 90° multiplied by the number of poles one decade or more below this point. For a low-pass filter, the phase shift one decade or more below this point is 0° and 90° multiplied by the number of poles one decade or more above this point. Some example Bode plots for different filters are given in Figure 11.28.

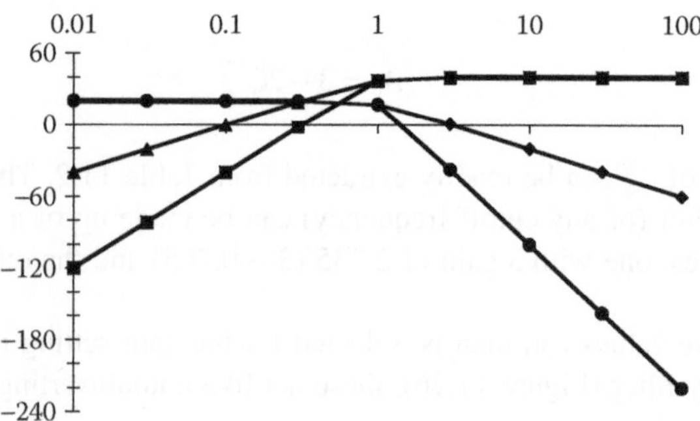

FIGURE 11.28 Example Bode plots for high- and low-pass Butterworth filters; gain in dB plotted against frequency (normalized) in radians per second, low-pass filters have a pass-band gain of 10 (20 dB) and two poles (diamonds) and six poles (circles); high-pass filters have a gain of 100 (40 dB) and two poles (triangles) and four poles (squares).

Group Delay

Group delay is a measure of the time that a signal of a particular frequency takes to traverse a filter or another circuit element. It has a bearing on distortion of the signal and is subtly different from phase shift. A phase shift of 45° in a 1-kHz signal represents a delay of 0.125 msec, whereas the same phase shift in a 10-kHz signal represents a delay of 0.0125 msec. Filters that apply an appropriate phase shift to correct for group delay can be designed; this is particularly important for signals that have traveled along long communication lines.

11.5.3 Switched-Capacitor Filters

Dedicated ICs are available for continuous-time active filters; however, there are problems with these, particularly for low-frequency operation. The cutoff frequency is tuned by selection of a capacitor, which gives rise to two problems. First high-tolerance capacitors cannot be easily fabricated in ICs, and the large capacitors required for low-frequency (audio range) filters take up a lot of area on the silicon die. Second, the cutoff frequency of the filter cannot be easily tuned (as may be required in some microprocessor-controlled situations).

As was hinted at in the previous section, filters can be constructed using integrator circuit elements. Consider the circuit in Figure 11.29. It can be shown that the circuit has the following characteristic:

$$v_o = f_0 \frac{C_1}{C_2} \int v_{in} dt \tag{11.46}$$

This is termed a *switched-capacitor* filter because of the manner in which it operates — switching between two capacitors. It also chops up the input signal in two, thus it is not a continuous-time filter (it is a discrete-time filter).The first thing to notice is that this depends upon the ratio of two capacitors and not their absolute values. Although it is difficult to fabricate a capacitor of a specific value

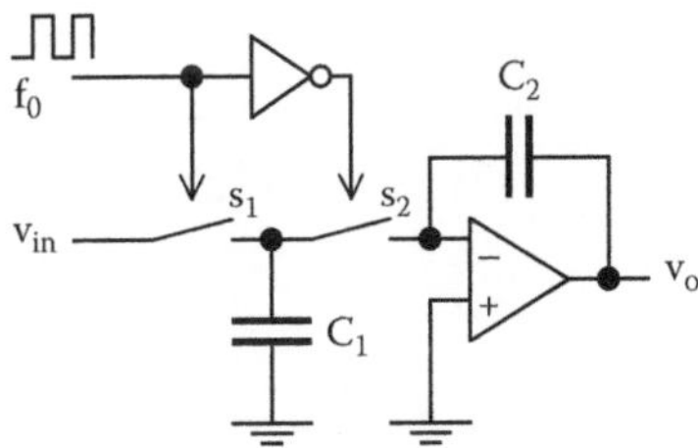

FIGURE 11.29 Switched-capacitor filter; the switches, s_1 and s_2, are digitally controlled by a signal f_0; the triangle with a circle on one end is the symbol for a digital inverter, which means that when s_1 is closed, s_2 is open, and vice versa.

on an IC, it is relatively easy to match components. Thus, it is relatively easy to fabricate two capacitors in a fixed ratio with a good degree of reproducibility. Secondly, the operation is dependent on the frequency with which the switches are opened and closed, so it is tunable.

Switched-capacitor filter ICs are available in a variety of configurations, including high- and low-pass. There are, however, some problems with switched-capacitor filters.

Switched-capacitor filters suffer from clock feedthrough. Noise appears at the output signal at the same frequency as the clock signal. Ideally, this could be removed by a simple continuous-time filter, but this may not always be possible. Also, the filter suffers from aliasing: signals near (above) the switching frequency are reflected down, below the switching frequency, and appear in the output signal without being attenuated. Again, this has to be solved by a simple continuous-time filter at the input. Finally, switched-capacitor filters tend to be noisier than continuous-time filters.

REFERENCES

1. Bogart, T.F., *Electronic Devices and Circuits*, Merrill Publishing Company, Columbus, OH, 1986.
2. Horowitz, P. and Hill, W., *The Art of Electronics*, 2nd ed., Cambridge University Press, Cambridge, New York, and Melbourne, 1989.

12 Computer Interfacing

12.1 INTRODUCTION

During the device development stage, it soon becomes necessary to provide microengineered devices with relatively complex signal-conditioning circuitry to drive them or process the signals that they produce. These will normally be produced by a computer equipped with analog-to-digital and digital-to-analog converters. For commercial devices, at least some of the signal processing circuitry may be provided on chip, and often you will want to interface them to a microcontroller for integration into a larger system.

This chapter aims to provide an outline of analog-to-digital and digital-to-analog conversion, which will form the basis of such interfaces. Practical schemes for posting the output levels of digital-to-analog converters and other circuits are considered separately in Chapter 13.

12.1.1 Number Representation

Most used for representing numbers is the decimal system, which has ten digits (0…9) arranged in columns, the rightmost column being the number of units one has, the next column to the left being the number of tens, the next the number of hundreds, and so on. So the number 143 has three units, four tens, and one hundred. Adding these up, one hundred plus four tens plus three units, gives the number one hundred and forty three. This may seem obvious, but computers and digital circuits do not use the same number system; so it helps to state the obvious before getting into the more complex. Notice, also, as one moves a column to the left the significance of the digit written there is 10 times that of the previous one, or 10^n, where n is the column number (starting from 0). Figure 12.1 shows this.

Digital computers have only two digits available, 0 (0 V, off or low), and 1 (+5 V, on or high). These are termed binary digits, or bits; the number system is binary (base 2), compared to the familiar decimal (or base 10) system. A group of 8 bits is termed a byte. Proceeding from right (least significant bit) to left (most significant bit), the value of each digit place doubles: units, twos, fours, eights, etc., or 2^0, 2^1, 2^2, 2^3, etc. (Figure 12.2). So the number 10001111 in binary represents: one one hundred and twenty eight, no sixty fours, no thirty twos, no sixteens, one eight, one four, one two, and one unit. Adding these up, one hundred and twenty eight plus eight plus four plus two plus one (128 + 8 + 4 + 2 + 1) is one hundred and forty three (143) in the decimal system.

Column number	n	....	3	2	1	0
Represents	10^n	....	10^3	10^2	10^1	10^0
	....	....	1000	100	10	1
Name	....	....	Thousands	Hundreds	Tens	Units

FIGURE 12.1 Representation of decimal numbers.

Writing everything out in binary gets a bit long winded; even quite small numbers take up a lot of room. For this reason, hexadecimal (base 16) is often used. This requires 16 digits, so 0…9 are supplemented with A to F (A being ten, B being 11, …, and F being fifteen). The reason that hexadecimal is used is that four binary digits can be quickly converted to one hexadecimal digit (Figure 12.3).

12.2 DRIVING ANALOG DEVICES FROM DIGITAL SOURCES

Analog devices require a voltage (or current) level that can be varied according to a digital demand signal. Ultimately, it is impossible to obtain a continuously variable signal level from a digital demand signal; the output will be quantized. The key to design is to ensure that the level of quantization is negligible and that the rate at which the signal can be varied is high enough for the desired application.

Column	7	6	5	4	3	2	1	0
2^n	128	64	32	16	8	4	2	1
	One hundred and twenty eights	Sixty fours	Thirty twos	Sixteens	Eights	Fours	Twos	Units
e.g.	1	0	0	0	1	1	1	1

$$
\begin{aligned}
10001111 \text{ is } \quad & 1 \times 128 = 128 \\
+ & \; 0 \times 64 = 0 \\
+ & \; 0 \times 32 = 0 \\
+ & \; 0 \times 16 = 0 \\
+ & \; 1 \times 8 = 8 \\
+ & \; 1 \times 4 = 4 \\
+ & \; 1 \times 2 = 2 \\
+ & \; 1 \times 1 = 1 \\
\hline
\text{Total in decimal: } \quad & 143
\end{aligned}
$$

FIGURE 12.2 Binary place values, with (10001111) converted to decimal (143).

Column	3	2	1	0
16^n	4096	256	16	1
	Four thousand and ninety sixes	Two hundred and fifty sixes	Sixteens	Units
e.g.			8	F

8F converted to decimal is $8 \times 16 = 128$
$+ \quad F \times 1 = 15$
Total in decimal 143

8F converted to binary

8				F			
(=1 × 8 + 0 × 4 + 0 × 2 + 0 × 1)				(=1 × 8 + 1 × 4 + 1 × 2 + 1 × 1)			
1	0	0	0	1	1	1	1

8F in binary is 10001111

FIGURE 12.3 Hexadecimal place values; 8F converted to decimal and binary. Note that one hexadecimal digit converts exactly to four binary digits.

12.2.1 Pulse-Width Modulation (PWM)

One of the easiest ways to implement digital-to-analog conversion is the PWM approach, because this only requires that the digital system produce two voltage levels. As such, it can be relatively easily implemented, although its use is restricted without additional analog-signal-conditioning circuitry.

Consider the signal displayed in Figure 12.4a. This is a square wave signal, which could quite easily be produced from a digital circuit. The period for which

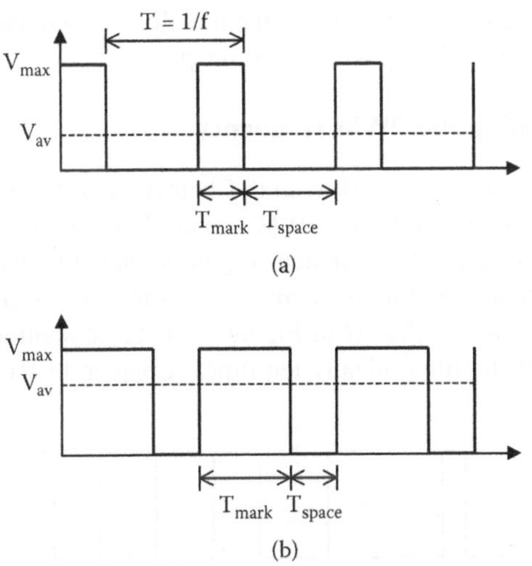

FIGURE 12.4 PWM signal, frequency *f*, period *T* (*1/f*), amplitude V_{max}, average signal level V_{av}, mark–space ratio (T_{mark}, T_{space}): (a) 1:2, (b) 2:1.

the signal is high is referred to as the *mark* and the period for which it is low as the *space*. It is possible to calculate the average signal level from this:

$$V_{av} = V_{\max} \frac{T_{mark}}{T} \tag{12.1}$$

It is normal to describe such signals in terms of mark–space ratio, because the average signal level is independent of frequency (i.e., T does not need to be known in order to find V_{av}). Equation 12.1 can be rewritten as:

$$V_{av} = V_{\max} \frac{T_{mark}}{T_{mark} + T_{space}} \tag{12.2}$$

In Equation 12.2, T_{mark} and T_{space} can be dimensionless. In Figure 12.4a, the mark–space ratio, $\frac{T_{mark}}{T_{space}}$, is 1:2, so the average signal level is $0.33V_{max}$. Compare this with Figure 12.4b, in which the mark–space ratio is reversed and V_{av} will be $0.66V_{max}$. By continuously varying the mark–space ratio, the average signal level can be varied to represent a variety of different signal shapes (Figure 12.5).

The obvious problem is that the analog circuit that is being driven by such a signal must be susceptible to the average signal level and not the instantaneous signal level. This implies that it must have some integrating or filtering function. From the design point of view, it is irrelevant if this occurs as part of the electrical design of the system (e.g., it incorporates large inductors or capacitors), the thermal design of the system (thermal masses that take time to heat or cool), the mechanical design of the system (inertia, damping), or some part of the function outside the system (e.g., the eye has an integrating function; a motion picture looks like a moving scene rather than what it really is, a sequence of still images). The system has to be a low-pass filter or integrator.

12.2.1.1 Estimating the PWM Frequency

As with the electronic filters dealt with in Chapter 11, we need to ensure that the system that is being driven by the PWM signal filters out the unwanted part and leaves us with the desired component of the signal: the average signal level. Because the system is a low-pass filter, we want to design the fundamental frequency of our PWM signal (f in Figure 12.4) to be significantly higher than the cutoff point of the filter, ideally ten times. Chapter 11 has already equipped

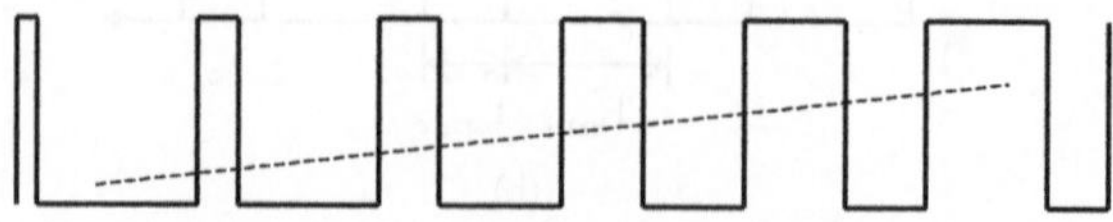

FIGURE 12.5 Varying the mark–space ratio of a PWM signal changes the average signal level; this can be used to mimic complex waveforms.

you to deal with electronic filter design; without going into detailed system analysis, it is possible to estimate the frequency required to make PWM successful by estimating the time constant of the system.

Given a step input to the system, the time constant is the time taken for the system to reach 63% of its final output (final position, final temperature, etc.). If you can determine this for your system, then a good rule of thumb is to select a PWM frequency that is ten times higher than this.

An alternative approach is to place an electronic filter on the output of the system. If you know what the maximum frequency of the output signal is to be, then an electronic filter can be designed with a –3-dB point at, or above, this frequency and suitable attenuation at the PWM frequency to bring it below noise level. One pulse from the PWM output (0 to V_{max}) should not be more than about 10% of a quantization step in the output, ideally (see the following subsections).

12.2.1.2 Digital Implementation and Quantization

Many modern microcontrollers are equipped with PWM generators implemented in hardware, although it is equally possible to implement them in software. PWM chips are also available for purchase. However, understanding the process elucidates some of the design constraints on PWM.

Figure 12.6 outlines how PWM may be implemented in hardware. Counter A acts to time the period of the PWM signal, which ultimately derives from a clock signal of a fixed frequency. For flexibility, counter A is provided with a preload register. When the counter reaches its maximum value, instead of starting back from 0, it is loaded with whatever value the preload register contains. The frequency of the PWM signal can thus be adjusted:

$$f = \frac{1}{clockperiod \times (counterA\,\text{max} - preload)} \tag{12.3}$$

The value in counter A is continuously compared with the value in the mark–space register. When counter A is greater than the value in the mark–space

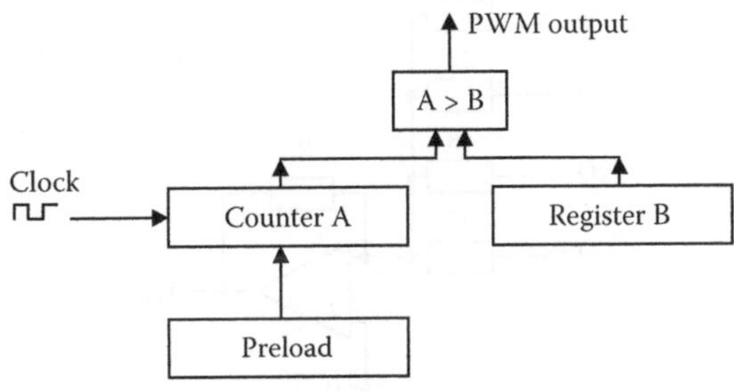

FIGURE 12.6 PWM hardware implementation; counter A is the period timer, and register B is the mark–space register.

register, the output is turned on; otherwise the output is turned off. The smallest mark–space ratio that can thus be obtained is:

$$1:(counterA\,\max - preload - 1) \tag{12.4}$$

This gives you a quantized (stepped) output, where the minimum step size can be computed using Equation 12.4 and Equation 12.2.

In this design, quantization of the output is dependent on the value in the preload register, and hence on the frequency of the signal. It is obviously possible to leave the preload register set at 0 and change the frequency of the PWM signal by altering the clock frequency. This can be done digitally with an additional counter and preload register. Ultimately, the upper limit on the PWM frequency depends on the maximum frequency to which the electronic logic circuitry can be driven, which determines the maximum count rate of the counters. It is also limited by the maximum quantization step with which you can cope. For a minimum step size of $0.01V_{max}$, you need 100 counts (1/0.01). The maximum PWM frequency that can be achieved is then:

$$f = \frac{maximum\_count\_rate\_in\_Hz}{100} \tag{12.5}$$

12.2.1.3 Reproducing Complex Signals with PWM

Figure 12.6 showed how varying the mark–space ratio could be used to create an average signal level that varies in a sinusoidal manner. The number of samples required to accurately reproduce a particular signal is dealt with in more detail in subsequent sections on analog-to-digital conversion. As a rule of thumb, at least five samples are required, preferably ten.

12.2.2 R-2R Ladder Digital-to-Analog Converter (DAC)

In Chapter 11, you came across the summing amplifier. This can be adapted to form a DAC (Figure 12.7). However, its use is limited. As the number of bits

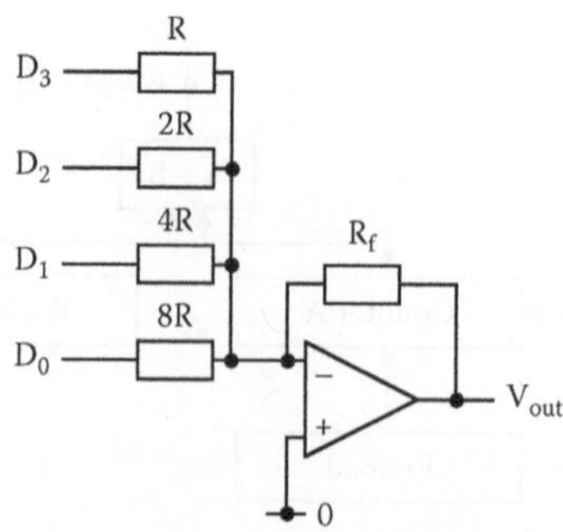

FIGURE 12.7 Summing amplifier DAC (4-bit).

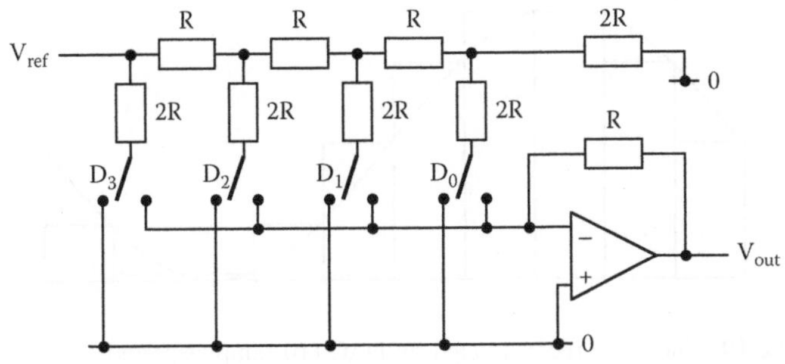

FIGURE 12.8 4-bit R-2R ladder DAC.

involved increases, so the size of the resistors doubles and very soon it becomes impractical. The R-2R ladder DAC (Figure 12.8) relies only on the relative ratios of resistors, and it is a lot easier to match ratios than absolute values when fabricating devices on ICs.

The R-2R ladder DAC is so called because the resistors are arranged in a ladder-like manner. It works because of the virtual earth point at the inverting input of the op-amp. Analysis of the circuit is left to the reader as an exercise.

12.2.3 Current Output DAC

The final form of DAC that will be considered here is the current output DAC. This can be constructed entirely from transistors and is, therefore, ideal for IC applications. If a voltage output is required, then some kind of current-to-voltage converter is required (e.g., a resistor and buffer amplifier).

Current output DACs can be constructed with either bipolar transistors (see p. 617 of Reference 1) or MOSFETS (Figure 12.9). In Figure 12.9, a p-channel MOSFET as an active load creates a reference current, and the n-channel MOSFET below it has its gate connected to the gates of eight other MOSFETs in a current mirror fashion. If these MOSFETs had the same dimensions as the first, then, having the same gate voltage applied, the current flowing through them would be the same.

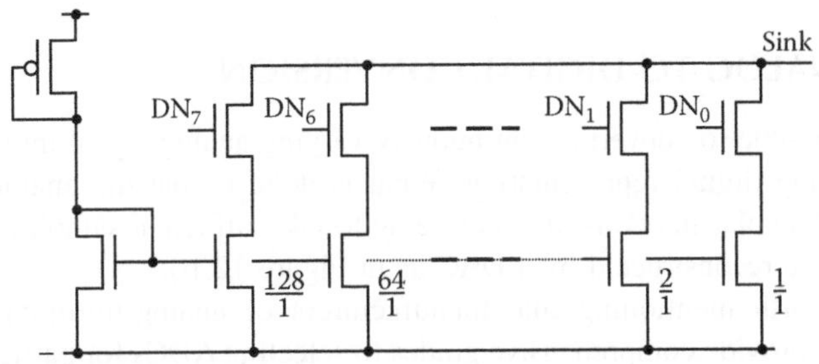

FIGURE 12.9 8-bit current sink DAC. Not all transistors are shown.

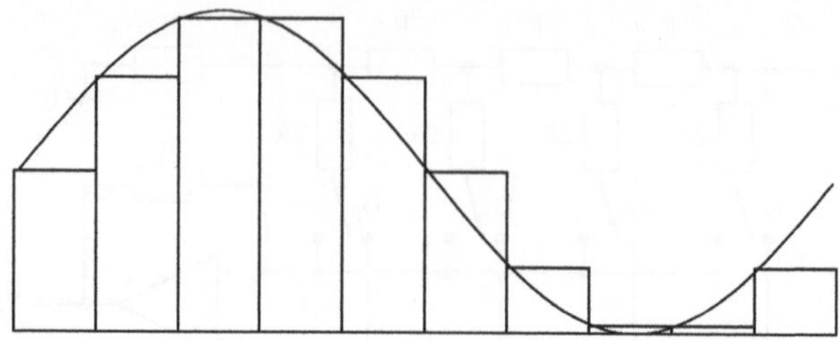

FIGURE 12.10 Sine wave reconstructed by DAC (10 samples/period).

The channel ratios of these MOSFETs are scaled, however, so the current flowing through the one on the far right is half that flowing through the next one, and so on.

A second row of MOSFETS, above the scaled ones, can be turned on or off to allow current to flow as required. A basic treatment of this topic is given elsewhere [2].

12.2.4 Reproducing Complex Signals with Voltage Output DACs

The voltage output DACs described earlier produce a discrete output voltage that corresponds with the digital number presented. The simplest way to create a time-varying signal, then, is to have a computer or microcontroller load the DAC with a sequence of numbers. Given that these can be presented to the DAC fast enough (and the DAC can keep up), then a waveform of arbitrary shape and frequency can be produced (Figure 12.10).

The rate at which data are presented to the DAC is the reconstruction clock frequency. For very high frequencies, a microprocessor may not be able to keep up; in such cases, the signal pattern would be preloaded into memory, and dedicated digital logic would present it to the DAC at the required reconstruction rate.

In Figure 12.10, the stepped nature of the signal is apparent. To remove these, a filter needs to be constructed. As stated before, in order to ease filter design, the reconstruction frequency should be selected such that five to ten samples fit into the period of the highest frequency signal that you wish to reconstruct (i.e., the reconstruction frequency should be five to ten times higher than the highest frequency in the signal that you wish to produce). This is a general rule of thumb, and the following section looks in more detail at this.

12.3 ANALOG-TO-DIGITAL CONVERSION

It is not possible to convert a continuously varying analog signal into a continuously varying digital representation. What is done is that the analog signal is sampled at regular intervals, the aim being to take sufficient samples so that the signal can be reconstructed by a DAC as in Figure 12.10.

It is worth mentioning that manufacturers of analog-to-digital-converter (ADC) ICs provide comprehensive guides to selecting ADCs for particular applications. National Semiconductor also publishes comprehensive guides to ADC specifications [3, 4].

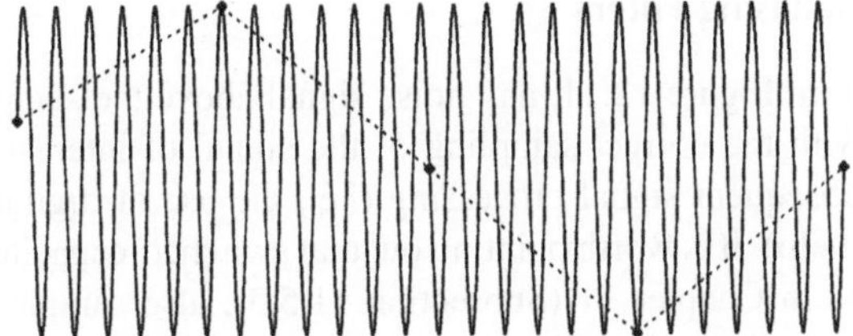

FIGURE 12.11 Illustration of the effects of undersampling. The frequency of the signal (continuous line) is approximately 6π times the sampling frequency. Samples (black diamonds) make the signal appear to be of a much lower frequency than it actually is.

12.3.1 Sample Rate

As Figure 12.11 shows, if the sampling frequency selected is too low, a signal will still be seen, but it will not accurately represent the actual signal. This effect, in which a signal above the sample rate appears at a lower frequency in the sampled signal, is known as *aliasing*. When aliasing occurs, information is lost, and there is no way to reconstruct the original signal.

The minimum sample frequency required to reproduce a signal is determined from the Nyquist criterion. This states that the sample frequency must be at least twice the maximum frequency of the signal to be reproduced. This is, however, an idealized situation; Figure 12.12 shows that $f_s = 2f_{max}$ can work provided that the samples are taken at the maxima and minima of the signal (Figure 12.12a). If, however, the samples are taken at the zero-crossing points, then the samples will give an erroneous impression of the actual signal (Figure 12.12b).

Compared to the maximum significant frequency expected in the signal, f_{max}, the sampling frequency, f_s, should be (as a general guideline):

- $f_s = 2f_{max}$: Nyquist criterion
- $f_s = 3f_{max}$: absolute minimum practical
- $f_s = 5f_{max}$: a practical value
- $f_s \geq 10f_{max}$: ideal situation

The sampling frequency will be limited by the technology employed by the ADC. High-precision converters, working to 24 bits, will be slower than 8-bit converters.

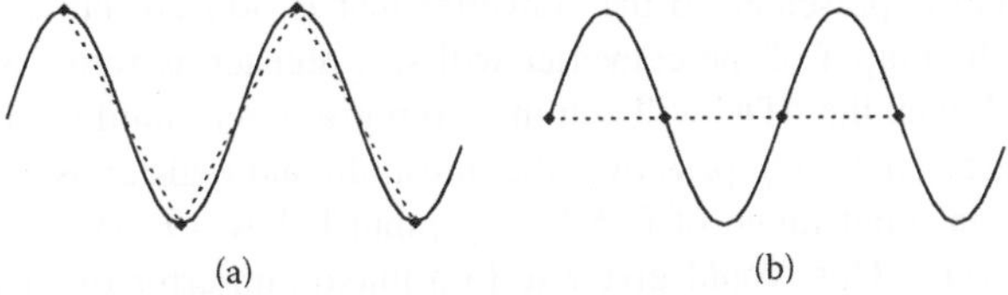

FIGURE 12.12 (a) Sine sampled at twice its frequency, at maxima and minima, (b) sine sampled at twice its frequency, at zero-crossing points.

12.3.1.1 Antialiasing Filters

As is evident from Figure 12.11, any noise signal above the sampling frequency will interfere with the correct sampling of the signal of interest and cannot be extracted by subsequent signal processing (i.e., the use of digital filtering techniques will not help). It is worth pointing out that switched-capacitor filters, which were introduced in Chapter 11 (Subsection 11.5.3), also sample the signal that they are filtering and so are subject to aliasing as well.

Consequently, it is necessary to employ analog antialiasing filters prior to any converters to remove noise above the sampling frequency. These can be simple RC filters, or more complex filters (see Chapter 11 for filter designs). Ideally, the filter would have a 3-dB point below the sampling frequency but above the maximum significant frequency of the signal.

Such filters are not normally included in the input stages of ADCs. It is common to find the bandwidth of an ADC specified as higher than the maximum sampling frequency. This allows them to be used more flexibly; they can reconstruct square waves quite effectively; for instance, if the bandwidth were limited, then square waves would be somewhat rounded when reconstructed. It does mean, however, that in such applications, care has to be taken to control sources of high frequency noise.

12.3.2 Resolution

The resolution chosen for the converter will depend on the signal that you wish to sample. The signal should be amplified so that the maximum amplitude signal that is expected almost fills the complete input range of the ADC. For example, if the input range of the ADC were ±1 V (this is also known as the *full-scale deflection*, or FSD) and the maximum signal amplitude expected was less than ±0.5 V, then the signal should be amplified with a gain of 2 prior to its being presented to the ADC.

The resolution of the ADC can then be computed based on the acceptable measurement error. So, for instance, if 1% of FSD is an acceptable error (this would be ±0.01 V in our original ±0.5 V signal), then this would imply that a converter with a 7-bit resolution would be acceptable.

Quantization Error

This error is an inherent part of the analog-to-digital conversion process. If an analog signal is presented to the converter that is nonzero but very close to zero, then the output of the converter will still register as zero. As the signal is increased, then the ADC will continue to register zero until the input signal becomes large enough to pass over the threshold and register as 1. In an 8-bit ADC with an input range of 0–5 V, any input below 19 mV (5/255) would register as zero. This would give rise to a maximum error of –1 LSB (least significant bit). Most converters are designed for an error of ± ∫ LSB, although this can give rise to other complications.

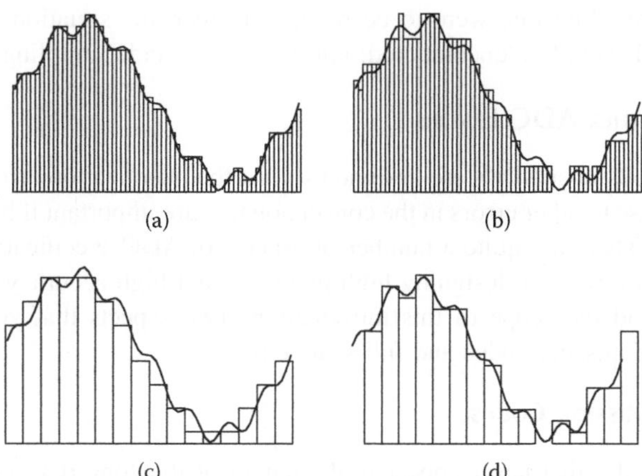

FIGURE 12.13 (a) Reconstructing a signal, 4 bits, 8 times the sample frequency, (b) 3 bits, 8 times the sample frequency, (c) 4 bits, 2 times the sample frequency, (d) 4 bits, 2 times the sample frequency, shifted to align samples with maxima and minima.

12.3.3 Signal Reconstruction: Sampling Rate and Resolution Effects

Figure 12.13 illustrates how the sample rate and resolution of the DAC combine when reconstructing an analog signal. This signal consists of two sine waves added together; the smaller of the two is eight times the frequency of the larger and one eighth the amplitude. The converters are presented such that the full-scale deflection matches the maximum amplitude of the signal.

Figure 12.13a shows the reconstructed signal sampled at eight times the maximum frequency of the signal (i.e., eight times the frequency of the smaller amplitude sine wave, or 64 times the amplitude of the larger), using a 4-bit converter. The 4-bit converter is only just enough to reproduce the smaller sine wave (the whole amplitude of this being less than 2 units of the output of the converter). Figure 12.13b shows what happens when a 3-bit converter, with only 8 discrete levels, is employed instead — even more information is lost.

Returning to the 4-bit converter, Figure 12.13c shows what happens when the sampling rate is reduced to the Nyquist minimum: twice the frequency of the highest frequency in the signal. The smaller amplitude component of the sine wave appears to have disappeared. As mentioned previously, this is not a problem with the Nyquist criterion but with exactly where the samples are taken. Figure 12.13d shows that the information is available if it can be contrived to ensure that the samples are taken at the maxima and minima of the smaller component of the waveform. Obviously, it is only possible to do this when using computer software to generate examples, but in most practical applications it is not possible to arrange this.

If the sampling rate were to be reduced further, the situation illustrated in Figure 12.11 would be encountered: aliasing due to undersampling.

12.3.4 Other ADC Errors

As mentioned, most ADCs are designed so that the quantization error is ± ∫ LSB. This gives rise to other errors in the conversion that are important if high precision is required. There are quite a number of aspects of ADC specifications that you need to be aware of if designing high-precision and high-accuracy systems, but this is beyond the scope of this introduction. Two aspects that you should be aware of are missing codes and full-scale error.

12.3.4.1 Missing Codes

These are codes that never appear in the output of the converter, and arise from various aspects of converter design. A converter with missing codes is no less accurate than an equivalent resolution converter with no missing codes, but they can be inconvenient. If this should prove the case for a particular design, then a converter has to be selected that has "no missing codes" specified on the data sheet.

12.3.4.2 Full-Scale Error

To achieve a ±∫ LSB quantization error, the threshold at which a signal moves from one output code to the next has to be offset by ∫ a bit. This results in an error in the last code (full-scale input) of 1∫ LSB. This is not normally very important, and it is the quantization error that is usually of most significance.

12.3.5 Companding

Throughout the earlier discussion, it has been assumed that changes in the output code of the converter (from $0 \rightarrow 1 \rightarrow 2 \rightarrow 3$, …, etc.) occur at specific threshold inputs that are linearly spaced (i.e., at ΔV, 2 ΔV, 3 ΔV, …, etc.). This leads to larger signals being measured with less overall error than small signals.

For instance, with an 8-bit converter, ±∫ LSB in a measurement of 128, e.g., represents an error of approximately 0.4% ($100 \times 0.5/128$). If the measurement were 16, then a ± ∫ LSB error would represent an error of more than 3%.

Companding is a method that employs nonlinear translation between input signal and output code; i.e., the change signal amplitude required to move from one code to the next depends on the current output, being larger if the present signal amplitude is already large. This means that the absolute error in measurement is evened out over the range.

12.4 ANALOG-TO-DIGITAL CONVERTERS

A number of different ADC architectures can be implemented. This section will consider PWM output ADCs, successive approximation ADCs, flash ADCs, and sigma-delta ADCs.

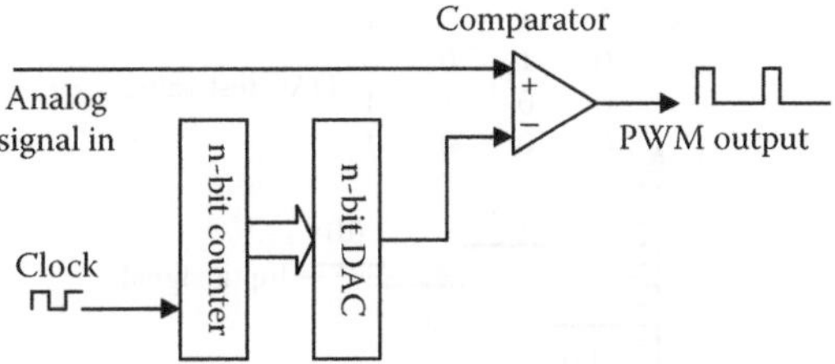

FIGURE 12.14 Schematic showing the operation of an ADC with PWM output. A digital circuit (counter and DAC) is used to generate the ramping signal against which the input analog signal is compared. The input will frequently need a sample and hold circuit to hold it constant during the conversion.

12.4.1 Sample-and-Hold Circuit

Some of the converters discussed in the following text take a long time to perform the conversion. It may, therefore, be necessary to employ a circuit known as a *sample-and-hold* circuit. This samples a signal quickly, then presents an unchanging signal to the converter on which for it to work. Some converters incorporate these circuits, whereas for others it may be necessary to obtain a separate component. Dedicated sample and hold circuits can be readily purchased.

12.4.2 PWM Output ADCs

The simplest form of PWM output ADC is illustrated in Figure12.14. It consists of an n-bit counter and an n-bit DAC, where n is the resolution of the converter. Counter A presents its output to the DAC. The output of the DAC, V_{DAC}, is compared to the analog signal, V_{in}. While $V_{DAC} < V_{in}$, the output of the converter is maintained high (digital 1). While $V_{DAC} > V_{in}$, the output of the converter is 0. The result of the conversion can be determined by measuring the mark–space ratio of the output signal; this would be performed with a second counter synchronized to the first, and the PWM signal would enable (when 1) or disable (when 0) counting on this counter.

The advantage of this approach is that high-resolution conversions can be performed and transmitted to a microcontroller using very few wires, which is often a restriction on MEMS devices.

12.4.2.1 Integrating ADC

The PWM signal can be generated without resort to digital circuitry. In this case, an integrating ADC converter would be employed.

A sample-and-hold circuit is used to present the input voltage to the converter. The counter and DAC are replaced by a capacitor and a constant-current source, which will charge the capacitor linearly (see Equation 11.2, Chapter 11). The voltage on this capacitor is compared to the sampled voltage, and when it exceeds the sampled voltage, the output of the converter is switched from 1 to 0. When

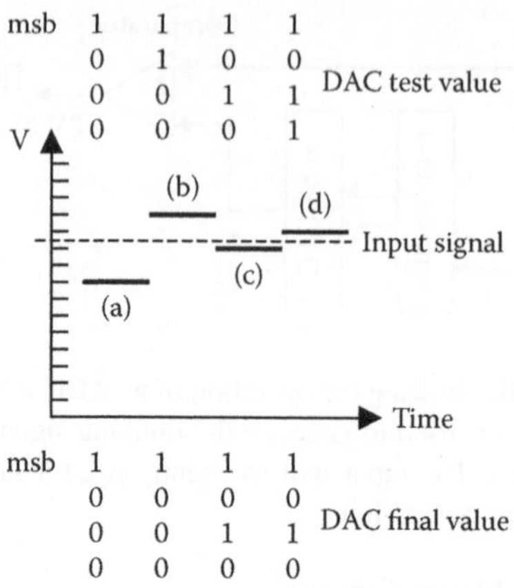

FIGURE 12.15 Illustration of the operation of a 4-bit successive approximation ADC. The input signal is shown as a dashed line, and the test value of the DAC is shown as a heavy line. In the first cycle (a) the most significant bit (MSB) of the DAC is set and the output compared to the input signal. The input signal is greater, therefore the bit remains set at the end of the step. Then (b) the next bit of the DAC is set. The input signal is less than the output of the DAC so this bit is reset. This is repeated in steps (c) and (d) to give a final value of 1010 after 4 steps.

the capacitor is fully charged, it is rapidly discharged, and the cycle repeats itself, thus creating a PWM signal.

In this case, the resolution of the converter is determined by the resolution of the counter used to decode the PWM signal.

This only presents the technique in outline. Further refinements are possible.

12.4.2.2 Conversion Time

For an n-bit converter, driven by a clock frequency, f_c, the time taken to perform a conversion will be:

$$T_c = \frac{2^n}{f_c} \tag{12.6}$$

(In other words, the time t takes for the counter to count from 0 to maximum).

12.4.3 Successive Approximation

The successive approximation ADC uses the following algorithm to perform a conversion (illustrated in Figure 12.15 for a 4-bit converter). It, too, uses a DAC, the output of which is compared with the input signal:

1. Set input to DAC to 0
2. Set most significant bit of DAC to 1

3. If the output of the DAC is greater than the input signal, then clear the bit that has just been set (i.e., reset it back to 0).
4. If you have not run out of bits, set the next most significant bit and go back to step 3.
5. Finished; the output of the ADC is the same as the current input to the DAC.

The algorithm basically works by dividing the input range of the ADC in two and seeing if the signal lies in the top or bottom half. Whichever half it lies in, this is divided in two and the test repeated until it runs out of bits to test. This is much faster than the PWM output converter described earlier. The conversion time for an n-bit successive approximation ADC is:

$$T_c = \frac{n}{f_c} \tag{12.7}$$

So the conversion time increases linearly with the resolution. The disadvantage of this converter compared to the PWM output version is that it requires n output lines (unless additional circuitry is included to implement a serial communications port), and a DAC is essential.

12.4.4 Flash ADC

The flash ADC is, as its name suggests, the fastest possible converter. For an n-bit converter, a potential divider chain is used to divide a reference (full scale) voltage into 2^n different voltages. Then, 2^n comparators are used to compare the input signal with each point on the potential divider chain. This is then converted to an n-bit output code by digital logic circuitry.

The conversion time is limited only by the time it takes the signal to propagate through the circuitry, but the converter has several disadvantages.

The large number of component parts means that it is the least accurate of the available ADC architectures. Also, because the area of silicon required increases with 2^n, only relatively low-resolution converters are possible.

It is possible to achieve some compromise by combining successive approximation and ADCs flash ADCs to produce reasonably fast high-resolution converters, but these are becoming less common with improving technology.

12.4.5 Sigma-Delta Converter

The sigma-delta has recently become more popular because of the increasing speed of CMOS technology. There are various approaches, and its operation is reminiscent of that described for integrating PWM output converters (earlier text).

The sigma-delta converter consists of an integrator and a single-bit DAC (i.e., a comparator). A number of pulses are applied to the input of the integrator until its output exceeds the level of the input signal, as determined by the comparator. The number of pulses required is counted, and this gives the output code of the converter.

The time taken to perform a conversion is, therefore, dependent on the input signal, but the maximum conversion time can be computed using Equation 12.6.

Sigma-delta converters offer the highest accuracy and resolutions of all converter types discussed here, but they are also one of the slowest types.

12.5 CONVERTER SUMMARY

The following are the main features of the DAC and ADC:

- DAC:
  - PWM is simple and can be used to drive integrating loads directly.
  - R-2R ladder DAC is, generally speaking, most useful.
  - Reconstruction frequency should be at least 5 times, preferably 10 times the highest frequency of the reconstructed signal.
  - Reconstruction filters should be employed as required.
- ADC:
  - Nyquist criterion: $f_s = 2f_{max}$.
  - Sample rate should be 5 to 10 times the maximum significant frequency in the signal to be measured, and at least 3 times.
  - Antialiasing filters must be employed.
  - Converter resolution should be selected to ensure an acceptable measurement error for the expected signal.
  - PWM output converters are slow, but give a serial output.
  - The successive approximation ADC offers good compromise design.
  - The flash ADC is fast, but less accurate.
  - The sigma-delta ADC is slow but very accurate and precise.

REFERENCES

1. Horowitz, P. and Hill, W., *The Art of Electronics*, 2nd ed., Cambridge University Press, Cambridge, New York, and Melbourne, 1989.
2. Banks, D., Neurotechnology: microelectronics, in Finn, W.E. and LoPresit, P., Eds., *Handbook of Neuroprosthetic Methods*, CRC Press, Boca Raton, FL, 2003.
3. Gray, N., ABCs of ADCs, Revision 2, August 2004, National Semiconductor Corporation (www.national.com).
4. National Semiconductor, A/D Converter Definition of Terms, January 2000, National Semiconductor Corporation (www.national.com).

13 Output Drivers

13.1 INTRODUCTION

The previous chapters of Part III introduced basic electronics and how this may be used to interface microengineered sensors or generate signals to control microengineered transducers. This chapter starts by showing how operational amplifiers (op-amps) can be used to drive electrodes in biosensors, although the same configurations can be used in other applications. However, basic op-amps have only limited capability to drive large currents or voltages, although most microactuators are not very demanding in this respect, it may be necessary to employ different output stages when driving electrostatic, piezoelectric or thermal actuators, or in μTAS applications. Power op-amps may be used in some situations, but these are limited and some alternative strategies, mainly based on the employment of discrete power transistors, are introduced here.

13.2 CONTROLLING CURRENTS AND VOLTAGES WITH OP-AMPS

The common voltage feedback op-amp can be used in a variety of configurations to control and measure both voltages and currents. For brevity and generality of application, this section does not feature complete circuit diagrams, and readers are referred to the material introducing op-amp and differential amplifiers in Chapter 11 if they wish to develop application-specific circuit designs.

13.2.1 Op-Amp Current Control

In many situations it is desirable to generate a current between two electrodes that is controlled by a voltage. This can be done quite simply by reference to the ideal inverting amplifier configuration. Figure 13.1 illustrates a simple circuit by which current flowing through a measurement electrode (M) may be controlled.

Ignoring, for the moment, the reference electrode shown in Figure 13.1, the current flowing from the control electrode (C) to the measurement electrode (M) must be equal to the current flowing through the resistor, R. Thus:

$$i_M = \frac{-v_i}{R} \tag{13.1}$$

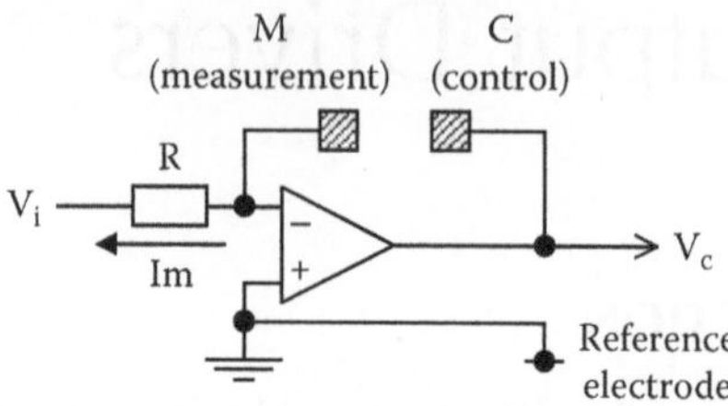

FIGURE 13.1 An op-amp controlling the current through two electrodes. The virtual earth means that the current is defined by the applied voltage: $i_M = \frac{-v_i}{R}$. This circuit has limited applications, more effective circuits are dealt with later in this chapter.

Clearly, this is a very convenient way by which current can be controlled. The system enables the resistance or impedance between the electrodes to be measured by monitoring the voltage of the control electrode, v_c. The reference electrode, however, requires some explanation.

In many situations the reference electrode will not be required. If, however, the measurement is being made in an aqueous environment (as is common with many biosensor applications), the reference electrode should be used to stabilize the measurement. If the measurement and control electrode are immersed in a volume of liquid in order to effect the measurement without the presence of a reference electrode, then errors will be introduced because the potential of the environment will drift with respect to the measurement system. A reference of some sort is required in all such situations.

For the reference electrode to work, it must be placed a long way away from the control and measurement electrodes at a distance greater than that between these two electrodes. Furthermore, the path between the reference electrode and the control and measurement electrodes must be resistive so that the control electrode can work (otherwise the entire measurement system would remain at 0 V because all current sourced or sunk by the control electrode would flow to Earth via the reference electrode rather than to the virtual earth of the measurement electrode). For a referenced system to be effective, the control electrode has to be able to raise or lower the potential in the vicinity of the measurement electrode without the reference electrode sourcing or sinking a large amount of current.

13.2.1.1 Four-Electrode Configuration

An alternative approach to measure change in impedance or resistance of a biosensing medium is the four-electrode configuration, illustrated in Figure 13.2. Here, the functions of current generation and voltage measurement have been separated. If it can be assumed that all the current flowing between the two outer electrodes passes through the two inner electrodes, then an absolute measurement can be made. At the least, however, changes in impedance can be registered.

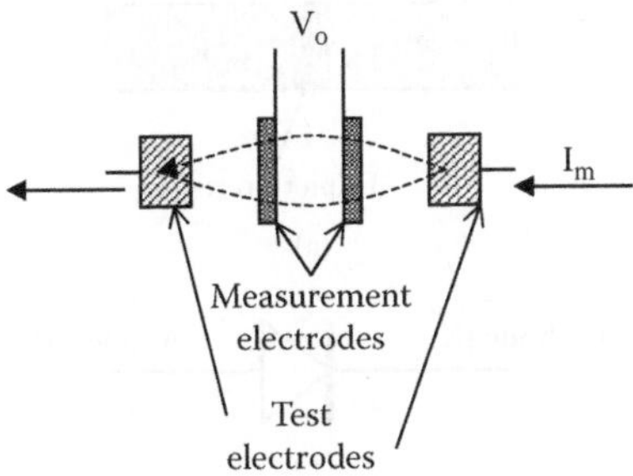

FIGURE 13.2 Four-electrode impedance measurement. Current is applied through two test electrodes, passing through the medium under test across two measurement electrodes. A high-impedance differential voltage amplifier is used to determine the voltage difference between the two measurement electrodes, which are a known distance apart. In this way, problems in applying the current are separated from measurement problems.

The advantage of this approach is that passing even relatively small electric currents through electrodes can introduce errors in voltage measurements in some situations. In this situation, the only currents passing through the voltage measurement electrodes will be the bias currents of the differential amplifier (these will need to flow to Earth, so a reference electrode may again be necessary to stabilize the system). The use of AC signals rather than DC, to make impedance measurements, will also help if irreversible electrochemical activity caused by the latter introduce errors.

13.2.2 Op-Amp Voltage Control

The use of an op-amp with negative feedback to control an output voltage is shown in Figure 13.3. A small amount of the quantity being controlled is fed back via V_{sense}. Assuming that V_{sense} has the same polarity as V_o (i.e., as V_o increases, so does V_{sense}) then, because the ideal op-amp attempts to maintain its inputs at the same potential, V_o will follow V_{in}. V_o may be larger or smaller than V_{in}, depending on whether V_{sense}/V_o is less than or greater than unity.

The applications of this approach will become apparent in subsequent sections, which consider the use of transistors to boost the output of an op-amp, and optoisolators. It should be noted that if V_{sense}/V_o is negative, or if V_{sense} lags V_o by a significant time delay, then the system will become unstable and oscillate.

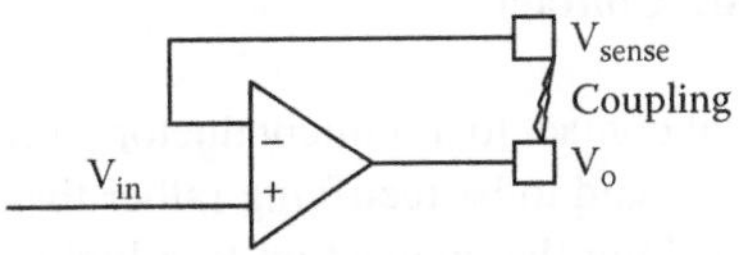

FIGURE 13.3 An op-amp controlling a voltage.

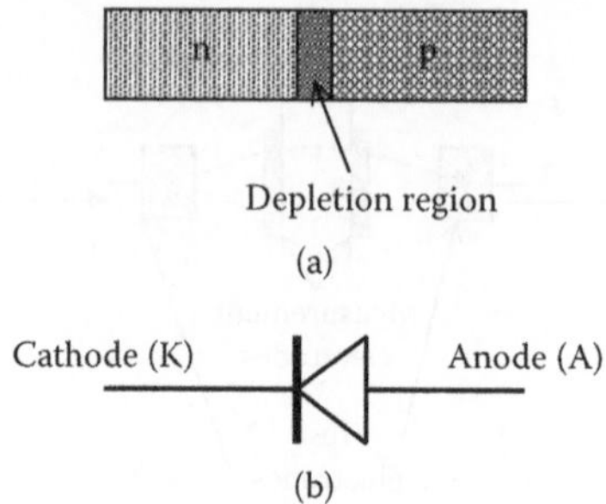

FIGURE 13.4 (a) Schematic of pn junction, (b) diode symbol.

13.3 TRANSISTORS

Of the several kinds of transistors, only four or rather two pairs, will be considered in this section, the npn and pnp bipolar junction transistor (BJT) and the n- and p-channel enhancement-mode metal-oxide-semiconductor field effect transistor (MOSFET). These are the most useful transistors for the sort of applications likely to be encountered by anyone developing instrumentation for MEMS.

13.3.1 THE BJT

The bipolar junction transistor is so called because it is formed by two pn junctions. The doping used to create pn junctions for electronic devices is generally lower than that required for concentration-dependent etching. The pn junction is shown schematically in Figure 13.4a, along with the circuit symbol for a pn junction diode (Figure 13.4b). The ideal diode acts as a one-way switch: current will flow through the device from the p-doped semiconductor to the n-doped semiconductor (in the same direction as the "arrow" of the diode symbol), but will not flow in the other direction.

The real diode is, of course, nonideal. The most important difference is that a small voltage must be dropped across the component before any current will flow. In the case of a silicon diode this voltage drop is about 0.7 V. In the case of a metal semiconductor (Schottky) diode, it is 0.2 V, and in the case of a light-emitting diode, 2 V. Figure 13.5 shows the idealized current–voltage (I–V) characteristic of a silicon diode.

Rectifying and Ohmic Contacts

When making electrical contact to a semiconductor with a metal, diode effects can occur; the contact is said to be rectifying rather than ohmic. The situation can be improved by making the connection to a highly doped, and therefore conductive, part of the device.

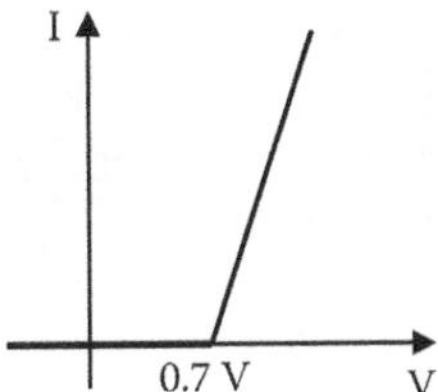

FIGURE 13.5 Idealized I–V (current–voltage) characteristic for a silicon diode. The graph is very steep, indicating a low resistance.

The BJT is a three-layer device with a very thin layer of p-doped material (the base) sandwiched between two n-doped layers, the collector and the emitter, (npn) or a thin n-doped base with p-doped collector and emitter (PNP). Figure 13.6 schematically shows the structure and circuit symbols of npn and pnp transistors. The direction of the arrow on the symbol indicates the direction of the base–emitter junction diode. Although the devices look symmetrical, they are normally fabricated in an asymmetrical manner, so attention should be paid to the package labeling.

The BJT is a current-controlled current amplifier. Figure 13.7a shows a configuration common for an npn transistor. The collector is the most positive part of the circuit, and the emitter is nominally the most negative part of the circuit; if necessary, the base can be driven to a more negative voltage than the emitter, and both diode junctions will then be reverse biased and no current will flow. The npn transistor is used more often than the pnp, so this treatment will proceed with reference to the npn. The currents in pnp circuits are reversed (Figure 13.7b).

For large signals, the BJT acts as if it has a current gain of β:

$$I_c = \beta I_b \tag{13.2}$$

If a base current is flowing such that the base–emitter junction is forward biased, then:

$$V_b = V_e + 0.7 \tag{13.3}$$

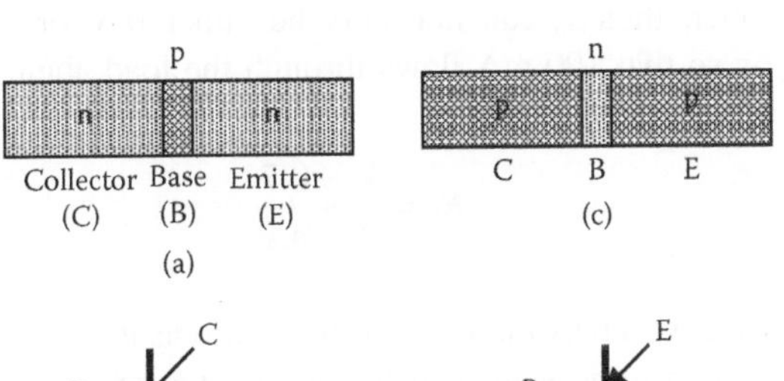

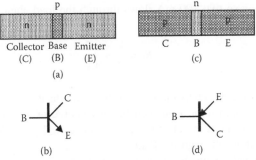

FIGURE 13.6 (a) npn sandwich, (b) npn symbol, (c) pnp sandwich, (d) pnp symbol.

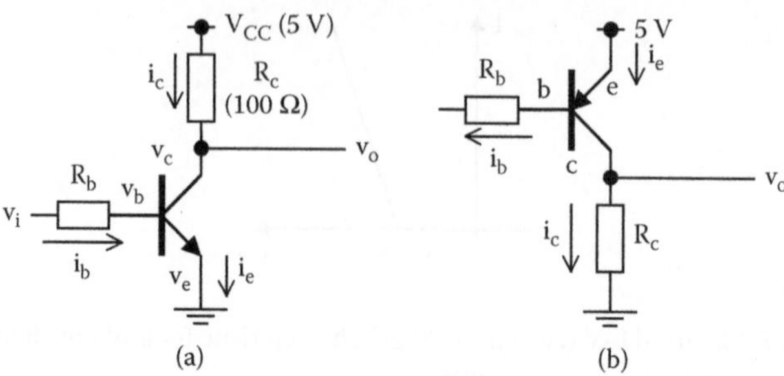

FIGURE 13.7 BJT as a switch: (a) npn, (b) pnp.

Armed with this knowledge and the tools introduced in Chapter 11, it is possible to analyze various simple transistor circuits.

It should be noted that for a single type of commercially available transistor, there is a wide tolerance range for values of β — over a factor of three or four is typical (e.g., from 60 to 240). Additional circuitry is required to compensate for this when the transistor is used as an amplifier; however, in its most common application as a switch, this is not a concern — the designer simply allows for the lowest possible value of β.

For example, in Figure 13.7, a BJT is used to control current through a 100 Ω load. The BJT has a β of at least 40, and is rated at 250 mA for maximum values of I_c to. If V_i is 0 V, then no base current will flow because V_b must be greater than or equal to V_e + 0.7 in order to forward bias the base–emitter junction. If I_b is 0, then from Equation 13.2, I_c must be 0 and no current will flow through the load. This also means that V_c must be 5 V (i.e., the transistor is an open circuit).

To turn the load on, 10 V must be dropped across it, so at least 100 mA must flow through it (from Ohm's Law). This implies that I_b must be at least 2.5 mA (100 ÷ 40).

If the system is being controlled from a logic gate, for example, the output of a microcontroller, then V_i can normally be either 0 V or 5 V. To turn the transistor fully on so that 100 mA flows through the load, then:

$$R_b = \frac{V_i - 0.7}{2.5\ \text{mA}} \tag{13.4}$$

Because V_i will have a maximum of 5 V, this suggests that R_b should be 1.72 kΩ or less. Many modern logic gates can happily source or sink up to 10 mA, so there would be no harm in selecting a value of, e.g., 1 k for R_b to ensure that enough current flows through the load.

Here, the transistor has been used to interface a low-current system with a limited supply voltage to drive a high-current load at double the supply voltage.

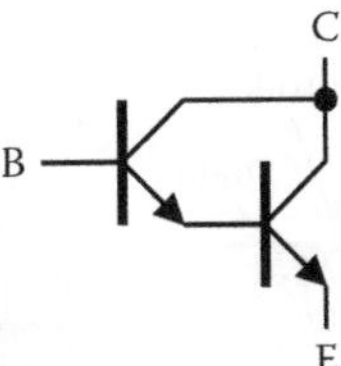

FIGURE 13.8 Darlington pair (npn).

The voltage dropped across the load will not be exactly 10 V because even when the transistor is fully turned on (saturated) some voltage is still dropped across it (V_{CESAT}). Additionally, should the load be capacitive, as is the case when driving an electrostatic actuator for example, then a large amount of current will need to be available to ensure that the capacitor charges rapidly enough.

Additional current amplification can be achieved by employing two transistors together, as in a *Darlington* configuration (Figure 13.8). Darlington transistors are available in a variety of formats and packages.

13.3.2 The MOSFET

The name metal-oxide-semiconductor field effect transistor (MOSFET) is derived from its structure. It has three terminals: drain, source, and gate. Current flows between the drain and source through a semiconducting channel in a silicon substrate. This channel is controlled by the metal gate electrode, which is separated from the channel by a thin oxide layer. MOSFETs can be found in one of two modes: enhancement or depletion, and may be either n-channel or p-channel. Enhancement-mode MOSFETs are more commonly employed than the depletion-mode, and for simplicity, these will be the focus of this section.

Figure 13.9 shows the structure and symbols for n-channel (Figure 13.9a and Figure 13.9b) and p-channel (Figure 13.9c and Figure 13.9d) enhancement-mode MOSFETs.

Note the substrate connections in Figure 13.9. For correct operation, this has to be connected to the most negative voltage for n-type and to the most positive voltage for p-type transistors. This is required to ensure that the pn junctions between substrate and the active parts of the device (channel and source or drain implants) are reverse biased. Although the structure of the device looks symmetrical, discretely packaged transistors are usually supplied with the substrate internally connected to source, which will normally be the more negative (n-channel) or more positive (p-channel) than the drain during normal operation of the device. (In some cases, asymmetrical doping may be employed as well).

CMOS processes employ both n-channel and p-channel MOSFETs, and so will be referred to as either n-well or p-well processes. In the n-well process, a p-type wafer is specified and deep n-type diffusions or implants are introduced into this at points where p-channel MOSFETs will be fabricated, and vice versa for the p-well process.

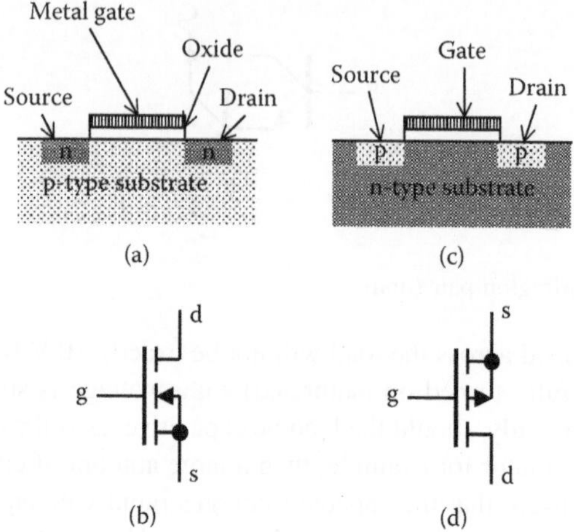

FIGURE 13.9 (a) n-channel MOSFET structure, (b) symbol, (c) p-channel, (d) symbol. All are in the enhancement mode. It is common to abbreviate the circuit symbols by joining the broken line and omitting the substrate connection. In many cases the substrate is internally connected, as shown in the symbols, to the most negative part of the circuit for the n-channel MOSFET and to the most positive part for the p-channel.

The operation of an n-channel enhancement-mode MOSFET is shown schematically in Figure 13.10. As the gate voltage is increased, electrons are attracted towards it. Eventually, there will be sufficient electrons beneath the gate to locally convert the silicon to an n-type semiconductor. At this point, current will be able to

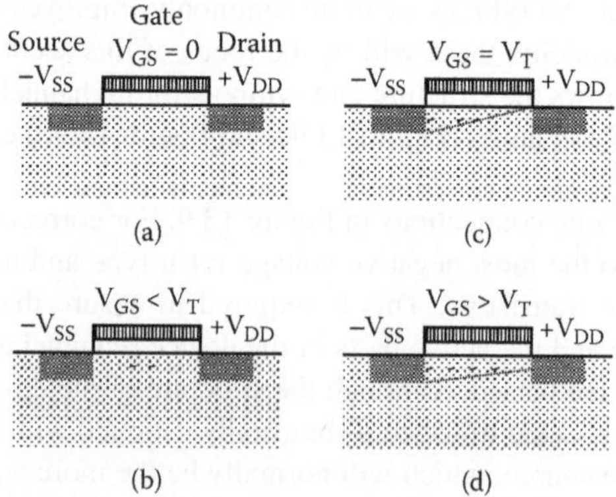

FIGURE 13.10 Operation of n-channel enhancement-mode MOSFET: (a) with no gate bias; (b) as the potential difference between the gate and source, V_{GS}, is increased, electrons are attracted towards the gate; (c) at a certain threshold value, V_T, sufficient electrons are beneath the gate region to make it an n-type semiconductor, and a conducting channel opens between the drain and source. As V_{GS} is increased, the channel opens further, and its equivalent resistance is reduced.

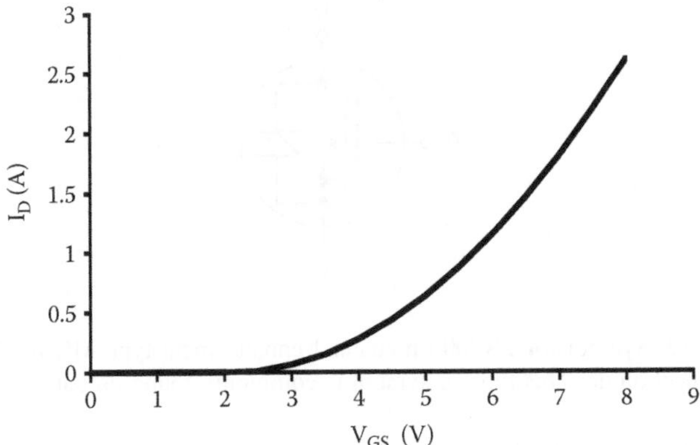

FIGURE 13.11 Graph of I_D vs. V_{GS} for an n-channel enhancement-type MOSFET with $V_T = 2.1$V and $k = 0.075$ S; this is similar to the readily available 2N7000.

flow from drain to source. The voltage between gate and source (V_{GS}) required to open up the channel is the threshold voltage, or V_T. As V_{GS} is increased, more current will be able to flow.

The current flow will be limited by the drain–source voltage if:

$$V_{DS} > V_{GS} - V_T \tag{13.5}$$

In this case, the current flowing, I_D, is given by:

$$I_D = k(V_{GS} - V_T)^2 \tag{13.6}$$

The value of k depends on the channel dimensions and the process, and will not be pursued any further in this chapter. Figure 13.11 shows the characteristics of an n-channel MOSFET with $k = 0.075$ S and $V_T = 2.1$ V. This sort of characteristic is typical of a discrete MOSFET used for low-power switching, such as the 2N7000.

From Equation 13.6, it is apparent that, unlike the BJT, the MOSFET is a transconductance amplifier: a change in gate voltage produces a corresponding change in drain current. So its transfer characteristics are given in terms of a current divided by a voltage (I_D/V_{GS}) — which is a conductance.

MOSFETs make very good switches, because virtually no gate current is required to switch large drain currents. Additionally, the channel behaves like a resistor, and this can be made very low when the transistor is turned on fully (data sheets will list this as $R_{DS(ON)}$ — the drain–source on-resistance). For low-power applications, some MOSFETs are designed with low-threshold voltages so that they can be driven by logic level signals. (Obviously, MOSFETs on ICs

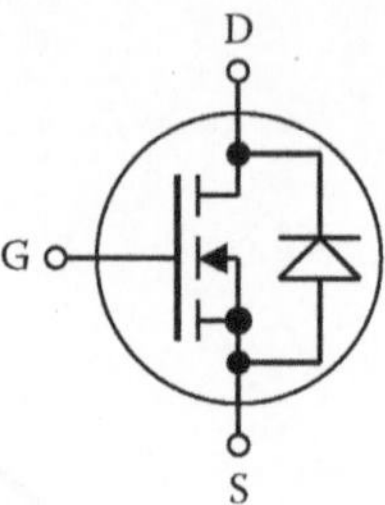

FIGURE 13.12 Symbol for 2N7000 n-channel enhancement-type MOSFET showing a built-in flywheel diode. This type of symbol is commonly found on data sheets.

are designed specifically for low-voltage operation, but will not be discussed as this chapter deals with output drivers for power applications).

As a switch, the MOSFET has the following characteristics:

- Negligible steady-state gate current required.
- Gate current must be sufficient to charge the gate capacitance (between the gate and the channel) quickly when turning the transistor on (or discharge it when turning it off).
- Gate capacitance and available current to the gate limit the switching time.
- V_{GS} must be sufficient to overcome the threshold voltage and ensure a low R_{DS} (not always possible with a logic level drive).

An additional caveat when employing discrete MOSFETs is that they are frequently intended to be used as switches, and to this end many incorporate flywheel diodes that will carry reverse currents that appear when inductive loads are being switched. Normally, these can be ignored, but sometimes their incorporation can cause problems. Figure 13.12 shows the symbol for a 2N7000 n-channel enhancement-mode MOSFET, which incorporates such a diode.

13.4 RELAYS

The relay is an electromechanical switch. An electromagnet is used to close a switch element. When no current is flowing through the electromagnet, a spring holds the switch in one position. When current flows the electromagnet holds the switch in a second position. Figure 13.13a shows the circuit symbol for a single-pole double-throw (SPDT) relay, and Figure 13.13b shows how this relay can be driven by a MOSFET.

The relay coil is inductive, so a diode is required to protect the transistor during switching. The resistance of relay coils is normally in the range of 500 to 1000 Ω, so they cannot normally be switched directly by digital outputs. Heavy-duty relays will require higher currents.

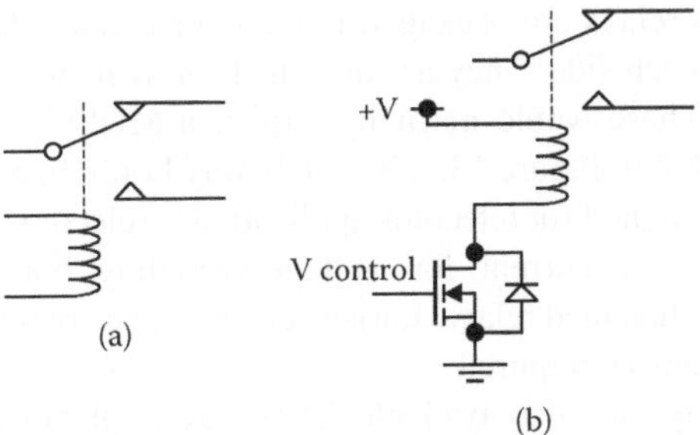

FIGURE 13.13 (a) SPDT relay symbol, (b) the same relay driven by a MOSFET.

Relays are commonly described by their switching arrangement. The number of poles is effectively the number of switches, and the number of throws is the number of contacts that each switch (pole) can make (normally only one or two in a relay: single or double throw).

13.4.1 Relay Characteristics

As mechanical switches, relays have several desirable characteristics:

- Low resistance
- High-current and high-voltage operation
- Can carry high-frequency signals
- Good isolation (high impedance when in the "off" state)
- Bidirectional current flow

These are counteracted by the following disadvantages:

- Physically large, generally
- High-power operation (electromagnets require high currents)
- Take a long time (ms) to switch over
- Contacts bounce: contact is partially made and broken several times before the switch settles into its final state (this can be eliminated by mercury-wetted contacts, but these are being withdrawn because of environmental concerns)

13.4.2 Relay Types

There are several different types of relay available, with different characteristics and applications.

The reed relay is the most compact conventional relay design. The switching element is a very fine leaf spring that can be magnetically actuated. These are designed for switching small currents, and can sometimes be driven directly by logic outputs. Reed relays are normally only SPST.

Most common relays are packaged in cubic or rectangular packages, a centimeter or two on each side. They are designed for switching several amperes at up to 250 V AC. These would normally require a MOSFET drive if controlled automatically, as that in Figure 13.13b. A half-way house between these and reed relays are relays designed for telecoms applications. Telecoms relays are compact relays designed for low-current, low-voltage switching, but are more mechanically sophisticated than reed relays. Larger relays can be obtained to switch higher currents and voltages as required.

Solid-state relays are also available. These are implemented in silicon technology and have no moving parts, and they are generally designed using CMOS transistors. Optically isolated versions are available (see the following text). These generally have fast switching times and no contact bounce, but high-power devices are at present quite costly.

Owing to the advantages of electromechanical switches, MEMS relays have been developed, and some are commercially available. These are usually based on cantilever structures, actuated by a variety of methods. MEMS relays have generally been designed for communications applications: they are more compact than any other relay design, and offer good isolation and bidirectional current flow. Power requirements for actuation depend on the strategy used, but they are generally portable-device-friendly.

13.5 BJT OUTPUT BOOST FOR OP-AMPS

An arbitrary scheme for op-amp voltage control was indicated in Figure 13.3. Figure 13.14 shows how this and a BJT can be combined to boost the current-handling capacity of an op-amp. The current available to drive the load will be magnified by the current gain (β) of the transistor. Another way of looking at it is that the load resistance will appear as a resistor of value βR_L to the output of the op-amp. This circuit cannot sink current, but it is a useful circuit for driving resistive heaters or coils on MEMS devices.

It may be desirable to control the current flowing through the load in Figure 13.14. One scheme for this is shown in Figure 13.15, where a sense

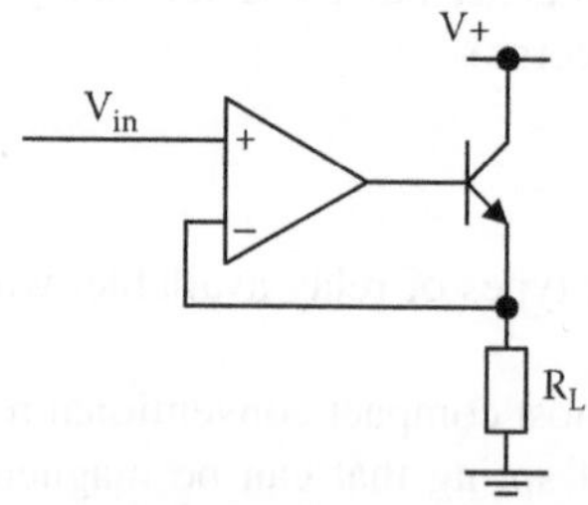

FIGURE 13.14 A BJT used to boost the output of an op-amp output.

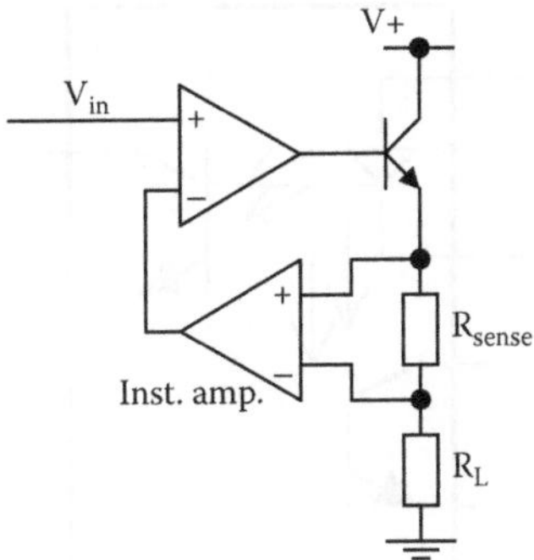

FIGURE 13.15 A BJT used to boost the output of an op-amp output as in Figure 13.14, but current, not voltage, is being controlled.

resistor is incorporated into the circuit. Note that R_{sense} must be much smaller than R_L, otherwise it will dissipate power. This is why an instrumentation amplifier has been incorporated into the circuit.

13.6 OPTOISOLATORS

When working with high voltages, such as in some MEMS situations: electrostatic drives or electroosmotic flow generation, it is advisable to isolate the high-voltage side of the equipment from the low-voltage control circuitry for reasons of safety if not concern about damaging sensitive equipment. Relays, offer one mechanism for this. A more flexible approach is optical isolation (optoisolation).

Optoioslators comprise a light-emitting diode and a light-sensitive diode or transistor in a single package. The simplest devices are designed to isolate digital signals, and these can effectively be employed for PWM-type signals (Figure 13.16).

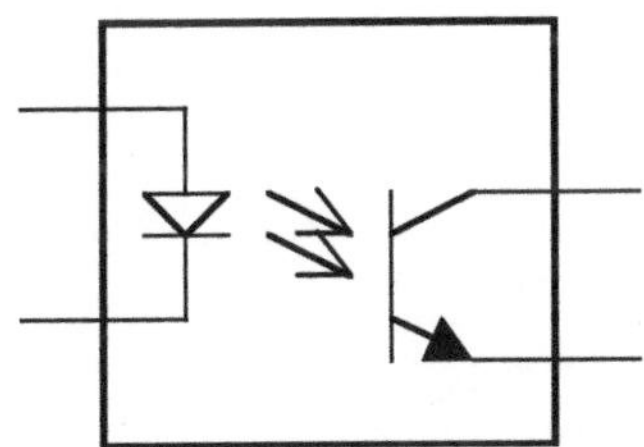

FIGURE 13.16 Optoisolator suitable for digital signals. The anode of the LED is connected to a positive power rail through a resistor and the cathode to a digital line. The transistor would be operated as a switch (collector pulled up to 5 V through a resistor, and the emitter connected to ground).

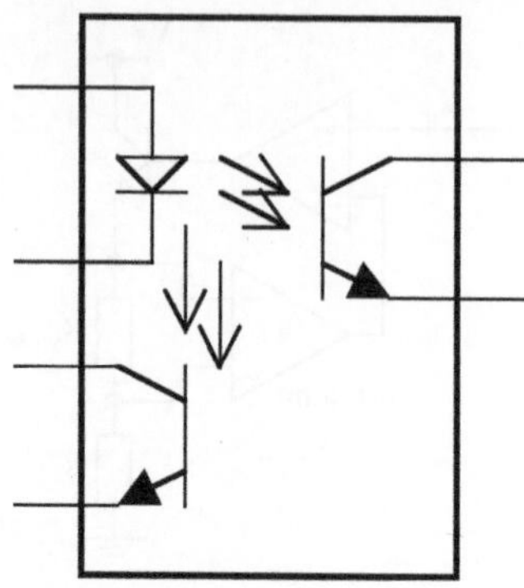

FIGURE 13.17 This optoisolator incorporates two matched transistors so that analogue signals can be reproduced accurately. The transistor on the LED side is incorporated into a feedback loop in the circuit that drives the LED.

Optical isolation of analog signals is more problematic because of nonlinearities in the optical components and variability from device to device. This is overcome by employing matched receivers on either side of the isolation (Figure 13.17). In this case, the signal on one side can be monitored, and the duplicate circuitry on the other side duplicates the signal correctly.

Index

A

Aberrations, optical, 19
Absolute values, *264*
Accelerometers, 146
AC-coupled amplifiers, *259,* 260–261
Acetone, 25, 66
Actin filaments, 230
Actuators, 131, *see also* Microactuators
ADC, *see* Analog-to-digital converters (ADCs)
Adenosine diphosphate (ADP), 229
Adenosine monophosphate (AMP), 229
Adenosine triphosphate (ATP), 229
Adhesion, 215–216, *216*
ADP, *see* Adenosine diphosphate (ADP)
AFM, *see* Atomic force microscope (AFM)
Alarms, intruder, 137
Aliasing, 289, *see also* Antialiasing filters
Aligners, mask, 12–15, *13–15*
Alignment
  crystal planes, *61*
  marks, mask design, 105–107, *106,* 111, *112–113*
  sequence, 121
Alignmentframe structure, 111, *112–113,* 117
Align structure, 108, *111*
Alternating current (AC), *241*
Aluminum films, 206
Amino acids, 168–170
Amorphous carbon, 211
AMP, *see* Adenosine monophosphate (AMP)
Amplifiers and filtering
  AC-coupled amplifiers, *259,* 260–261
  applications, 258–263, *259*
  bandwidth limitations, 254–255
  basics, 239
  bias currents, 255–256, *256*
  Butterworth filters, 273–279, *274–275*
  capacitor bridge, 266, 268
  capacitors, ideal, *240,* 242
  common-mode rejection ratio, 256–257
  conductors, ideal, *240,* 241
  controlled sources, *240,* 243–244
  current conventions, 239–241, *240–241*
  current sources, ideal, *240,* 243
  differentiators, *259,* 261–262
  electronics, 239–247, *240*
  filtering, *259,* 268–280
  frequency response, Butterworth filters, *278,* 278–279
  functions, 263
  inductors, ideal, *240,* 242–243
  input impedance, 255–256
  instrumentation amplifiers, 263–265, *264*
  insulators, ideal, *240,* 241
  integrators, *259,* 261–262
  inverting op-amp configurations, *249,* 251–252, *252–253*
  Kirchoff's laws, 246–247, *246–247*
  noise, 257–258
  nonideal amps, 253–257
  nonideal sources, *249,* 251–252, *252–253*
  noninverting op-amp configurations, *249,* 251–252, *252–253*
  op-amps, 247–263, *248*
  parallel components, 245, *245*
  power calculations, 244
  power supply rejection ratio, 256–257
  RC filters, 268–271, *269, 271–272,* 273
  resistors, ideal, *240,* 241
  series components, 245, *245*
  slew rate, 254–255
  summing amplifiers, *259,* 261
  switched-capacitor filters, *279,* 279–280
  thermal noise, 258, *258*
  unity-gain bugger amplifier, 258–260, *259*
  voltage conventions, 239–241, *240–241*
  voltage sources, ideal, *240,* 243
  Wheatstone Bridge, 265–268, *266–268*
  white noise source combination, 257, *257*
Analog devices, 282–288
Analogies, strange, *243*
Analog-to-digital conversion, *288,* 288–292
Analog-to-digital converters (ADCs), 292–296
Angell, Terry, Jerman and, studies, 155
Anistropic etchants, 53, *see also* Potassium hydroxide (KOH)
Anistropic reactive-ion etching, 66

Annealing, *51*
Anodic bonding, 67
Antialiasing filters, *289,* 290, *see also* Aliasing
Antireflective coating, 21
Applications
  amplifiers and filtering, 258–263, *259*
  biochemistry, 177–178
  biology, 178–183
  chemistry, 177
  devices, 176–183
  integrated optics, 205–208
  micro total analysis systems, 176–183
  photoresists, *23,* 27–28
Approximation, successive, *294,* 294–295
Argon plasma, 57
Arrow structure, 111, *112–113*
Artifacts, SPM, *221,* 221–222
Assembly and packaging
  assembly, 209–211
  auto-alignment, *210,* 210–211
  basics, 7, 209
  conventional IC packaging, *213,* 213–214
  design for assembly, 209–211
  flip-chip bonding, 215
  future possibilities, *210*
  materials, prototypes, 215–216, *216*
  multichip modules, 214
  packaging, 212–214
  passivation, 211–212
  prepackage testing, 212
  prototype materials, 215–216, *216*
  self-alignment, *210,* 210–211
  self-assembly, *210,* 210–211
  thermocompression bonding, 214
  ultrasonic bonding, 214–215
  wire bonding, 214–215
Astigmatism, 19
Atomic force microscope (AFM)
  artifacts and calibration, 222
  biology, 178–179
  nanotechnology, 219–220, *220*
  SPM pens, 224
  surface roughness, 93
Atomic number, 157
ATP, *see* Adenosine triphosphate (ATP)
ATPase, 231–232
Auto-alignment, *210,* 210–211
AutoCAD, 97
Automatic stirrers, 26

B

Bandwidth limitations, *254,* 254–255
Beams, 59, 61–63
BESOI, *see* Bonded and etched-back SOI (BESOI)
Bessel filters, 273–274
BHF, photostructurable glasses, 90
Bias currents, 255–256, *256*
Binary format, 97
Biochemistry
  basics, 167–168
  carbohydrates, 175, *176*
  cholesterol, 174–175, *175*
  devices applications, 176–178
  fats, 173, *174*
  lipids, 172–173
  nucleic acids, 170–172, *170–173*
  phospholipids, 173–174, *174–175*
  proteins, 168–170
Biology
  basics, 178
  chromatography, 180–181, *180–181*
  electrophoresis, 181–182, *182*
  mass spectrometry, 182
  microscopy, 178–179
  nuclear magnetic resonance, 182–183
  radioactive labeling, 179
  x-ray crystallography, 182–183
Bionanotechnology, 228–233, *229*
Biosensors, 138–141
Bipolar junction transistors (BJTs)
  AC-coupled amplifiers, 261
  input bias currents, 256
  output boost, op-amps, *307–309,* 308–309
  output drivers, 300–303, *300–303*
BJT, *see* Bipolar junction transistors (BJTs)
Black wax, *70,* 215
Blood glucose sensors, 138
Bode plot, 270–271, 278
Bogart, T., 243
Boltzmann's constant, 258
Bonded and etched-back SOI (BESOI), 43
Bond formation, *156, 158,* 159–161, *160–161*
Bonding pads, 111, 114
Borosilicate glass (BSG), *57*
Bosch process, 57
Breaking design rules, 120
Bridges, *54,* 61–63
British government, 1
BSG, *see* Borosilicate glass (BSG)
Buckminsterfullerene, 234
Bulk silicon micromachining, 59–64
Business, incorporation of microengineering, *7*
Butane, *162*
Butterworth filters, 273–279, *274–275*

C

Cabling, *265*
Calibration
- blood glucose sensors, 138
- exposure time, *13*
- nanotechnology, *221*, 221–222

CalTech Intermediate Form (CIF) file format, 97, *97*, 123
Capacitive sensors, 144–145
Capacitor bridge, 266, 268
Capacitors, ideal, *240*, 242
Capillary drilling (CD), 94
Carbohydrates, 168, 175, *176*
Carbon dioxide dryers, 66
Carbon dioxide lasers, 76
Caveats, 100
Cavities, *56*, 71
CCDs, *see* Charge-coupled devices (CCDs)
CD, *see* Capillary drilling (CD)
Cell membranes, 229
Charge-coupled devices (CCDs), 135, *136*
Chebyshev filters, 273–274
Chemical-mechanical polishing (CMP), 73
Chemical sensors, 138–141
Chemical vapor deposition (CVD)
- deposition, 47, *48*, 49, *51*
- RIE changeover, 57

Chemistry
- applications, devices, 176–183
- basics, 156
- biochemistry, 167–175
- bond formation, *156*, *158*, 159–161, *160–161*
- carbohydrates, 175, *176*
- devices applications, 176–183
- environmental monitoring, 177
- inorganic chemistry, *156*, 157–161, *158*
- lipids, 172–175
- microengineered devices applications, 176–183
- micro total analysis systems, 156–175
- nucleic acids, 170–172, *170–173*
- organic chemistry, *162–166*, 162–167
- pH, 161
- polymers, 164
- process monitoring, 177
- proteins, 167–170
- silicones, *166*, 166–167
- synthesis, 177

Chlorine, *159*
Cholesterol, 174–175, *175*
Chromatic aberrations, 19
Chromatography, 180–181, *180–181*
Chrome-on-quartz mask, 223
Chromium-coated mask blanks, 33
CIF, *see* CalTech Intermediate Form (CIF) file format
Cleaning, substrate, 25–26
CMP, *see* Chemical-mechanical polishing (CMP)
CMRR, *see* Common-mode rejection ratio (CMRR)
Coaxial cable, *265*
Codons, 172, *173*
CO_2 lasers, 76
Comb drives, *148*, 148–149
Combinatorial chemistry, 178
Common-mode rejection ratio (CMRR), 256–257
Companding, 292
Complex signal reproduction, *285*, 286, 288, *288*
Components, integrated optics, 204–205, *205*
Computer interfacing
- analog devices, 282–288
- analog-to-digital conversion, *288*, 288–292
- analog-to-digital converters, 292–296
- antialiasing filters, *289*, 290
- basics, 281
- companding, 292
- complex signal reproduction, *285*, 286, 288, *288*
- conversion time, 294
- current output DAC, *287*, 287–288
- digital implementation, *285*, 285–286
- digital sources, 282–288
- digital-to-analog converters, *286–287*, 286–288
- flash ADCs, 295
- full-scale error, 292
- integrating ADCs, 293
- missing codes, 292
- number representation, 281–282, *282–283*
- pulse-width modulation, *283–284*, 283–286
- PWM output ADCs, *293*, 293–294
- quantization, *285*, 285–286
- resolution, *289*, 290–292, *291*
- R-2R ladder DAC, 286–287, *286–287*
- sample-and-hold circuits, 293
- sample rate, *289*, 289–292, *291*
- sigma-delta converters, 295–296
- signal reconstruction, *289*, *291*, 291–292
- signal reproduction, *285*, 286
- successive approximation, *294*, 294–295

Concentration-dependent etching, 55
Conducting polymers, 197–198, *198*
Conductors, ideal, *240*, 241
Confocal microscopes, 34

Contact aligners, *13*
Contact photolithography, 12–14, *15*
Contact printing, 16–17, *17,* 18
Controlled sources, *240,* 243–244
Conventional IC packaging, *213,* 213–214
Conventional stereolithography, 88
Conversion time, 294
Corners, rounding, 20
Costs
 assembly and packaging, 7
 low-cost photolithography, 32–34
 masks, 32
 shadow masks, 21
 x-ray lithography, 30
Covalent bond, 158, 160
Critical-point-drying equipment, 66
Cross-linking, *164*
Crystals
 growth, 39–40, *40*
 orientation, 42
 planes, *61*
Current control, 297–299, *298*
Current conventions, 239–241, *240–241*
Current output DAC, *287,* 287–288
Current sources, ideal, *240,* 243, *243*
CVD, *see* Chemical vapor deposition (CVD)
Cyanoacrylate adhesives, 216
Cyclic voltammetry, *193,* 193–194
Cytoskeleton, 230
Czochralski process, 39

D

DAC, *see* Digital-to-analog converters (DACs)
Daltons (Da), *170*
Dark current, 134
Darlington configuration, 303
de Broglie hypothesis, 30
Decibels, *251,* 270
Deckert, Kern and, studies, 52
Deep RIE (DRIE) processes, 57, 117
Deionized (DI) water, 25
Dendrimers, 234, *234–235*
Dental wax, 215
Deoxyribonucleic acid (DNA)
 associated molecular machines, 232
 basics, 170–172, *170–173*
 chips, 196–197
 chromatography, 181
 engineering, 233
 strand direction, *171*
Deposition, 45–51, *46, 48, 51*
Derivatization, 190, *191*
Design, 120, 209–211
Design, masks
 alignment marks, 105–107, *106,* 111, *112–113*
 basics, 103, *103*
 devices, 111–117, *113–114, 116*
 file details, *116*
 frame, 104–111
 layers, 108, *108–109*
 layout hints, *115*
 rules, 117–118, *118–122,* 120–121
 scribe lane, 104–105, *105*
 set identification marks, 108, *108–109*
 test structures, *107,* 107–108
Design rule checker (DRC), *118*
Design structure, 111, *113–114*
Desirable properties, photoresists, *22*
Detection
 basics, 190
 electrochemical detection, *192,* 192–194
 laser-induced fluorescence, 190–191
 mass spectrometry, 194–195, *195*
 nuclear magnetic resonance, 195
 radioactive labeling, 194
 ultraviolet absorbance, 191–192
Detlefs and Pisano studies, 4–5
Devices, mask design, 111–117, *113–114, 116*
Devices applications, 176–178
Diamond-like carbon (DLC), 211
Diamond scribe, 69
Dicing saw, 68–69, *69*
Differential values, *264*
Differentiators, *259,* 261–262
Digital implementation, *285,* 285–286
Digital sources, 282–288
Digital-to-analog converters (DACs), *286–287,* 286–288, 296
Dimensional reproduction, 20
Direct current (DC), *241*
Direct silicon bonding, 68
Direct-write lithography, 30–32, *31–32*
Displays, integrated optics, 206, *206*
Distortions, 19
Divider, potential, *250*
DLC, *see* Diamond-like carbon (DLC)
DNA (deoxyribonucleic acid)
 associated molecular machines, 232
 basics, 170–172, *170–173*
 chips, 196–197
 chromatography, 181
 engineering, 233
 strand direction, *171*
Dopant levels, *40*
Doping, 40–42
Double-sided polishing, 42

Drawing hints, *103*
DRC, *see* Design rule checker (DRC)
Drexler, K. Eric, 233
Drilling processes, 94
Dry etching, 56–58, *58,* 63, *63*
Dry film resists, 81–82, *82*
Dyenins, 230

E

E. coli, 228
Eagle software, *125*
Easy PC software, *125*
E-beam lithography, 30–32, *31–32*
EDM, *see* Electrical discharge machining (EDM) and related processes
Effenhauser studies, 86
Electrical discharge machining (EDM) and related processes, *89,* 89–90, 94
Electric software (Static Free Software), 95, *96*
Electrochemical detection, *192,* 192–194
Electrochemical etching, 67
Electroless plating, 93–94
Electronics
  basics, 239, *240*
  controlled sources, *240,* 243–244
  current conventions, 239–241, *240–241*
  ideal electronics, *240,* 241–243
  ideal voltage and current sources, 243, *243*
  Kirchoff's laws, 246–247, *246–247*
  parallel components, 245, *245*
  power calculations, 244
  series components, 245, *245*
  switching losses, 244
  voltage conventions, 239–241, *240–241*
Electroosmotic flow, *185,* 185–186
Electrophoresis, 181–182, *182*
Electroplating, *74,* 74–75
Electrostatic actuators, 147–150, *148*
Electrostatic bonding, 67
Electrostream drilling (ESD), 94
Elemental metals, pure, 158
Elements similarities, *159*
Embossing, 82–83, *83–85*
Engineering, science comparison, 2–3
Environmental monitoring, 177
Enzyme-based biosensors, *140,* 140–141
Epoxy glues, 216
Equations, *29*
Error voltage, 248
Escheria coli, 228
ESD, *see* Electrostream drilling (ESD)
Etching, *56–57, 70, see also specific type of etching*
Ethane, *162*
Ethylene diamine, *54*
Eukaryotic cells, 230
Europe, 1
Evaporation, 50
EV Group, 27
Excimer laser micromachining, 77–79, *77–79*
Exploitation, 5, *6*
Exposure, low-cost photolithography, 33
Exposure systems, 11–21

F

Fabrication, optical fibers, 202–203, *203*
Face-centered cubic (FCC) unit cell, 37–38
Facets, 97
Fats, 173, *174*
Fatty acids, 173
Feature size, minimum, 95
Feynman, Richard, 3
Fiber coupling, 205, *205*
Fiber-optic cross-point switches, 206, *207*
Fiber-optics cable, *265*
FIBM, *see* Focused ion beam milling (FIBM)
Field effect transistors, 136, 139
Files
  details, *116,* 123
  formats, *97,* 97–100
  Gerber, 124, *125,* 126
  technology, 98–100
Film mask parameters, *126*
Filtering, *see also* Amplifiers and filtering
  basics, 268
  Butterworth filters, 273–279, *274–275*
  RC filters, 268–271, *269, 271–272,* 273
  switched-capacitor filters, *279,* 279–280
Fine points, 63, *64*
Flagella motors, 231
Flash ADCs, 295
Flip-chip bonding, 215
Fly-eye homogenizers, 19
Focused ion beam milling (FIBM), 57–58, 225–226
Four-electrode configuration, 298–299, *299*
Frame, 104–111
Frame structure, 108, *109–111,* 111
Frequency response, Butterworth filters, *278,* 278–279
FSD, *see* Full-scale deflection (FSD)
Full-scale deflection (FSD), 290
Full-scale error, 292
Functions, amplifiers and filtering, 263
Future possibilities, assembly and packaging, *210*

G

Gain, *248*
Gain-bandwidth product, *254*
GC-Prevue software, *125*
GDSII file format, 97, *97,* 123
Gel electrophoresis, 181
Gerber files
  mask design, 124, *125,* 126
  printed circuit board industry, 34
Glass filters, 93, *94*
Glass micropipettes, *195*
Glass wafers, 44
Graded change, 202
Graphics, 100–101, *103*
Grass, *58*
Grayscale masks, 14, *14*
Grid, 101, *101*
Group delay, *279*
Growth potential, microengineering, 5–7

H

Hall effect sensors, 137
Hard masks, 18
H-bonds, 161
Hexane, *162*
HF, *see* Hydrofluoric acid (HF)
High-pressure liquid chromatography (HPLC), 181
Hill, Horowitz and, studies, 257, 263
Historical perspectives, 3–4
Horowitz and Hill studies, 257, 263
HPLC, *see* High-pressure liquid chromatography (HPLC)
Hydraulic actuators, 147, 152, *152*
Hydrazine, *54*
Hydrocarbons, *163*
Hydrofluoric acid (HF), 53
Hydrogels, 197–198
Hydrogen bonds, 161

I

IBM, *see* Ion beam milling (IBM)
IC packaging, *213,* 213–214
Ideal electronics
  capacitors, *240,* 242
  conductors, *240,* 241–243
  current sources, *240,* 243, *243*
  inductors, *240,* 242–243
  insulators, *240,* 241
  op-amps, 248–525, *249–250*
  resistors, *240,* 241
  voltage sources, *240,* 243, *243*
Ikuta studies, 88
Image-reversal process, 59
Impedance, *243*
Imported files, *98*
Impurities, *see* Doping
Incomplete etching, *57*
Incorporation, microengineering, *7*
Inductors, ideal, *240,* 242–243
Infrared lasers, 76, *76*
Inorganic chemistry, *156,* 157–161, *158*
Input impedance, 255–256
Instrumentation amplifiers, 263–265, *264*
Insulators, ideal, *240,* 241
Integrated optics
  applications, 205–208
  basics, 201
  components, 204–205, *205*
  displays, 206, *206*
  fabrication, optical fibers, 202–203, *203*
  fiber coupling, 205, *205*
  fiber-optic cross-point switches, 206, *207*
  lenses, 205–206
  optical fiber waveguides, 201–203, *202–203*
  planar waveguides, 204, *204*
  tunable optical cavities, 206–208, *207*
  waveguides, 201–204
Integration, ADCs, 293
Integrators, op-amp application, *259,* 261–262
Interfacing
  amplifiers and filtering, 239–280
  basics, 237–238
  computer interfacing, 281–296
  output drivers, 297–310
Internal units, 99, *100, see also* Units
Internet, *see* Web sites
Intruder alarms, 137
Inverting amplifier, 249
Inverting op-amp configurations, *249,* 251–252, *252–253*
Ion beam milling (IBM), 57–58, 225–226, *226*
Ionic bond, 158
Ion implantation, 41–42, *45*
Ions, 157
Ion-sensitive field effect transistors (ISFETs), *138–139,* 138–140
IPA, *see* Isopropyl alcohol (IPA)

ISFETs, *see* Ion-sensitive field effect transistors (ISFETs)
Isopropyl alcohol (IPA), 25, 66
Isotropic etchants, 52

J

Jackman studies, 87
Japan, 1
Jerman and Angell, Terry, studies, 155
Johnson noise, *see* White noise
Junction field effect transistors (JFETs), 256, 261

K

Kern, Vossen and, studies, 33, 50, 56
Kern and Deckert studies, 52
Kinesins, 230
Kirchoff's laws, 246–247, *246–247,* 249
KOH, *see* Potassium hydroxide (KOH)

L

Lambda units, 99–100, *100*
Laminar flow, *184,* 184–185
Langmuir-Blodgett films, 224, 227–228, *227–228*
Laser-cut stencils, *21,* 34
Laser-induced fluorescence (LIF), 190–191
Laser machining, 75–79
Lasers, 76, *76*
Laser scribe, 69
Layernames structure, 108, *111*
Layers, mask design, 108, *108–109*
Layout hints, *115*
Layout software
- basics, 95, *96,* 97
- caveats, 100
- file formats, 97–100
- graphics, 100–101
- grid, 101, *101*
- layers, 102
- Manhattan geometry, *102,* 102–103
- technology files, 98–100
- text, 101
- units, *99,* 99–100

Lead zirconate titanate (PZT-PbZiTiO3), 144
L-Edit Pro software (Tanner), 95, *96*
Lenses, integrated optics, 205–206
LIF, *see* Laser-induced fluorescence (LIF)
Liftoff, *58,* 58–59
LIGA (Lithographie, Galvanoformung, Abformung)
- hydraulic actuators, 152
- magnetic actuators, 150
- mask plate details, 123
- nonsilicon processes, *74,* 74–75
- wobble motors, 150

Light sources, 15, *16*
Linear molecular motors, 231
Lipids, 172–173
Lithographie, Galvanoformung, Abformung, *see* LIGA (Lithographie, Galvanoformung, Abformung)
Low-cost photolithography, 32–34
Low-pressure CVD (LPCVD), 47, *48,* 49, *51–52*
Low-reflection coating (LRC), 122
LRC, *see* Low-reflection coating (LRC)

M

Magnetic actuators, 147, 150, *150*
Magnetic sensors, 137, *137*
Magnetic stirrers, 26
Manhattan geometry, *102,* 102–103
Mark, high signal, 284
Market and opportunities, microengineering, 5–7
Maskname structure, 108, *111*
Mask plate details, 122–123, *123*
Masks
- aligners, 12–15, *13–15*
- cavities, *56*
- costs, 32
- hard masks, 18
- set details, 123
- wear and tear, 13
- x-ray lithography, 29–30, *30*

Masks, design
- alignment marks, 105–107, *106,* 111, *112–113*
- basics, 95, 126
- caveats, 100
- design, *103,* 103–118, *118–120,* 120–121
- details, mask plate, 122–123, *123*
- devices, 111–117, *113–114, 116*
- file details, *116,* 123
- file formats, 97–100
- frame, 104–111
- Gerber files, 124, *125,* 126
- graphics, 100–101
- grid, 101, *101*
- layers, 108, *108–109*
- layout software, 95, *96,* 97–103
- Manhattan geometry, *102,* 102–103
- mask plate details, 122–123, *123*
- mask set details, 123
- minimum feature size, 95

placement requirements, 124
production, 122–124
scribe lane, 104–105, *105*
set identification marks, 108, *108–109*
step and repeat, *13,* 124
technology files, 98–100
test structures, *107,* 107–108
text, 101
Mass spectrometry, 182, 194–195, *195*
Materials
prototypes, 215–216, *216*
silicon micromachining, 45–47, *46,* 49–51, *51*
Mechanical sensors
accelerometers, 146
basics, 143
capacitive sensors, 144–145
optical sensors, 145
piezoelectric sensors, 144
piezoresistors, 143–144, *143–144*
pressure sensors, 146
resonant sensors, *144,* 145
Membranes, *54,* 59, *60–63,* 61–63
MEMS, *see* Microelectromechanical systems (MEMS)
MEMS Industry Group, 4–5
Mesas, 59, *60,* 61–63
Metalorganic CVD (MOCVD), 49
Metal-oxide semiconductor field effect transistors (MOSFETs)
AC-coupled amplifiers, 261
chemical sensors, 137
current output DAC, 287–288
input bias currents, 256
ISFET sensors, 139
output drivers, 303–306, *304–306*
relays, 308
self-aligned gate, *45*
Methane, *162*
Microactuators
basics, 147
comb drives, *148,* 148–149
electrostatic actuators, 147–150, *148*
hydraulic actuators, 152, *152*
magnetic actuators, 150, *150*
microstimulators, 153–154
multilayer bonded devices, 153, *153*
piezoelectric actuators, 151, *151*
thermal actuators, 151–152, *152*
wobble motors, *149,* 149–150
Microchannel electrophoresis, 186–189, *188*
MicroChem Corp., 24
Microcontact printing, 87, *87*
Microelectrodes, 141, *142*
Microelectromechanical systems (MEMS)
advantages, *4*
basics, 2
BJT output boost, 308
conducting polymers, 197
conventional stereolithography, 88
deep RIE processes, 57
defining characteristics, 1
dry etching, 56
electronic circuit fabrication, 41
glass wafers, 44
goals, 28
hydrogels, 197
interfacing, 237
Manhattan geometry, 102
modeling software, 98
oxygen RIE, 57
photolithography features, *9*
polygons, 101
porous silicon, 67
relays, 308
resists, 18
sensors, 133
silicon, 37
silicon-on-insulator wafers, 42–43
tunable optical cavities, 207–208
wafer size, 31, 42
Microengineering
advantages, 4
basics, 1–3
exploitation, 5, *6*
historical perspectives, 3–4
importance, 3–5
incorporation into business, *7*
market and opportunities, *5,* 5–7
packaging, 6–7
photoresist properties, *22*
Microfluidic chips, 183–184, *184*
Micromachined structures, 66
Micromachining
advantages, 4
basics, 1–3
exploitation, 5, *6*
goals, 28
historical perspectives, 3–4
importance, 3–5
incorporation into business, *7*
market and opportunities, *5,* 5–7
mask design, 95–126
nonsilicon processes, 73–94
packaging, 6–7
photolithography, 9–35
silicon micromachining, 37–71

Micrometers, 100
Microscopy, 178–179, *see also specific type of microscope*
Microsensors
accelerometers, 146
basics, 131
biosensors, 138–141
capacitive sensors, 144–145
charge-coupled devices, 135, *136*
chemical sensors, 138–141
enzyme-based biosensors, *140,* 140–141
ISFET sensors, *138–139,* 138–140
magnetic sensors, 137, *137*
mechanical sensors, 143–146
microelectrodes, 141, *142*
neurophysiology, 141, *142*
optical sensors, 145
photodiodes, 134–135
phototransistors, 135, *135*
piezoelectric sensors, 144
piezoresistors, 143–144, *143–144*
pressure sensors, 146
pyroelectric sensors, 136–137
radiation sensors, 134–137
resonant sensors, *144,* 145
thermal flow-rate sensors, 133–134, *134*
thermal sensors, 131–134
thermocouples, 131–132, *132*
thermoresistors, 132–133
Microstereolithography, 87–88, *88*
Microstimulators, 153–154
Microsystems
assembly and packaging, 209–216
basics, 127–128
components, *128,* 128–129
integrated optics, 201–208
microactuators, 147–154
microsensors, 131–146
micro total analysis systems, 155–198
nanotechnology, 217–234
Microsystems technology (MST), 1
Micro total analysis systems (μTAS)
applications, 176–183
basics, 155–156, *156,* 183
biochemistry, 167–175, 177–178
biology, 178–183
bond formation, *156, 158,* 159–161, *160–161*
carbohydrates, 175, *176*
chemistry, 156–175
chromatography, 180–181, *180–181*
conducting polymers, 197–198, *198*
detection, 190–195
DNA chips, 196–197
electrochemical detection, *192,* 192–194
electroosmotic flow, *185,* 185–186
electrophoresis, 181–182, *182*
environmental monitoring, 177
hydrogels, 197–198
inorganic chemistry, *156,* 157–161, *158*
laminar flow, *184,* 184–185
laser-induced fluorescence, *190,* 190–191
lipids, 172–175
mass spectrometry, 182, 194–195, *195*
microchannel electrophoresis, 186–189, *188*
microfluidic chips, 183–184, *184*
microscopy, 178–179
nuclear magnetic resonance, 182–183, 195
nucleic acids, 170–172, *170–173*
organic chemistry, *162–166,* 162–167
pH, 161
polymerase chain reaction, 183, 197
polymers, 164
process monitoring, 177
proteins, 167–170
radioactive labeling, 179, 194
sample injection, 186, *186*
silicones, *166,* 166–167
surface tension, *184,* 184–185
synthesis, 177
ultraviolet absorbance, 191–192
x-ray crystallography, 182–183
Microtubules, 230
Miller indices, 38–39, *39*
Millimeter wave technologies (MMIC), 128
Miniaturization, speed, 181, 197
Minimum feature size, 95, 120
Missing codes, 292
MMIC, *see* Millimeter wave technologies (MMIC)
MOCVD, *see* Metalorganic CVD (MOCVD)
Molecular beam epitaxy (MBE), *51*
Molecular mass, *170*
Molecular motors, 230–232, *230–232*
Molecular nanotechnology, 233–234
Molybdenum silicide, 223
Monocrystalline forms, 37
MOSFETs (metal-oxide semiconductor field effect transistors)
AC-coupled amplifiers, 261
chemical sensors, 137
current output DAC, 287–288
input bias currents, 256
ISFET sensors, 139
output drivers, 303–306, *304–306*
relays, 308
self-aligned gate, *45*
MOSIS, 118, *118–120*
Motion detectors, 137

MST, *see* Microsystems technology (MST)
MTAS, *see* Micro total analysis systems (μTAS)
Muller, Williams and, studies, 52, 56
Multichip modules, 214
Multilayer bonded devices, 153, *153*
Multimode fibers, 202
Myosins, 230–231

N

Nanoelectromechanical systems (NEMS), 222–226
Nanolithography, 20, 222–224
Nanostructures, 222–225, *223–225*
Nanotechnology
  artifacts, *221*, 221–222
  atomic force microscope, 220, *220*
  basics, 217
  bionanotechnology, 228–233, *229*
  buckminsterfullerene, 234
  calibration, *221*, 221–222
  cell membranes, 229
  cytoskeleton, 230
  dendrimers, 234, *234–235*
  DNA-associated molecular machines, 232
  DNA engineering, 233
  ion beam milling, 225–226, *226*
  Langmuir-Blodgett films, 227–228, *227–228*
  molecular motors, 230–232, *230–232*
  molecular nanotechnology, 233–234
  nanoelectromechanical systems, 222–226
  nanolithography, 222–224
  nanostructures, 222–224, *223*, 224–225, *224–225*
  protein engineering, 233
  scanning electron microscope, 217–219, *218*
  scanning near-field optical microscope, *220*, 221
  scanning probe microscopy, *219*, 219–222
  scanning tunneling electron microscope, 219–220, *220*
  silicon micromachining, 224–225, *224–225*
  SPM pens, 224
  UV photolithography, 222–224, *223*
National Institute of Standards and Technology (NIST), 29
Native oxide layer, 44
Nd:YAG lasers, 76
Negative resists, 19, *19*
NEMS, *see* Nanoelectromechanical systems (NEMS)
Neodymium YAG (Nd:YAG) lasers, 76
Network of Excellence for Multifunctional Microsystems (NEXUS), 4–5
Neurophysiology, 141, *142*
NEXUS (Network of Excellence for Multifunctional Microsystems), 4–5
NIST (National Institute of Standards and Technology), 29
Nitride-oxide comparison, *49*
Noble metals, 45
Noise, 257–258
Nonideal electronics
  op-amps, 253–257
  sources, *249*, 251–252, *252–253*
Noninverting op-amp configurations, *249*, 251–252, *252–253*
Nonsilicon processes
  basics, 73
  chemical-mechanical polishing, 73
  drilling processes, 94
  dry film resists, 81–82, *82*
  electrical discharge machining and related processes, *89*, 89–90, 94
  electroless plating, 93–94
  electroplating, *74*, 74–75
  embossing, 82–83, *83–85*
  excimer laser micromachining, 77–79, *77–79*
  glass filters, 93, *94*
  infrared lasers, 76, *76*
  laser machining, 75–79
  LIGA, *74*, 74–75
  microcontact printing, 87, *87*
  microstereolithography, 87–88, *88*
  parylene, 81
  PDMS casting, 83, 86, *86*
  photochemical machining, 75
  photoformable epoxies, 80–81, *81*
  photostructural glasses, 90–91, *90–91*
  polyimides, 80
  polymer microforming, 79–88
  precision engineering, 91–93, *92*
  PTFE, 81
  roughness measurements, 92–93, *93*
  sand blasting, 93
  SU-8, 80–81, *81*
  ultrasound, 93
Nonsinusoidal signals, *273*
Nuclear magnetic resonance, 182–183, 195
Nucleic acids, 170–172, *170–173*, *see also specific type of nucleic acid*
Number representation, 281–282, *282–283*
Nyquest criterion, 289, 291

O

Ohmic contacts, *300*
Ohm's Law, 302

OLEDs, *see* Organic light-emitting diodes (OLEDs)
Operational amplifiers (op-amps)
AC-coupled amplifiers, *259,* 260–261
applications, 258–263, *259*
bandwidth limitations, 254–255
basics, 247–248, *248*
bias currents, 255–256, *256*
common-mode rejection ratio, 256–257
differentiators, *259,* 261–262
functions, 263
ideal op-amps, 248–252, *249–250*
input impedance, 255–256, *256*
integrators, *259,* 261–262
inverting configurations, *249,* 251–252, *252–253*
noise, 257–258
nonideal op-amps, 253–257
nonideal sources, *249,* 251–252, *252–253*
noninverting configurations, *249,* 251–252, *252–253*
output drivers, 297–299, *307–309,* 308–309
power supply rejection ratio, 256–257
slew rate, 254–255
summing amplifiers, *259,* 261
thermal noise, 258, *258*
unity-gain buffer amplifier, 258–260, *259*
white noise source combination, 257, *257*
Opportunities in microengineering, 5–7
Optical fiber waveguides, 201–203, *202–203*
Optical oddities, 19–21
Optical proximity effects, 20
Optical sensors, 145
Optical systems, 15–18
Optics, *see* Integrated optics
Optoisolators, 309–310, *309–310*
Organelles, 229
Organic chemistry, *162–166,* 162–167
Organic light-emitting diodes (OLEDs), 198
Output boost, *307–309,* 308–309
Output drivers
basics, 297
BJTs, 300–303, *300–303, 307–309,* 308–309
current control, 297–299, *298*
four-electrode configuration, 298–299, *299*
MOSFETs, 303–306, *304–306*
op-amps, 297–299, *307–309,* 308–309
optoisolators, 309–310, *309–310*
output boosts, *307–309,* 308–309
relays, 306–308, *307*
transistors, 300–306
voltage control, 299, *299*
Oxidation, 192
Oxides, *49, 57*
Oxygen RIE, 57

P

Packaging, 6–7, 212–214, *see also* Assembly and packaging
Parallel components, 245, *245*
Parameter extraction, 98
Parylene, 81
Passivation, 211–212
PbZiTiO3 (PZT-lead zirconate titanate), 144
PCM, *see* Photochemical machining (PCM)
PCR, *see* Polymerase chain reaction (PCR)
PDMS casting, *see* Polydimethylsiloxane (PDMS) and casting
Peckerar studies, 16–17
Pentane, *162*
Periodic Table, *156,* 158, *159*
Petersen, Kurt, 3, 37
pH, 161
Phase-shift masks, *223,* 223–224
Phospholipids, 173–174, *174–175*
Phosphosilicate glass (PSG), *57*
Photochemical machining (PCM), 75
Photodiodes, 134–135
Photoengraving, *11*
Photoetching, *11*
Photoformable epoxies, 80–81, *81*
Photolithography
aligners, mask, 12–15, *13–15*
basics, 9, *10,* 34–35
direct-write lithography, 30–32, *31–32*
e-beam lithography, 30–32, *31–32*
exposure systems, 11–21
light sources, 15, *16*
low-cost photolithography, 32–34
masks, 12–15, *13–15,* 29–30, *30*
optical oddities, 19–21
optical systems, 15–18
photoresists, 21–22, *22,* 24–28
resist processing, *24,* 24–28
shadow masks, 21
UV photolithography, *10,* 10–28
X-ray lithography, 28–30
Photoresists
adhesion, 216
application of, *23,* 27–28
basics, 11, 21–22, *22–24*
desirable properties, *22*
etching resistance, 45
popular, *23*
postexposure processing, 28
processing, *24,* 24–28

profiles, *22*
steps, *24*
substrate cleaning, 25–26
suppliers, *23*
thick resist processing, *27*
types, *23*
UV light source, 15
wafer bonding, 68
Photostructural glasses, 90–91, *90–91*
Phototransistors, 135, *135*
Piezoelectric actuators and sensors, 144, 147, 151, *151*
Piezoresistors, 143–144, *143–144*
Piranha clean and etch, 26, 53
Pisano, Detlefs and, studies, 4–5
Pits, *54,* 59, *60,* 61–63
Placement requirements, 124
Planarization, 73
Planar waveguides, 204, *204*
Planck's constant, 30
Plasma-enhanced CVD (PECVD)
basics, 47, *48,* 49, *51*
etch rates, *52*
passivation, 211–212
wafer protection, 71
PMMA-resists, 74
Points, fine, 63, *64*
Polycrystalline forms, 37
Polydimethylsiloxane (PDMS) and casting
microcontact printing, 87
polymer microforming, 80, 83, 86, *86*
silicones, 166
Polyethylene glycol, 216
Polygons, 97, 101
Polyimides, 80
Polymerase chain reaction (PCR), 183, 197
Polymer microforming
basics, 79–80
dry film resists, 81–82, *82*
embossing, 82–83, *83–85*
microcontact printing, 87, *87*
microstereolithography, 87–88, *88*
parylene, 81
PDMS casting, 83, 86, *86*
photoformable epoxies, 80–81, *81*
polyimides, 80
PTFE, 81
SU-8, 80–81, *81*
Polymers, 164
Polytetrafluoroethylene (PTFE)
polymer microforming, 81
reactive ion etching, 57
tape, *70,* 216
Porous silicon, 67
Positive resists, 19, *19*
Postetch rinsing, *53*
Postexposure processing, 28
PostScript output files, 34
Potassium hydroxide (KOH)
anisotropic etchant comparison, *54*
bond formation, 160
etching, wafers, *70*
pH, 161
releasing structures, etching, 70–71, *70–71*
silicon, electrochemical etching, 67
silicon micromachining, *54,* 59, *60–63,* 61–63
wafer damage, 117
wet etching, 53–55
Potential divider, *250*
Power calculations, 244
Power supply rejection ratio (PSRR), 256–257
Power transfer, *251*
Precision engineering, 91–93, *92*
Prepackage testing, 212
Presensitized mask blanks, 33
Pressure sensors, 146
Printed circuit boards and industry
dry film resists, 81–82
Gerber files, 124
laser-cut stencils, 34
resist technologies, 28
Printers, 33
Probe station, 114
Processing, photoresists, *24,* 24–28
Process monitoring, 177
Production, mask design, 122–124
Profiles, photoresists, *22*
Projection printing, *15,* 17–18, *18*
Propane, *162*
Prosthetics, visual, 153–154
Protective sleeving, 202
Protein engineering, 233
Proteins, 168–170
Protel software, *125*
Prototype materials, 215–216, *216*
Proximity effects, 20, *20,* 32
Proximity printing, 16–17, *18*
PSG, *see* Phosphosilicate glass (PSG)
PSRR, *see* Power supply rejection ratio (PSRR)
PTFE, *see* Polytetrafluoroethylene (PTFE)
Puers and Sansen studies, 61
Pulse-chase experiments, 179
Pulse-width modulation (PWM)
computer interfacing, *283–284,* 283–286
output ADCs, *293,* 293–294
Pure elemental metals, 158
PWM, *see* Pulse-width modulation (PWM)

Pyroelectric sensors, 136–137
PZT (PbZiTiO3-lead zirconate titanate), 144

Q

Quantization, *285,* 285–286, *290*
Quartz glass, 122

R

Radiation sensors, 134–137
Radioactive labeling, 179, 194
Radio Corporation of America (RCA) clean, 26
Radio frequency MEMS, 128
Ragged lines, *56*
Rayleigh criterion, 15–16
RCA clean, 26
RC filters, 268–271, *269, 271–272,* 273
Reactance, *243*
Reactive parts, impedance, *242*
Reactive/relative ion etching (RIE)
- dry etching, 56
- pattern transfer, 64
- photostructurable glasses, 90
- releasing structures, 71
- silicon micromachining, 56–57
- surface micromachining, 66

Reche studies, 16–17
Rectifying contacts, *300*
Redox potential, 192
Reduction, 192
Reed relays, 307
Reflection, substrate, *20,* 20–21
Reflective proximity effects, 21
Reflow, 64, *65*
Refractive index, 202–204
Relays, 306–308, *307*
Research and development (R&D), 5–7
Resistance, *243*
Resistive parts, *242*
Resistors, ideal, *240,* 241
Resist processing, 21, *24,* 24–28
Resolution, computer interfacing, *289,* 290–292, *291*
Resonant sensors, 6, *144,* 145
Restriction enzymes, *196*
Ribonucleic acid (RNA)
- basics, 170–172, *170–173*
- biochemistry, 167
- engineering, 233

RIE, *see* Reactive/relative ion etching (RIE)
Root mean square, *257*
Roughness measurements, 92–93, *93*
R-2R ladder DACs, 286–287, *286–287*
Rubber, *164*

S

Sallen and Key filter, 275, 277
Sample-and-hold circuits, 293
Sample injection, 186, *186*
Sample rate, *289,* 289–292, *291*
Sand blasting, 93
Sandmaier studies, 61
Sansen, Puers and, studies, 61
Scalable CMOS (SCMOS) process, 118, *118–120*
Scaling, 5, 598
Scanning electron microscope (SEM)
- direct-write e-beam system, 31
- low-cost photolithography, 34
- nanotechnology, 217–219, *218*
- prepackage testing, 212

Scanning near-field optical microscope (SNOM), 179, *220,* 221
Scanning probe microscopy (SPM)
- biology, 179
- nanotechnology, 217, *219,* 219–222
- pens, 224
- surface roughness, 93

Scanning tunneling electron microscope (STEM), 178, 219–220, *220*
Schmitt trigger, 263
Science, engineering comparison, 2–3
Screened cable, *265*
Scribelane structure, 104–105, *105,* 108, 111, *111*
Self-alignment, *210,* 210–211
Self-assembly, *210,* 210–211
SEM, *see* Scanning electron microscope (SEM)
Semiconducting thermoresistors/thermistors, 132
Sensors, *see* Microsensors
Series components, 245, *245*
Set identification marks, 108, *108–109*
Shadow masks, 21
Shaped tube electrolytic milling (STEM), 94
Showered ion beam milling (SIBM), 57–58, 225–226
SIBM, *see* Showered ion beam milling (SIBM)
Sigma-delta converters, 295–296
Signal reconstruction, *289, 291,* 291–292
Signal reproduction, *285,* 286
Silanes, *167*
Silicon, 37–39, *38*
Silicones, *166,* 166–167
Silicon micromachining

basics, 37
beams, 59, 61–63
bridges, *54,* 61–63
bulk silicon micromachining, 59–64
crystal growth, 39–40, *40*
deposition, 45–47, *46,* 49–51, *51*
diamond scribe, 69
dicing saw, 68–69, *69*
doping, 40–42
dry etching, 56–58, 63, *63*
electrochemical etching, 67
fine points, 63, *64*
ion-bean milling, 57–58
ion implantation, 41–42, *45*
laser scribe, 69
liftoff, *58,* 58–59
materials, 45–47, *46,* 49–51, *51*
membranes, *54,* 59, *60–63,* 61–63
mesas, 59, *60,* 61–63
Miller indices, 39, *39*
nanotechnology, 224–225, *224–225*
pits, *54,* 59, *60,* 61–63
porous silicon, 67
potassium hydroxide, *54,* 59, *60–63,* 61–63
potassium hydroxide etching, 70–71, *70–71*
reflow, 64, *65*
relative ion etching, 56–57
RIE pattern transfer, 64
silicon, 37–39, *38*
structures, 59–68, 70–71, *70–71*
surface micromachining, 64–66, *65–66*
thermal diffusion, 41
thin films, 45–47, 49–59
wafer bonding, 67–68, *68*
wafer dicing, 68–71
wafer specifications, 42–44, *43–44*
wet etching, *52,* 52–56, *54–55,* 63, *64*
Silicon-on-insulator (SOI) wafers, 42–43, 71
Silver-loaded epoxy, 216
Single mode fibers, 202
Single-point diamond turning, 92
Single-sided polishing, 42
Slew rate, *254,* 254–255
SNOM, *see* Scanning near-field optical microscope (SNOM)
Sodium, *159*
Software, 95, *96, 125, see also* Layout software
SOI, *see* Silicon-on-insulator (SOI) wafers
Solder reflowing, 64
Solid-state relays, 308
Space, low signal, 284
Spark erosion, 89
Speed, miniaturization, 181, 197
Spinning, 27, *48,* 50
Spin-on-glass, 68
SPM, *see* Scanning probe microscopy (SPM)
SPM pens, 224
Spot size, 31
Spraying, photoresist application, 27
Sputtering
deposition, *48,* 49–50, *51*
dry etching, *58*
shadow masks, 21
SQUIDs (superconducting quantum interference devices), 137
Static Free Software's Electric software, 95, *96*
STEM, *see* Scanning tunneling electron microscope (STEM); Shaped tube electrolytic milling (STEM)
Stencils, laser-cut, 21, 34
Step-and-repeat approach, *13,* 14, 124
Step change, 202
Stereolithography, 88
Stick-and-ball models, 163, *163*
Stress signs, *49*
Structures, silicon micromachining, 59–68, 70–71, *70–71*
SU-8, 24–25, 80–81, *81*
Substance, amount, *157*
Substituent groups, *165*
Substrate cleaning, 25–26
Substrate reflection, *20,* 20–21
Successive approximation, *294,* 294–295
Summing amplifiers, *259,* 261
Superconducting quantum interference devices (SQUIDs), 137
Super IH process, 89
Suppliers, photoresists, *23*
Surface charge formation, *186*
Surface micromachining, 64–66, *65–66*
Surface tension, *184,* 184–185
Switched-capacitor filters, *279,* 279–280
Switching losses, 244
Synthesis, micro total analysis systems, 177

T

Talysurf, 219
Tanner's L-Edit Pro software, 95, *96*
Target 3001 software, *125*
MTAS, *see* Micro total analysis systems (μTAS)
TCR, *see* Temperature coefficient of resistance (TCR)
Technology files, *98,* 98–100
Temperature coefficient of resistance (TCR), 132
TEOS-based CVD, *48,* 49, 64

Terry, Jerman and Angell studies, 155
Test structure, 108, *111*
Test structures, *107,* 107–108
Tetramethyl ammonium hydroxide (TMAH), 53, *54*
Text, 97, 101
Thermal actuators, 147, 151–152, *152*
Thermal diffusion, 41, *48*
Thermal flow-rate sensors, 133–134, *134*
Thermal noise, 258, *258*
Thermal oxidation, 47, *48,* 225
Thermal sensors, 131–134
Thermocompression bonding, 214
Thermocouples, 131–132, *132*
Thermoresistors, 132–133
Thick resist processing, *27*
Thin films, 45–47, 49–59
Three-mask process, 104
Tipping, 26
TMAH, *see* Tetramethyl ammonium hydroxide (TMAH)
Transconductance, *248*
Transducers, 131
Transimpedance, *248*
Transistors, 300–306
Tunable optical cavities, 206–208, *207*
Twisted pair cabling, *265*

U

Ultrasonic bonding, 214–215
Ultrasonic cleaning bath, 26
Ultrasound, 93
Ultraviolet (UV) absorbance, 191–192
Ultraviolet (UV) photolithography
  aberrations, optical, 19
  application, photoresists, *23,* 27–28
  basics, *10,* 10–11
  cleaning, substrate, 25–26
  contact printing, 16–17, *17,* 18
  distortions, 19
  exposure systems, 11–21
  light sources, 15, *16*
  mask aligners, 12–15, *13–15*
  nanotechnology, 222–224, *223*
  negative resists, 19, *19*
  optical oddities, 19–21
  optical systems, 15–18
  photoresists, 21–22, *22,* 24–28
  positive resists, 19, *19*
  postexposure processing, 28
  projection printing, 17–18, *18*
  proximity effects, 20, *20*
  proximity printing, 16–17, *18*
  resist processing, 21, *24,* 24–28
  shadow masks, 21
  substrate cleaning, 25–26
  substrate reflection, *20,* 20–21
Ultraviolet (UV) sources and wavelengths, *16*
Uneven etching, *56*
United Kingdom, 1
Units
  electronics, *240, 243*
  technology files, 99–100, *99–100*
Unity-gain bugger amplifier, 258–260, *259*
Unsaturated hydrocarbons, 164
User units, 99, *100, see also* Units
UV-curing adhesives, 216

V

van der Waals bonding, 160
Vector-scan approach, 32
Virtual earth/ground, 261
Viruses, 228
Visual prosthetics, 153–154
Voltage control, 299, *299*
Voltage conventions, 239–241, *240–241*
Voltage sources, ideal, *240,* 243, *243*
Vossen and Kern studies, 33, 50, 56

W

Wafers
  bonding, 67–68, *68*
  dicing, 68–71
  lapping, 73
  potassium hydroxide etching, *70*
  size, 31
  specifications, 42–44, *43–44*
Wafer structure, 115, *116*
Waveguides, 201–204
Web sites
  EV Group, 27
  MOSIS, 118
  PCB design packages, *125*
  photoresists, *23*
  Static Free Software's Electric software, *96*
  Tanner's L-Edit software, *96*
Wet etching
  micromachined structures, 66
  silicon micromachining, *52,* 52–56, *54–55,* 63, *64*
Wheatstone Bridge, 265–268, *266–268*
White noise, *257,* 257–258
Whiteside studies, 87
Williams and Muller studies, 52, 56

Wire bonding, 214–215
Wobble motors, *149*, 149–150
World Wide Web, *see* Web sites

X

X-ray crystallography, 182–183
X-ray lithography
basics, 28–29
LIGA, 74
mask plate detail, 123
masks, 29–30, *30*

Y

Yellow room, *25*

Z

Zinc oxide, 137, 144
Zone refining, 39

T - #0035 - 111024 - C0 - 229/152/20 - PB - 9780367391027 - Gloss Lamination